DIGITAL CIRCUITS AND MICROCOMPUTERS

DIGITAL CIRCUITS AND MICROCOMPUTERS

D. E. JOHNSON, J. L. HILBURN, and P. M. JULICH

Department of Electrical Engineering
Louisiana State University

PRENTICE-HALL, INC., *Englewood Cliffs, New Jersey 07632*

Library of Congress Cataloging in Publication Data

Johnson, David E
Digital circuits and microcomputers.

Includes index.
1. Digital electronics. 2. Electronic digital computers. 3. Microcomputers. I. Hilburn, John L., 1938– joint author. II. Julich, Paul M., 1938– joint author. III. Title.
TK7868.D5J63 621.3815'3 78-13244
ISBN 0-13-214015-2

Editorial production supervision and interior design by: JAMES M. CHEGE

Cover design by: SAIKI & SPRUNG DESIGN

Manufacturing buyer: GORDON OSBOURNE

10 9 8 7

Printed in the United States of America

PRENTICE-HALL INTERNATIONAL, INC., *London*
PRENTICE-HALL OF AUSTRALIA PTY. LIMITED, *Sydney*
PRENTICE-HALL OF CANADA, LTD., *Toronto*
PRENTICE-HALL OF INDIA PRIVATE LIMITED, *New Delhi*
PRENTICE-HALL OF JAPAN, INC., *Tokyo*
PRENTICE-HALL OF SOUTHEAST ASIA PTE. LTD., *Singapore*
WHITEHALL BOOKS LIMITED, *Wellington, New Zealand*

CONTENTS

PREFACE

Digital circuits have been important for many years, but their use dramatically accelerated in the 1960s with the development of integrated circuit technology. The circuits of this era were designed to implement the basic digital logic functions fundamental to all digital systems. These circuits, known as *random logic* circuits, are still used extensively and are of great importance. The most revolutionary recent development in integrated circuitry is the microprocessor, which performs the complex operations of a digital computer. The impact of this device on the field of electronics has been compared to that of the transistor in 1948.

The astonishing growth of integrated circuit technology has caused digital circuits to rival, in importance, the conventional analog circuits. In the foreseeable future, digital circuits will become of even greater importance. Thus there is a need for an elementary textbook which introduces fundamental digital circuit concepts, random logic design, microprocessors, and microcomputers. The primary purpose of this book is to fill this need in a manner that is easily understood by a beginning student with a background in high school algebra or its equivalent.

Before the microcomputer was in common use, random logic networks were required to be much more complex in many applications. Consequently, earlier textbooks on digital circuits presented elaborate design methods which are, in general, too difficult for the beginner. With the advent of the microcomputer, however, the emphasis on sophisticated circuitry is usually not necessary and the simpler design methods of this text are sufficient in most cases.

Chapter 1 presents a historical review of digital circuits and digital computers, including large-scale, mini, and microcomputers. Chapters 2, 3, 4, 8, and 10 contain the basic concepts of number systems, Boolean algebra, logic gates, codes, signed numbers, and complementary arithmetic. Chapters 5, 6, 7, and 9 are devoted to the design of random logic networks, such as combinational circuits, flip-flops, counters, and registers.

The final two chapters are devoted entirely to microprocessors and microcomputers. Chapter 11 deals with the architecture and hardware aspects of microcomputers. Chapter 12 concludes the book with microcomputer programming methods, known as software. An appendix for the Intel 8080 microprocessor is also included, as a supplement to Chapters 11 and 12.

To aid the reader in understanding the textual material, examples are liberally supplied and numerous exercises, with answers, are given at the end of virtually every section. Problems, some more difficult and some less difficult than the exercises, are also given at the end of every chapter.

There are many people who have provided invaluable assistance and advice concerning this book. We are indebted to our colleagues and our students for the form the book has taken and to Mrs. Marie Jines for the expert typing of the manuscript.

DAVID E. JOHNSON
JOHN L. HILBURN
PAUL M. JULICH

Louisiana State University
Baton Rouge, LA

DIGITAL CIRCUITS AND MICROCOMPUTERS

1

INTRODUCTION

Digital computers have brought about a revolution in our everyday lives since their development in the 1940s. Consequently digital, or logic, circuits, of which computers and all other digital devices are examples, have become more important with each passing year. In the last quarter of the twentieth century digital circuits are becoming as important as the conventional analog circuits that have dominated electrical engineering since its beginnings. Many experts believe that before the end of this century, digital circuits will dominate the field.

Recent history has seen a number of revolutionary scientific developments: the automobile and airplane in the early 1900s, the radio in the 1920s, atomic power in the 1940s, television in the 1950s, and space travel and astronauts on the moon in the 1960s. Similarly, in its short history the digital computer has passed through a number of dramatic stages, beginning with the advent of vacuum tube computers in the 1940s, and continuing with the new generation of the 1950s, which used transistors, and the succeeding generation of the 1960s, employing integrated circuits. The 1970s is the age of the microprocessor–microcomputer, and this may well be the most revolutionary of all the computer eras. The microprocessor, a tiny integrated circuit containing hundreds of transistors on a single silicon chip, is the central data-processing unit at the heart of a microcomputer. Its development in the late 1960s and early 1970s is undoubtedly the most exciting technological event in electronics since the appearance of the transistor, in 1948.

The hardware of digital devices becomes increasingly sophisticated as vacuum tubes give way to transistors and transistors to integrated circuits and microprocessors; however, the basis for the hardware, the digital circuit theory, changes very little. A primary purpose of this book is to present the basic theory of digital circuits and the related theory of numbers and codes. We do this in Chapts. 2 through 10. Fortunately, the mathematics of digital circuitry is relatively simple, and our presentation assumes only a knowledge of elementary algebra.

The last two chapters are an introduction to microprocessors and microcomputers. Chapter 11 deals with the hardware of these devices, and Chap. 12 considers *software*, which is the programming of a sequence of detailed instructions for the computer to carry out.

1.1 Digital Circuits

Data, or information, takes many forms when it is stored, communicated, or processed. We may classify a set of data, or a signal, in two broad categories, *analog* and *digital*. An analog signal is usually a *smooth*, *continuously varying* one such as a column of mercury in a liquid thermometer. The height of the column varies continuously with temperature, and its value is *analogous* to the temperature.

An *analog device* is one that can manipulate, or *measure*, analog signals. The thermometer is one such device, and another example is a slide-rule, which measures a distance analogous to a number, and adds two distances which are analogous to two numbers being multiplied. Still another example is an analog electric circuit, such as a filter in a television set, which accepts a continuous wave signal of the desired frequency and blocks, or filters out, signals of other frequencies.

A digital signal is one that is by nature *discrete*, or *discontinuous*, such as a voltage that can only be either "high" or "low." Another example is the time of day as displayed by a digital clock, which indicates the time in whole minutes but does not indicate any fractional time between, say, 3:58 and 3:59.

A digital device is then one that processes digital signals. It may detect whether a signal is present or not, whether it is true or false, whether there is a punched hole in a card or not, etc. Examples are digital clocks (already mentioned), adding machines, and, of course, digital computers. Numerical values are represented by such devices as digits, which may be stored or displayed. We may say that the major difference between analog and digital devices is that the former *measures* and the latter *counts*. For instance, the digital counterpart of the analog slide rule (which measures) is the hand calculator, which displays digital answers.

In this book we will be interested in digital data that can change discretely from one to the other of two distinct states. We may characterize the two states in many ways, such as true or false, high or low, present or absent, etc., but we shall represent them abstractly by the digits 0 and 1. The digital devices we consider are digital circuits whose elements are so-called logic gates (which are themselves digital devices). The best known case of a digital circuit is, of course, a digital computer.

1.2 History of Digital Computers

Since humans learned to count they have probably been searching for devices to help in the process. One of the earliest such devices is the abacus, a mechanical calculator that has been used for 5000 years. It consists of a frame containing columns of beads strung on wires. The beads and their positions on the wires determine values of digits in a number. A person skilled in the use of an abacus can add, subtract, multiply, and divide with it faster than most mechanical calculators of today can perform these operations.

As far as digital devices go, the abacus had no competition until 1642, when the great French philosopher and mathematician Blaise Pascal (1623–1662) invented a mechanical calculator. Pascal's calculator, which may still be viewed in a French museum, had rotating gears and could be used to add and subtract decimal numbers.

Charles Babbage, an English mathematician and Cambridge professor, developed the idea of a mechanical digital computer in the 1830s. He designed and tried to build his device, which he called an "analytical engine," but because the engineers of Babbage's day could not meet the tight tolerances his plans called for, his machine was never completed (Fig. 1.1). However, Babbage's concepts can be found, with minor variations, in every digital computer built today.

In 1890 Herman Hollerith, a U.S. Census Bureau employee, conceived the idea of using holes punched in a card for the purpose of storing the census data to be processed by machines. Hollerith's ideas were used extensively with card-processing equipment for many years prior to the appearance of electronic computers. In the case of the latter machines, punched cards are, of course, a fundamental means of entering and extracting information.

In the late 1930s George Stibitz of the Bell Telephone Laboratories developed a computer using relays. His machine, called the Complex Calculator, was the first of several other relay computers developed during the 1940s and 1950s.

The first general-purpose computer was a relay machine called the Mark I, built by Howard Aiken of Harvard University in cooperation with the IBM Corporation. Aiken's computer, which became operational in 1944, used punched cards and could perform division in about 60 seconds.

The first-generation vacuum tube computers began with the Electronic Numerical Integrator and Computer, or ENIAC, developed by a team of engineers and mathematicians headed by J. P. Eckert, Jr. and J. W. Mauchly of the Moore School of Electrical Engineering of the University of Pennsylvania. The ENIAC, completed in 1946, contained over 18,000 tubes and was about 30,000 times faster than the relay Mark I computer.

Figure 1.1 *Babbage's analytical engine (Courtesy, IBM)*

The most important computer advancement of the 1940s was perhaps the work of the American mathematician John von Neumann (1903–1957), who developed the idea of storing the computer program in the machine's memory. Earlier computers had used programs, but they were not stored in memory until the late 1940s, when von Neumann's idea was implemented. With stored program capabilities, the program may be changed at will without rewiring the computer; this, of course, is a great achievement in computer history.

We may divide computer history into roughly three generations, the first of which were the vacuum tube computers of the late 1940s and early 1950s. Some examples were the ERA 1101, built by Engineering Research Associates of St. Paul, Minnesota; the first UNIVAC (Universal Automatic Computer), built for the National Bureau of Standards by the Eckert–Mauchly Division of Remington Rand Corporation; the IBM 650; and the IBM 704.

The invention of the transistor in 1948 made possible the second generation of computers, in which transistors replaced vacuum tubes. These machines were smaller, faster, and much more reliable, the once common power and tube failures being things of the past. Some examples are the IBM 1401, the IBM 7090, and the Control Data Corporation's CDC 6600. As an example of the greatly increased speed and capabilities of the second-generation machines as compared to those of the first generation, we note that the CDC 6600 can perform more than 3 million instructions per second.

Around 1965 the production of third-generation computers began. These machines employ integrated circuits, which are miniature packages about the size of a typewriter letter containing hundreds of transistor circuits. As a consequence, third-generation computers are 100 times as fast as second-generation machines and are greatly reduced in size, with greater dependability. Examples are the IBM series of System 360 machines, General Electric's GE 600 series, RCA's Spectra 70 series, Burroughs Corporation's 5700, 6700, and 7700 machines, and IBM's System 370.

Minicomputers and microcomputers, which we discuss in the next section, are another new breed of computer that appeared in the late 1960s and early 1970s. They are integrated circuit computers and thus may be said to be of the third generation. However, they are radically different from their large ancestors and probably should be considered to be the fourth generation of computers.

1.3 Microcomputers

In the late 1960s and early 1970s the shrinkage in size of computers due to integrated circuit technology made it possible to have full-fledged computers that would fit on top of an office desk. The first of these computers, known as minicomputers, was the PDP 8, developed by Digital Electronics Corporation.

Minicomputers cannot perform the large tasks assigned to the big computers, such as the IBM 360 series, but they are faster and more powerful than many second-generation, and even third-generation, computers. To illustrate how far computer technology had come by this time, we might compare a minicomputer sitting on a desk with the ENIAC, mentioned in the previous section. The ENIAC had, as we have said, 18,000 vacuum tubes, it weighed 28 metric tons, and was the size of a six-room house; however, it did not have the capability or anything like the reliability of a typical minicomputer. The IBM Series/1 minicomputer (Fig. 1.2) is a good example.

Minicomputers, however, were just the beginning. The amazing decrease in size of the advanced integrated circuits, known as *large-scale integrated* (LSI) chips, started a new kind of computer revolution. The LSI chips were, in themselves, elementary computer components that served as subsystems for more complex computers. In particular, the basic ingredients of a computer, aside from the input–output devices (facilities for putting in and taking out data), are an arithmetic

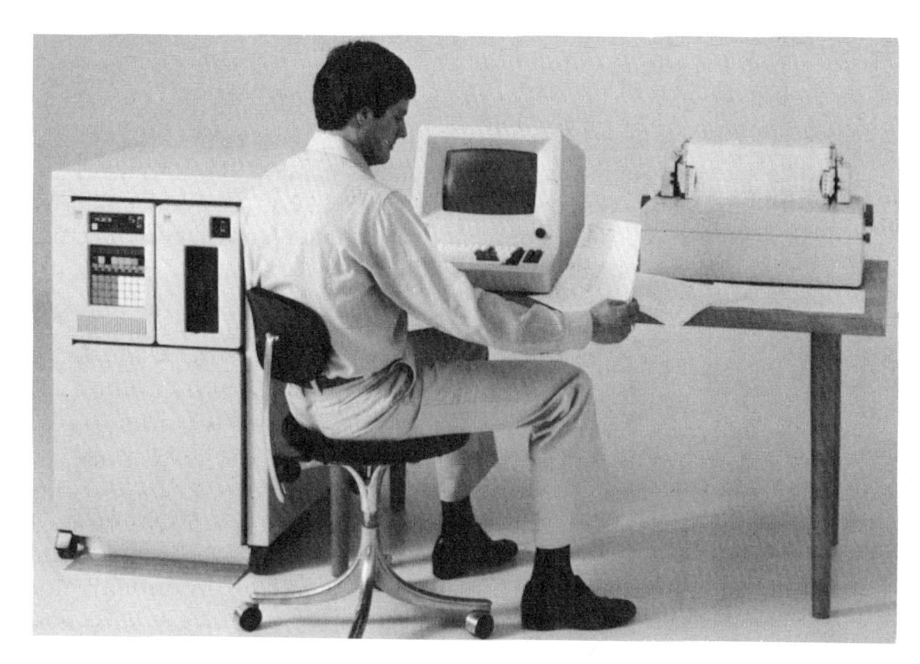

Figure 1.2 *The IBM Series/1 minicomputer (Courtesy, IBM)*

Figure 1.3 *The Intellec Series II microcomputer system (Courtesy, Intel Corp.)*

Figure 1.4 *The Intel 8748 microcomputer integrated circuit (Courtesy, Intel Corp.)*

system that performs the mathematical or logic operations, a memory system that stores the information being used, and the control system that guides the computer in manipulating the stored data. In 1969 Marcian E. Hoff, an engineer at Intel Corporation in Santa Clara, California, conceived a method of condensing the computer's arithmetic system so that it could be put on a single chip, which was called a *microprocessor*. He succeeded in putting the memory and control units on two other chips, which were attached to the microprocessor to form the first *microcomputer*. Hoff's microcomputer could perform 10,000 calculations a second, which is as many as ENIAC could do with its 18,000 vacuum tubes.

Intel thus was a pioneer in developing microprocessors and microcomputers. Its Intel 4004, introduced in 1971, was the first microprocessor to be marketed. This was followed by the Intel 8008 and later the 8080, both in 1972. Others introduced about that time were National Semiconductor's IMP-16 in 1973, Motorola's M6800 in 1974, and the LSI-16, developed by General Automation, Inc. of Anaheim, California, also in 1974. A typical microcomputer system is the Intellec Series II of Intel Corp. (Fig. 1.3).

Prior to 1977 several chips were required to constitute a microcomputer, but in that year Intel Corporation introduced a one-chip microcomputer, called the 8748 (Fig. 1.4). This device contains a central processing unit, program memory, data memory, connections between input and output signals, and clocks and timers, all on a single silicon chip. In the same year Texas Instruments, Inc. produced their TMS 9940, which is also an extremely powerful, one-chip microcomputer.

Microprocessors and microcomputers are being used in applications for which computers were never before considered. These include hand calculators, programmable tennis games, learning aids, computerized supermarket checkout counters, automobile fuel-saving devices, and a host of others. It is too early, of course, to predict the ultimate impact of these new components on our lives, but the microcomputer revolution may easily be the greatest of all the computer revolutions.

2

NUMBER SYSTEMS

A numeration system, or number system, is a way of counting and of naming numbers. Primitive humans undoubtedly counted on their fingers and had names or symbols representing one finger, two fingers, a hand of fingers, etc. For example, the Roman numeral I (one) probably evolved from a mark representing one finger, V (five) was perhaps a symbol for a hand, and X (ten) looks very much like two V's combined into one symbol.

In most early systems, such as those of the ancient Chinese, Egyptians, and Romans, numbers were formed by repeating basic symbols and adding their values. For example, the Roman numeral XXI (twenty-one) is evidently two tens and a one combined into a single number. The Romans added XI (eleven) to XXI by simply forming XXXII (thirty-two). Of course, they also had refinements, such as IX (nine), in which the smaller number is written first and is subtracted rather than added.

A revolutionary step occurred with the realization that it requires only a few basic numbers, or *digits*, inserted in different *places*, or *positions*, to form numbers of any size. This was first done around 300 B.C. by Hindu mathematicians, who used ten basic numbers. Their system was later adopted by the Arabs, who introduced it into Europe around 1200 A.D. The Hindu–Arab system, which is regarded by many mathematicians as one of the world's greatest inventions, is, of course, the forerunner of our modern decimal number system. (The word *decimal* comes from the Latin word *decem*, meaning *ten*.)

There have been systems developed throughout the world based on numbers other than ten. For example, some evidence of systems based on 12 and 60 still survives in our units of 12 inches in a foot and 60 minutes in an hour. Because we have ten fingers, however, the most natural and thus the most widely used system is the decimal system. Even the word *digit*, used for the basic numbers, comes from the Latin word *digitus*, meaning *finger*.

In this chapter we consider the general idea of number systems, with particular emphasis on two special cases, the decimal system, based on ten, and the *binary* system, based on two. The former is, as we have said, the familiar system that each of us uses everyday, and the latter is the system used in the calculations done by a digital computer. In a later chapter we shall consider systems based on other numbers.

2.1 Decimal System

All number systems modeled on the Hindu–Arab system are based on an ordered set of numbers, or *digits*. The total number of digits used in a system is called the *base*, or *radix*, of the system. For example, the *decimal* number system has a base of ten and uses the ten digits 0, 1, 2, 3, 4, 5, 6, 7, 8, and 9.

The beauty of the Hindu–Arab system, and the feature that has won it universal acceptance, is its use of *positional notation*, or *place value*. Positional notation enables any number, no matter how large or how small, to be expressed by means of the basic digits in this system. Because the position, or place, of the digit in the number determines the value, or *weight*, assigned to the digit, a relatively small number of digits is sufficient to represent any number.

As an example, let us consider the decimal number

$$N = 235 \tag{2.1}$$

which, as we all know, means

$$N = 200 + 30 + 5$$

In other words, we have

$$N = 2 \times 100 + 3 \times 10 + 5$$

or

$$N = 2 \times 10^2 + 3 \times 10^1 + 5 \times 10^0 \tag{2.2}$$

Evidently, the position of each digit determines the value it represents in the sum given by the number N. In Eq. (2.1), for example, the digit 2 represents 200, 3 represents 30, and 5 is 5.

Stated another way, the position of each digit in a decimal number determines the power of 10 by which the digit is multiplied to yield the value it represents in the number. In Eq. (2.2), 2 is in the *hundreds* (10^2) place, 3 is in the *tens* (10^1) place, and 5 is in the *units* (10^0) place.

The example we have just considered and our familiarity with the decimal system enables us to note a few important characteristics of number systems in general. Some of these are as follows:

1. The number of digits used in the system is equal to the base.
2. The largest digit is one less than the base.
3. In the sum represented by a number each digit is multiplied by the base raised to the appropriate power for the digit position.

For example, in the decimal system the number of digits is ten, which is equal to the base, the largest digit is 9, one less than the base, and the units digit is multiplied by 1 (10^0), the tens digit by 10 (10^1), the hundreds digit by 100 (10^2), etc.

Decimal numbers with fractional parts, such as

$$M = 764.28 \tag{2.3}$$

are handled in exactly the same manner as whole numbers, such as Eq. (2.1). For example, M may be written

$$\begin{aligned} M &= 764.28 \\ &= 700 + 60 + 4 + 0.2 + 0.08 \end{aligned}$$

or

$$M = 7 \times 10^2 + 6 \times 10^1 + 4 \times 10^0 + 2 \times 10^{-1} + 8 \times 10^{-2}$$

In this case, in addition to the units, tens, and hundreds places, we have the *tenths* (10^{-1}) place and the *hundredths* (10^{-2}) place. The *decimal point* separates the *integer* part 764 from the *fractional* part 0.28.

In general, if the base or radix of the number system is b, then a number may be expressed in positional notation as

$$N = (a_n a_{n-1} \ldots a_1 a_0 . a_{-1} \ldots a_{-m})_b \tag{2.4}$$

The point shown, which is called the *radix point*, corresponds to the decimal point of the decimal system. For example, in Eq. (2.3) we have $b = 10$, $a_2 = 7$, $a_1 = 6$, $a_0 = 4$, $a_{-1} = 2$, and $a_{-2} = 8$. In this case $n = 2$ and $m = 2$.

The representation that exhibits the various weights assigned to the digits, by the definition of positional notation, is

$$N = a_n b^n + a_{n-1} b^{n-1} + \cdots + a_1 b^1 + a_0 b^0 + a_{-1} b^{-1} + \cdots + a_{-m} b^{-m} \tag{2.5}$$

This, of course, is the number in Eq. (2.4) expressed in decimal form. Because of its form, this representation is sometimes called the *polynomial representation.*

As an example, suppose we have the number in base 5 given by

$$N = 1230.41_5$$

[We will not use the parentheses appearing in Eq. (2.4) except when they are needed for clarity. Moreover, we will omit the subscripts when the base is understood]. By the definition of positional notation, we have the decimal form

$$N = 1 \times 5^3 + 2 \times 5^2 + 3 \times 5^1 + 0 \times 5^0 + 4 \times 5^{-1} + 1 \times 5^{-2}$$

which is

$$N = 125 + 50 + 15 + 0 + 0.8 + 0.04$$

or

$$N = 190.84$$

That is,

$$1230.41_5 = 190.84_{10}$$

The decimal system, as we have said, is by far the most popular system in current use. Other systems that are very important because of their use in connection with digital computers are the *binary* system (base 2), the *octal* system (base 8), and the *hexadecimal* system (base 16). We shall consider the first of these in the next section, and the last two will be discussed in Chap. 8.

EXERCISES

2.1.1 Convert the following binary numbers to decimal numbers:

(a) 1011

(b) 11011

(c) 101110

Ans: (a) 11; (b) 27; (c) 46

2.1.2 Convert the following base 6 numbers to decimal numbers:

(a) 125

(b) 2054

(c) 12450

Ans: (a) 53; (b) 466; (c) 1902

2.1.3 Convert the following *octal* numbers (base 8) to decimal numbers:

(a) 231

(b) 3107

(c) 13765

Ans: (a) 153; (b) 1607; (c) 6133

2.2 Binary Numbers

The simplest number system using positional notation is the *binary*, or base 2, system. This system was developed during the late 1600s by the German mathematician and philosopher Gottfried Wilhelm Leibniz (1646–1716), but no practical use for the system was found until the 1940s, when computers were developed. The binary system is well suited for applications such as internal computer operations, where electrical signals are used to represent digits. In the binary system there are two digits, 0 and 1, and thus only two signal levels are required. In the case of decimal numbers, ten signal levels would be necessary.

The digits in the binary system are called *bits*, which is a contraction of "binary digits." As in the decimal system, the position of the bit in a binary number determines the value it represents. The only difference is that powers of 2 are used instead of powers of 10. For example, to convert the binary number

$$N = 1011.1101_2 \tag{2.6}$$

to a decimal number, we note that

$$\begin{aligned} N &= 1 \times 2^3 + 0 \times 2^2 + 1 \times 2^1 + 1 \times 2^0 \\ &\quad + 1 \times 2^{-1} + 1 \times 2^{-2} + 0 \times 2^{-3} + 1 \times 2^{-4} \\ &= 8 + 0 + 2 + 1 + 0.5 + 0.25 + 0 + 0.0625 \\ &= 11.8125_{10} \end{aligned} \tag{2.7}$$

The *binary point*, which separates the integer part and the fractional part of N in Eq. (2.6), corresponds to the decimal point in the base 10 system.

Table 2.1 shows the decimal numbers 0 through 15 and their binary equivalents.

Table 2.1 *Decimal and binary equivalents*

Decimal	*Binary*	*Decimal*	*Binary*
	$2^3\ 2^2\ 2^1\ 2^0$		$2^3\ 2^2\ 2^1\ 2^0$
0	0 0 0 0	8	1 0 0 0
1	0 0 0 1	9	1 0 0 1
2	0 0 1 0	10	1 0 1 0
3	0 0 1 1	11	1 0 1 1
4	0 1 0 0	12	1 1 0 0
5	0 1 0 1	13	1 1 0 1
6	0 1 1 0	14	1 1 1 0
7	0 1 1 1	15	1 1 1 1

The *least significant bit* (LSB) in a binary integer is the units bit (2^0), or rightmost bit, and the *most significant bit* (MSB) is the leftmost bit. That is, the bit that could be changed (from 0 to 1 or vice versa) with the least effect on the number is the LSB, whereas the bit whose change has the most effect is the MSB. For example, in the number 1010_2 (equivalent to 10 in decimal), the LSB is 0, the units bit, and the MSB is 1, the 2^3 or eights bit. Changing the LSB yields 1011_2 (the decimal number 11), and changing the MSB yields 0010_2 (the decimal number 2).

We may change a binary number to its decimal equivalent by means of the definition of the positional notation, as was done in the steps leading to Eq. (2.7). We may also reverse the process to obtain the binary equivalent of a decimal number. In giving an example of the latter procedure we will use the powers of 2 listed in Tab. 2.2.

Table 2.2 *Powers of 2*

i	2^i
0	1
1	2
2	4
3	8
4	16
5	32
6	64
7	128
8	256
9	512
10	1024

For example, let us convert the decimal number 405 to binary. From Tab. 2.2 we see that the highest power of 2 contained in 405 is $2^8 = 256$. Therefore we write

$$405 = 256 + 149$$

The highest power of 2 in 149 is $2^7 = 128$, and thus we have

$$149 = 128 + 21$$

Continuing, we write

$$21 = 16 + 5$$
$$5 = 4 + 1$$

Summing up these results, we have

$$405 = 256 + 128 + 16 + 4 + 1$$

or

$$405 = 1 \times 2^8 + 1 \times 2^7 + 0 \times 2^6 + 0 \times 2^5 + 1 \times 2^4 + 0 \times 2^3 + 1 \times 2^2 + 0 \times 2^1 + 1 \times 2^0$$

Therefore we have

$$405_{10} = 110010101_2$$

This method works very well for relatively small numbers, but it could become quite unwieldy for large numbers. In the next section we consider other methods of conversion that are applicable to more general cases.

EXERCISES

2.2.1 Convert the following decimal numbers to binary numbers. Check by converting back to decimal.

(a) 19

(b) 260

(c) 500

Ans: (a) 10011; (b) 100000100; (c) 111110100

2.2.2 **(a)** Note that the fractional decimal number 0.625 may be written as the sum $0.5 + 0.125 = 2^{-1} + 2^{-3}$. Use this fact to find its binary equivalent.

(b) Find the binary equivalent of 0.4375_{10}.

Ans: (a) 0.101; (b) 0.0111

2.3 Number Conversion

In Sec. 2.1 we saw how the definition of placement position, as displayed in the polynomial representation of Eq. (2.5), could be used to convert to decimal form a number given in any other base. If we know the digits a_n, a_{n-1}, etc. and the base b, then the number N is the decimal equivalent. We may also use this system in reverse, as explained in the previous section, to convert a decimal number to a binary number. This is done through discovering the binary digits by successively removing the various powers of 2 contained in the decimal number. Both of these procedures work well for relatively small numbers but become cumbersome as the numbers increase.

In this section we present alternative methods of converting from decimal to binary and vice versa, which are applicable to both large and small numbers. We will not prove that these methods work in the general case, but for the interested reader, proofs are outlined in the problems at the end of the chapter. (See Probs. 2.14–2.19.)

Let us begin by separating the number N to be converted into its integer and fractional parts, which we denote, respectively, by N_I and N_F. That is,

$$N = N_I + N_F \tag{2.8}$$

The radix point in N separates the two parts, of course. For example, if we have

$$N = 13.8125_{10} \tag{2.9}$$

then $N_I = 13$ and $N_F = 0.8125$, and evidently Eq. (2.8) holds. If we wish to convert N to a binary number, we may convert the two parts separately and combine them as in Eq. (2.8). The result will be a binary integer and a binary fractional part separated by a binary point.

A very popular method of converting a decimal integer to binary involves successive division of the decimal integer by the base 2. That is, each quotient obtained is divided again to obtain another quotient, the process continuing until a zero quotient is obtained. The remainders at each stage are the binary digits in the order from the LSB to the MSB.

To illustrate the procedure we convert to binary the integer part 13 of the decimal number in Eq. (2.9). First we divide 13 by 2, which yields a quotient of 6 with a remainder of 1. Next, dividing 6 by 2, we obtain 3 with remainder 0. Then 3 divided by 2 yields 1 with remainder 1, and finally, 1 divided by 2 is 0 with remainder 1. The process terminates at this point, because the zero quotient is obtained. We may tabulate the steps more compactly as follows:

	Remainders
$13 \div 2 = 6$	1 (LSB)
$6 \div 2 = 3$	0
$3 \div 2 = 1$	1
$1 \div 2 = 0$	1 (MSB)

Therefore the binary number is 1101_2. The MSB is the last remainder, the second most significant bit is the next to the last remainder, etc., and the LSB is the first remainder. Therefore, we have,

$$13_{10} = 1101_2 \tag{2.10}$$

This result may be checked by converting the binary number back to decimal.

As another example, let us convert 24_{10} to binary. The procedure is as follows:

	Remainders
$24 \div 2 = 12$	0 (LSB)
$12 \div 2 = 6$	0
$6 \div 2 = 3$	0
$3 \div 2 = 1$	1
$1 \div 2 = 0$	1 (MSB)

Therefore, we have,

$$24_{10} = 11000_2$$

To convert a fractional decimal number to a binary number, successive multiplications by the base 2 are performed. In this case, instead of remainders of 0 or 1 we have *overflows* of 0 or 1. That is, 2 times a decimal fraction is either a decimal fraction (an overflow of 0) or 1 plus a decimal fraction (an overflow of 1). The overflows are the binary digits, and they occur in the order from MSB to LSB. (In the case of fractional numbers, the leftmost bit after the binary point is the MSB and the rightmost bit is the LSB.)

We illustrate the process with the decimal part 0.8125 of Eq. (2.9):

	Overflows	
$0.8125 \times 2 = 1.6250 = 0.6250$ plus	1	(MSB)
$0.6250 \times 2 = 1.2500 = 0.2500$	1	
$0.2500 \times 2 = 0.5000 = 0.5000$	0	
$0.5000 \times 2 = 1.0000 = 0.0000$	1	(LSB)

Thus we have

$$0.8125_{10} = 0.1101_2 \tag{2.11}$$

The procedure terminates, as in this case, when the multiplication results in an overflow plus 0. This will not always occur, since the binary fraction may be one that repeats itself indefinitely. For example, in the case of the decimal number 0.1 we have

	Overflows
$0.1 \times 2 = 0.2 = 0.2$ plus	0
$0.2 \times 2 = 0.4 = 0.4$	0
$0.4 \times 2 = 0.8 = 0.8$	0
$0.8 \times 2 = 1.6 = 0.6$	1
$0.6 \times 2 = 1.2 = 0.2$	1

The multiplication process has produced 0.2, which is the product we began with. To continue, therefore, will produce the same overflows and thus a repeating binary fraction. The next step produces 0.4 plus an overflow of 0, which is the beginning of the repeated pattern 0011. Therefore the result is

$$0.1_{10} = (0.0001100110011\ldots)_2$$

To obtain the binary equivalent of the decimal number in Eq. (2.9), we combine its equivalent integer and fractional parts, obtained in Eqs. (2.10) and (2.11). The result is

$$13.8125_{10} = 1101.1101_2$$

Let us now consider a method of converting a binary integer to a decimal.

Starting with the leftmost bit (MSB) and working from left to right, multiply the first digit by the base 2, add the product to the next digit, multiply the resulting sum by 2, add the product to the next digit, and so on, until the rightmost bit (LSB) is added. The last sum obtained is the decimal number.

As an example, let us convert the binary number 1101 to decimal. Multiplying the first (leftmost) bit by 2 and adding the result to the second, we have

$$2 \times 1 = 2, \qquad 2 + 1 = 3$$

Multiplying the sum 3 by 2 and adding to the next bit results in

$$2 \times 3 = 6, \qquad 6 + 0 = 6$$

Multiplying the sum 6 by 2 and adding to the next (rightmost) bit yields

$$2 \times 6 = 12, \qquad 12 + 1 = 13$$

Therefore the decimal equivalent is 13.

The work may be put in the following compact form:

$$\begin{array}{cccc} 1 & 1 & 0 & 1 \\ & 2 & 6 & 12 \\ \hline 1 & 3 & 6 & 13 \end{array}$$

The first row consists of the bits in the binary number to be converted. The leftmost bit is brought down to the third row, multiplied by 2 and added (in the second row) to the next bit, etc. The decimal equivalent (13) is the rightmost sum. (Readers who have had a second course in high school algebra may recognize this process as *synthetic division.*)

As another example, let us consider

$$\begin{array}{cccccc} 1 & 1 & 0 & 0 & 1 & 1 \\ & 2 & 6 & 12 & 24 & 50 \\ \hline 1 & 3 & 6 & 12 & 25 & 51 \end{array}$$

Therefore we have

$$110011_2 = 51_{10}$$

A method of converting binary fractions to decimal will be outlined in Probs. 2.18 and 2.19. It consists in dividing the LSB by 2 and adding the result to the next bit (moving leftward). This result is then divided by 2 and added to the next bit, and so on, until the MSB is added and the result divided by 2. The number at this point is the decimal fraction.

For example, consider the binary fraction 0.1101. The steps are as follows:

$$Q_1 = \frac{1}{2} = 0.5$$

$$Q_2 = \frac{0.5 + 0}{2} = 0.25$$

$$Q_3 = \frac{0.25 + 1}{2} = 0.625$$

$$Q_4 = N = \frac{0.625 + 1}{2} = 0.8125$$

Thus we have

$$0.1101_2 = 0.8125_{10}$$

EXERCISES

2.3.1 Using the method of this section, convert the following decimal numbers to binary numbers:

(a) 14

(b) 26

(c) 53

Ans: (a) 1110; (b) 11010; (c) 110101

2.3.2 Repeat Ex. 2.3.1 for the decimal numbers

(a) 11.25

(b) 18.71875

(c) 3.6

Ans: (a) 1011.01; (b) 10010.10111; (c) 11.10011001100...

2.3.3 Using the method of this section, convert the following binary numbers to decimal numbers:

(a) 10111

(b) 110111

(c) 1010011

Ans: (a) 23; (b) 55; (c) 83

2.3.4 Using the method of this section, convert the following binary numbers to decimals:

(a) 0.101

(b) 0.1011

(c) 0.10111

Ans: (a) 0.625; (b) 0.6875; (c) 0.71875

2.4 Binary Addition and Subtraction

Arithmetic operations on binary numbers are performed in exactly the same way as on decimal numbers, the only difference being in the bases and the basic digits. In decimal addition, for example, 6 and 4 added in a column result in 0 with a *carry* of 1 to the next column. In the same manner, in a column of binary numbers, $1 + 1 = 0$ with a carry of 1 to the next column. Thus in binary we have

$$1 + 1 = 10$$

which is one entry in a binary addition table. (The binary number 10 is, of course, the decimal 2.) The other entries are

$$0 + 0 = 0$$
$$0 + 1 = 1 + 0 = 1$$

To illustrate binary addition let us find the sum of the binary numbers 1101 and 1001. As in decimal addition we place the numbers in columns:

$$\begin{array}{r} 1\ 1\ 0\ 1 \\ +1\ 0\ 0\ 1 \\ \hline \end{array}$$

The addition is then performed with appropriate carries exactly as in the decimal system:

$$\begin{array}{rl} 1\quad\ \ 1\ \ \ \ & \text{Carries} \\ 1\ 1\ 0\ 1 & \\ +\ \ 1\ 0\ 0\ 1 & \\ \hline 1\ 0\ 1\ 1\ 0 & \end{array}$$

The two ones at the top are the carries resulting from adding $1 + 1 = 0$, carry 1, in the first (rightmost) column and in the fourth (leftmost) column.

As another example consider

$$\begin{array}{rl} 1\ 1\ 1\ \ \ \ & \text{Carries} \\ 1\ 1\ 0\ 1\ 0\ 1 & \\ +\qquad 1\ 1\ 1 & \\ \hline 1\ 1\ 1\ 1\ 0\ 0 & \end{array}$$

In this case the third column has two 1 bits plus a carry 1 bit which have to be added. The result is

$$\begin{aligned} 1 + 1 + 1 &= (1 + 1) + 1 \\ &= 10 + 1 \\ &= 11 \end{aligned}$$

In other words, the sum is 1 with a carry of 1, as indicated.

Binary numbers with fractional parts are handled in the same way as integers. For example, consider

$$\begin{array}{rr} & 1101.11 \\ + & 1010.10 \\ \hline & 11000.01 \end{array}$$

Columns with more than two digits may generate several carry bits. These are all added to the next column as in decimal addition. In some cases it may be easier to add several numbers by adding them in pairs to avoid more than one carry at a time.

To subtract two binary numbers we make use of the rules

$$0 - 0 = 0$$
$$1 - 0 = 1$$
$$1 - 1 = 0$$
$$10 - 1 = 1$$

The last rule illustrates the use of *borrowing*, which is done in the same way as in decimal subtraction. For example, if a column of decimal numbers is $5 - 6$, we borrow 1 from the next column and perform $15 - 6 = 9$. The borrowed 1 is subtracted from the next column. In the same manner, $0 - 1$ in binary requires borrowing 1 from the next column to produce $10 - 1$. Alternatively, we could say $0 - 1 = 1$, borrow 1.

An example that requires no borrowing is

$$\begin{array}{rr} & 11011 \\ - & 1001 \\ \hline & 10010 \end{array}$$

As another example let us subtract 1101 from 10110:

$$\begin{array}{rccccl} & 0 & 10 & 0 & 10 & \text{Borrows} \\ & 1 & 0\,1 & 1 & 0 & \\ - & & 1\,1 & 1 & 1 & \\ \hline & & 1\,0 & 0 & 1 & \end{array}$$

The top row indicates the borrows. In the first (rightmost) column we must borrow 1 from the second column, resulting in 10 in the first and 0 in the second column. Another such borrow is required in the fourth column.

Because of the borrowing procedure, subtraction in binary is more difficult to perform than addition. In a digital computer a number is subtracted by adding its *complement*, a concept we will consider in Chap. 10.

EXERCISES

2.4.1 Perform the following binary number additions. Check by changing to decimal numbers.

(a) 1101 + 1011

(b) 110111 + 11001

(c) 1110 + 1001 + 11011

Ans: (a) 11000; (b) 1010000; (c) 110010

2.4.2 Perform the following binary number subtractions. Check by changing to decimal numbers. (Note that in (c) the first borrow requires a second borrow, which requires a third borrow from the 2^3 position.)

(a) 1101 − 1011

(b) 110110 − 11001

(c) 11000 − 1111

Ans: (a) 10; (b) 11101; (c) 1001

2.5 Binary Multiplication and Division

Binary multiplication is done in exactly the same manner as decimal multiplication. However, the work, is easier because of the simplicity of the binary system. To illustrate the procedure let us multiply the binary numbers 1101 and 101. The work may be put in the following compact form:

$$
\begin{array}{ccccccc}
 & & & 1 & 1 & 0 & 1 \\
 & & \times & & 1 & 0 & 1 \\
\cline{4-7}
 & & & 1 & 1 & 0 & 1 \\
 & & 0 & 0 & 0 & 0 & \\
 & 1 & 1 & 0 & 1 & & \\
\cline{1-7}
1 & 0 & 0 & 0 & 0 & 0 & 1
\end{array}
$$

The three *partial products* 1101, 0000, and 1101 are obtained and added as in decimal multiplication. Evidently, also as in decimal multiplication, we may omit the listing of a zero partial product and simply shift the next partial product one entry to the left:

$$
\begin{array}{ccccccc}
 & & & 1 & 1 & 0 & 1 \\
 & & \times & & 1 & 0 & 1 \\
\cline{4-7}
 & & & 1 & 1 & 0 & 1 \\
 & 1 & 1 & 0 & 1 & 0 & \\
\cline{1-7}
1 & 0 & 0 & 0 & 0 & 0 & 1
\end{array}
$$

As a check, the result we have just obtained,

$$1101 \times 101 = 1000001$$

when converted to decimal gives us

$$13 \times 5 = 65$$

Likewise, division in binary is done in the same way as in decimal. Let us illustrate by dividing 110111 (55_{10}) by 101 (5_{10}):

```
          1 0 1 1
1 0 1 ⟌ 1 1 0 1 1 1
        1 0 1
        ---------
          1 1 1
          1 0 1
          -------
            1 0 1
            1 0 1
            -----
```

The result 1011 is, of course, 11_{10}. As in the decimal case, the procedure is a trial division followed by a subtraction, a trial division, a subtraction, etc. In binary the trials are easier to discover because there are only two possibilities, 0 and 1.

In the case of fractional numbers the process again is like in the decimal system. To illustrate, let us obtain the decimal result $1.625 \times 1.25 = 2.03125$ in binary:

```
      1.1 0 1
        1.0 1
      -------
      1 1 0 1
  1 1 0 1 0
  -----------
1 0.0 0 0 0 1
```

As in decimal multiplication, the number of places in the fractional part of the product is the sum of those in the multiplicand and the multiplier.

An example of division ($2.03125 \div 1.625 = 1.25$ in decimal) is

```
                      1.0 1
1.1 0 1‸ ⟌ 1 0.0 0 0‸0 1
           1 1 0 1
           -------------
           0 0 1 1 0 1
               1 1 0 1
           -------------
```

The binary point is found, as in decimal division, by moving the binary points of both the dividend and the divisor an equal number of places. Also, as in the decimal system, there may or may not be a remainder.

EXERCISES

2.5.1 Perform the following binary multiplications:

(a) 11011×110

(b) 1101.01×11.01

Ans: (a) 10100010; (b) 101011.0001

2.5.2 Perform the following binary divisions:

(a) $10101 \div 111$

(b) $10110 \div 1001$

(c) $100111.111 \div 101.1$

Ans: (a) 11; (b) 10 with remainder 100; (c) 111.01

PROBLEMS

2.1 Using the definition of positional notation, convert the following numbers to decimal:

(a) 110110_2

(b) 1100111_2

(c) 2012_3

(d) 215_6

(e) 317_8

2.2 Repeat Prob. 2.1 for the following binary numbers:

(a) 11001

(b) 11101

(c) 11110

(d) 101111

(e) 111110

(f) 1100111

2.3 Note that in Table 2.1 we have $11_2 = (2^2 - 1)_{10}$, $111_2 = (2^3 - 1)_{10}$, and $1111_2 = (2^4 - 1)_{10}$. These are special cases of the general result

$$111\ldots1_2 = (2^n - 1)_{10}$$

where n is the number of 1's in the binary number. Check this result by converting to decimal the binary numbers

(a) 11111

(b) 111111

(c) 11111111

2.4 Find the decimal equivalents of the numbers of Prob. 2.2 if

(a) the LSB is changed

(b) the MSB is changed

Compare these results to the decimal equivalents of the original numbers.

2.5 Convert the following binary numbers to decimal:

(a) 0.111

(b) 11.1001

(c) 110.11001

(d) 1001.10011

2.6 Convert the following decimal numbers to binary by using the definition of positional notation:

(a) 29

(b) 317

(c) 1125

(d) 2051

2.7 Solve Prob. 2.2 using the method of Sec. 2.3.

2.8 Solve Prob. 2.6 using the method of Sec. 2.3.

2.9 Using the method of Sec. 2.3, convert the following decimal numbers to binary:

(a) 0.78125

(b) 0.84375

(c) 23.4375

(d) 0.7

2.10 Convert

(a) 1320_4 to base 8

(b) 1212_3 to base 7

(c) 625_7 to base 5

(*Suggestion:* Convert to decimal first.)

2.11 Perform the following binary additions:

(a) $1101 + 1011$

(b) $1111 + 1011$

(c) $11111 + 1$

(d) 11011 + 1101

(e) 1011 + 111 + 10101

(f) 1111 + 111 + 11101

2.12 Perform the following binary additions:

(a) 1101.11 + 110.01

(b) 10.0111 + 1.1111

(c) 0.10101 + 10.10101

(d) 1.11111 + 0.11111

2.13 Perform the following binary subtractions:

(a) 1011 − 110

(b) 11010 − 10111

(c) 10000 − 1

(d) 1.1011 − 1.0101

(e) 1010.1101 − 101.101

2.14 Consider the decimal form

$$N = a_n b^n + a_{n-1} b^{n-1} + \cdots + a_2 b^2 + a_1 b + a_0$$

given in Eq. (2.5) of the base b integer

$$N = (a_n a_{n-1} \ldots a_2 a_1 a_0)_b$$

Show that $N/b = Q_1$ with remainder a_0, $Q_1/b = Q_2$, with remainder a_1, $Q_2/b = Q_3$ with remainder a_2, and so on, where $Q_1, Q_2, Q_3, \ldots$ are quotients that are integers. Thus the remainders are the digits in the base b form of N in the order from LSB to MSB. Let $b = 2$ to obtain the procedure in Sec. 2.3 for changing decimal numbers to binary numbers.

2.15 Consider the decimal form

$$N = a_{-1} b^{-1} + a_{-2} b^{-2} + a_{-3} b^{-3} + \cdots + a_{-m} b^{-m}$$

given in Eq. (2.5) of the fractional base b number

$$N = (0.a_{-1} a_{-2} a_{-3} \ldots a_{-m})_b$$

Show that $bN = P_1 + a_{-1}$, $bP_1 = P_2 + a_{-2}$, $bP_2 = P_3 + a_{-3}$, and so on, where $P_1, P_2, P_3, \ldots$ are products that are fractions and $a_{-1}, a_{-2}, a_{-3}, \ldots$ are the *overflows* which are the base b digits. Note that the digits occur in the order MSB to LSB and that if $b = 2$, we have the decimal to binary conversion procedure of Sec. 2.3. (*Suggestion:* Recall that each of the digits is less than b.)

2.16 In the decimal form given in Prob. 2.14 of a number N in base b, define the sums, $S_n, S_{n-1}, S_{n-2}, \ldots$ by

$$\begin{aligned} S_n &= a_n \\ S_{n-1} &= bS_n + a_{n-1} \\ S_{n-2} &= bS_{n-1} + a_{n-2} \\ S_{n-3} &= bS_{n-2} + a_{n-3} \end{aligned}$$

or in general,

$$S_{n-i} = bS_{n-i+1} + a_{n-i}; \quad i = 1, 2, 3, \ldots, n$$

From these results show that

$$\begin{aligned} S_{n-1} &= ba_n + a_{n-1} \\ S_{n-2} &= b^2a_n + ba_{n-1} + a_{n-2} \\ S_{n-3} &= b^3a_n + b^2a_{n-1} + ba_{n-2} + a_{n-3} \\ &\ldots \\ S_{n-i} &= b^ia_n + b^{i-1}a_{n-1} + \cdots + a_{n-i} \end{aligned}$$

and therefore, for $i = n$,

$$\begin{aligned} S_0 &= b^na_n + b^{n-1}a_{n-1} + \cdots + a_0 \\ &= N \end{aligned}$$

Finally, note that for $b = 2$, this procedure becomes that of Sec. 2.3 for converting binary integers to decimals.

2.17 Use the procedure of Prob. 2.16 to solve

(a) Prob. 2.1(c) with $b = 3$

(b) Prob. 2.1(d) with $b = 6$

(c) Prob. 2.1(e) with $b = 8$

(*Suggestion:* The procedure is identical to that described in Sec. 2.3 except that the multiplications are by b instead of by 2.)

2.18 Consider the fractional base b number

$$N = (0.a_{-1}a_{-2} \ldots a_{-m+1}a_{-m})_b$$

which in decimal form is

$$N = a_{-1}b^{-1} + a_{-2}b^{-2} + \cdots + a_{-m+1}b^{-m+1} + a_{-m}b^{-m}$$

Divide the least significant digit by b, obtaining the quotient

$$Q_1 = \frac{a_{-m}}{b}$$

Add to Q_1 the next digit (moving leftward) and divide the result by b, obtaining

$$Q_2 = \frac{a_{-m+1} + \dfrac{a_{-m}}{b}}{b} = \frac{ba_{-m+1} + a_{-m}}{b^2}$$

Add to Q_2 the next digit and divide the result by b, and continue the process. Show that in general we have

$$Q_i = \frac{b^{i-1}a_{-m+i-1} + b^{i-2}a_{-m+i-2} + \cdots + a_{-m}}{b^i}$$

When the most significant digit a_{-1} is added and the result is divided by b, the result is Q_m ($i = m$), or

$$Q_m = \frac{b^{m-1}a_{-1} + b^{m-2}a_{-2} + \cdots + a_{-m}}{b^n}$$
$$= N$$

2.19 Use the result of Prob. 2.18 with $b = 2$ to obtain

$$Q_1 = \frac{a_{-m}}{2}$$

$$Q_2 = \frac{Q_1 + a_{-m-1}}{2}$$

$$Q_3 = \frac{Q_2 + a_{-m-2}}{2}$$

$$\cdots$$

$$Q_m = N$$

By this procedure, convert to decimal the binary numbers

(a) 0.111

(b) 0.1001

(c) 0.11001

(d) 0.10011

2.20 Use the methods of Probs. 2.16 and 2.19 to solve Prob. 2.5.

2.21 Perform the following binary operations:

(a) 10011×101

(b) 101.11×1.011

(c) $1011111 \div 10011$

(d) $111.11101 \div 1.011$

3

BOOLEAN ALGEBRA

The binary digits 0 and 1, discussed in Chapter 2, are especially useful in describing theoretical or physical systems for which the associated quantities can take on exactly two values, or *states*. For example, in a system of logic a statement is either TRUE or FALSE, which could be represented mathematically by 1 and 0, respectively. Also, in an electrical system a light could be ON or OFF, or a switch could be OPEN or CLOSED. In a digital computer the numbers 1 and 0 could be represented by a HIGH and a LOW voltage, or by the PRESENCE and ABSENCE of a pulse. On a punched card we may have a HOLE or NO HOLE, and in a winner-take-all situation we have the states of ALL or NOTHING. The reader doubtless can think of many other two-state variables, such as MALE and FEMALE or GO and NO GO.

The mathematics of logic systems, known as *Boolean algebra*, was developed in the 1840s by the English mathematician George Boole (1815–1864). Boole's work, like Leibniz's on binary systems, was little noted for many years, but in the 1930s Claude E. Shannon, an American mathematician, made a major application of Boolean algebra to switching circuits. The algebra Shannon used is the *two-element* Boolean algebra, the special case that is applicable to two-state systems.

In this chapter we consider in some detail the two-state Boolean algebra, which we shall refer to simply as Boolean algebra. We will define Boolean variables and functions and show how they are used in switching circuits and logic. The ideas presented in this chapter will be the basis for the discussion of the electronic circuits called *gates*, which are considered in Chap. 4 and which constitute the basic hardware of digital computers.

3.1 Definitions

We define a *Boolean algebra* as a set of two elements, denoted 0 and 1, together with two operations + and · , called addition and multiplication, for which the following rules hold. The addition table is given by

$$\begin{aligned} 0 + 0 &= 0 \\ 0 + 1 &= 1 + 0 = 1 \\ 1 + 1 &= 1 \end{aligned} \tag{3.1}$$

and the multiplication table by

$$\begin{aligned} 0 \cdot 0 &= 0 \\ 0 \cdot 1 &= 1 \cdot 0 = 0 \\ 1 \cdot 1 &= 1 \end{aligned} \tag{3.2}$$

All of these arithmetic rules are exactly like those of ordinary addition and multiplication except for the addition rule in the last equation of (3.1). As we shall see, this rule is quite as plausible as the others, but it may help at this stage to illustrate it with an example. Suppose we have a winner-take-all system in which the elements 1 and 0 are ALL and NOTHING, respectively. Then it is certainly reasonable to write, by Eqs. (3.1),

$$\begin{aligned} \text{NOTHING} + \text{NOTHING} &= \text{NOTHING} \\ \text{NOTHING} + \text{ALL} &= \text{ALL} + \text{NOTHING} = \text{ALL} \\ \text{ALL} + \text{ALL} &= \text{ALL} \end{aligned}$$

A *Boolean variable* is a quantity such as A, B, C, etc., which may take on the values 0 or 1, and which satisfies the rules of Eqs. (3.1) and (3.2). A *Boolean function* is a function of Boolean variables, such as

$$f(A, B) = A + B \tag{3.3}$$

The only operations allowed in constructing a Boolean function are, of course, Boolean operations. (Besides addition and multiplication, there is a third Boolean operation, that of *complementation*, which we consider in Sec. 3.4.)

Since each Boolean variable in a function f may assume two values, there are often many *combinations* of values that may be assigned at any given time to the variables of f. For example, there are four sets of values, or *inputs*, that may be assigned to the variables in $f(A, B)$ given by Eq. (3.3). Since there are two variables,

A and B, and each may assume two values, 0 and 1, there are $2^2 = 4$ ways to assign values to *both* A and B. These ways are the inputs $A = B = 0$; $A = 0, B = 1$; $A = 1, B = 0$; and $A = B = 1$. For these cases we have, by Eqs. (3.1),

$$f(0, 0) = 0 + 0 = 0$$
$$f(0, 1) = 0 + 1 = 1$$
$$f(1, 0) = 1 + 0 = 1$$
$$f(1, 1) = 1 + 1 = 1$$

We may tabulate these results as shown in Tab. 3.1. Such a table, which displays all the combinations the variables of a function may assume, as well as the corresponding values of the function, is called a *truth table* for the function. In this example the truth table is the addition table of Eqs. (3.1).

Table 3.1 *Truth table for A + B*

A	B	$A + B$
0	0	0
0	1	1
1	0	1
1	1	1

As another example, the multiplication table of Eqs. (3.2) is the truth table for

$$f(A, B) = AB \tag{3.4}$$

(We use the shorthand notation AB for $A \cdot B$.) The resulting truth table is Tab. 3.2.

Table 3.2 *Truth table for AB*

A	B	AB
0	0	0
0	1	0
1	0	0
1	1	1

In constructing a truth table for any Boolean function of n variables, we note that since each variable can assume two values, there are 2^n possible inputs (ways of assigning 0 and 1 to the n variables). That is, the first variable may be assigned two ways, after which the second may be assigned two ways, and so on. The product of these n 2's is $2 \cdot 2 \cdot 2 \cdots 2$ (n factors), or 2^n. A systematic way of enumerating all the possible inputs is to list them as base 2 integers. For example, in the truth tables of Tabs. 3.1 and 3.2, if we think of the numbers listed under A and B as a column of two-digit binary numbers, then their decimal equivalents are simply 0, 1, 2, and 3.

As another example let us consider the Boolean function

$$f = AB + C$$

There are three variables and thus $2^3 = 8$ inputs. Listing the inputs as the binary equivalents of the decimals 0, 1, 2, . . . , 7 readily yields all the possibilities. The truth table is Tab. 3.3, where for convenience in computing f, an extra column AB is shown.

Table 3.3 *Truth table for AB + C*

A	B	C	AB	$AB + C$
0	0	0	0	0
0	0	1	0	1
0	1	0	0	0
0	1	1	0	1
1	0	0	0	0
1	0	1	0	1
1	1	0	1	1
1	1	1	1	1

EXERCISES

3.1.1 Find the value of the Boolean function

$$f = A + AB$$

and thus construct its truth table, for the inputs A, B given by

(a) 0, 0

(b) 0, 1

(c) 1, 0

(d) 1, 1

Ans: (a) 0; (b) 0; (c) 1; (d) 1

3.1.2 Repeat Ex. 3.1.1 for the function

$$g = A(A + B)$$

(Note that the truth tables for f in Ex. 3.1.1 and g here are identical, and indeed, every value of f or g is the same as the value used for A. As we shall see later, this means that $f = g = A$.)

Ans: (a) 0; (b) 0; (c) 1; (d) 1

3.2 Perfect Induction

Two Boolean functions f and g are said to be equal, that is, $f = g$, if f and g assume the same value for any set of inputs. This is the same definition that is used for equality of functions in ordinary mathematics, but in the latter case there is an infinite number of values to be checked, whereas in Boolean algebra each variable can take on only two values, 0 and 1. Thus the number of cases (2^n if there are n variables) is finite, and theoretically all of them could be checked.

The method of proving or disproving that two Boolean functions are equivalent by exhausting all the possibilities of the input variable combinations is called *perfect induction*. The most compact and direct way to use perfect induction is to compile the truth tables for the two functions. If the tables are identical, then the two functions are equal, and conversely, if the tables differ at all, then the functions are not equal.

As an example, let us consider the functions

$$f = A + AB$$

and

$$g = A(A + B)$$

given earlier, along with their truth tables, in Exs. 3.1.1 and 3.1.2. Combining the two truth tables into Tab. 3.4, we see that the f and g columns are identical. Indeed, these columns are also identical to the A column. Therefore, not only does $f = g$, but $f = g = A$.

Table 3.4 *Truth tables for f and g*

A	B	AB	$f = A + AB$	$A + B$	$g = A(A + B)$
0	0	0	0	0	0
0	1	0	0	1	0
1	0	0	1	1	1
1	1	1	1	1	1

As another illustration, suppose we wish to prove the statement

$$(A + B)(A + C) = A + BC \tag{3.5}$$

The truth table is shown in Tab. 3.5, where we see that the column $(A + B)(B + C)$ is identical to the column $A + BC$. Therefore, by perfect induction, Eq. (3.5) is true in general.

Table 3.5 *Truth table illustrating Eq. (3.5)*

A	B	C	$A+B$	$A+C$	$(A+B)(A+C)$	BC	$A+BC$
0	0	0	0	0	0	0	0
0	0	1	0	1	0	0	0
0	1	0	1	0	0	0	0
0	1	1	1	1	1	1	1
1	0	0	1	1	1	0	1
1	0	1	1	1	1	0	1
1	1	0	1	1	1	0	1
1	1	1	1	1	1	1	1

EXERCISE

3.2.1 Show by perfect induction that the following statements are true in general:

(a) $A + 1 = 1$

(b) $A + A = A$

(c) $A(B + C) = AB + AC$

(d) $A(A + BC) = A$

3.3 The AND and OR Operations

In this section we compare the Boolean algebra of 0 and 1, having the operations of addition and multiplication, with a system of logic, having operations that we shall call AND and OR. As we shall see, both systems are essentially identical, and thus we may represent the logic system by a simple system of numbers using rules of arithmetic.

Suppose A and B represent statements in logic that are either true (T) or false (F). There are four possibilities that can arise: (1) F, F (A and B both false), (2) F, T (A false, B true), (3) T, F (A true, B false), and (4) T, T (A and B both true). Let us now consider in each of these cases whether the statement A OR B is true (T) or false (F). Certainly in case (1) A OR B is a false statement, since it is not true that either A is true or B is true. In the other cases, however, A OR B is a true statement, since obviously either A is true or B is true [or both, as in case (4)]. These results may be compiled in a truth table, such as Tab. 3.6.

Next let us consider the statement A AND B for each of the four possible cases. Evidently A AND B is true only if both A and B are true statements. (Both cannot be true if either one is false or if both are false.) The truth table for this statement will have T as an entry in case (4) and F entries elsewhere. The result is shown in Tab. 3.7.

Table 3.6 *Truth table for A OR B*

A	B	A OR B
F	F	F
F	T	T
T	F	T
T	T	T

Table 3.7 *Truth table for A AND B*

A	B	A AND B
F	F	F
F	T	F
T	F	F
T	T	T

Let us compare the truth tables for the OR and AND operations with the truth tables for addition and multiplication, given previously in Tabs. 3.1 and 3.2. If we represent F by 0 and T by 1, Tabs. 3.1 and 3.6, the truth tables for the addition and OR operations, are identical. Similarly, Tabs. 3.2 and 3.7, the truth tables for the multiplication and AND operations, are identical. Therefore we may use the OR and addition operations, and the AND and multiplication operations, interchangeably.

Symbols in common usage for the OR operation are illustrated in the expressions

$$A + B = A \cup B = A \vee B \tag{3.6}$$

We will use the first notation exclusively and read it A plus B or A OR B, depending on the application. In the case of the AND operation, common expressions are

$$A \cdot B = AB = A \cap B = A \wedge B \tag{3.7}$$

We will use the first two notations and say A times B or A AND B, as is appropriate.

As we have seen in this section, there is a close analogy between the Boolean numbers 1 and 0 and the logic states TRUE and FALSE. Indeed, if the operations OR and AND are read as addition and multiplication, the logic system and the Boolean algebra are identical. For this reason, Boolean variables are sometimes called *logic variables*.

As an example, let us consider the statement, "Both Allen and Bob were at the game or else Charlie was mistaken." To put this statement in symbolic form, let A represent the statement, "Allen was at the game," let B denote "Bob was at the game," and let C denote "Charlie was mistaken." Then if D denotes the original statement, we may write

$$D = (A \text{ AND } B) \text{ OR } C$$

or

$$D = AB + C$$

The truth table for this function is Tab. 3.3, given earlier in Sec. 3.1. If we interpret 1 as true and 0 as false, the right column of the truth table shows clearly the conditions under which D is a true or a false statement. For instance, from the fourth row, we see that D is true when A is false, B is true, and C is true. That is, Allen was not at the game, Bob was at the game, and Charlie was mistaken. (He thought he saw both Allen and Bob there, but he did not.)

EXERCISES

3.3.1 Consider the logic statement D: Both A and B are true or else both B and C are true.

(a) Express D as a Boolean function, and determine whether D is T (true) or F (false) when A, B, and C are, respectively,

(b) T, F, T

(c) F, T, T

(d) T, T, F

Ans: (a) $AB + BC$; (b) F; (c) T; (d) T

3.3.2 Given the logic function

$$f = (AB + D)(B + CD) + ABC$$

Determine whether f is T (true) or F (false) when A, B, C, and D are, respectively,

(a) T, F, T, F

(b) T, T, F, F

(c) F, T, F, T

(d) F, F, T, T

Ans: (a) F; (b) T; (c) T; (d) T

3.4 Complementation and Duality

Thus far we have considered two Boolean operations, addition and multiplication, or equivalently, OR and AND. There is one other Boolean operation known as the *complementation,* or NOT, operation, which when applied to a Boolean variable A, changes it to NOT A. That is, since A can take on only two values, when

A has one value (0 or 1), then NOT A has the other value (1 or 0). Common symbols for the complement of A (NOT A) are

$$\text{NOT } A = \bar{A} = A' = \sim A \tag{3.8}$$

We shall use the symbol $\bar{A}$ exclusively. For example, we may write

$$\text{NOT } (A + B) = (\overline{A + B}) = \overline{A + B}$$

in which case, if $A = 0$ and $B = 1$, we have

$$\overline{0 + 1} = \bar{1} = 0$$

The truth table for $\bar{A}$ is Tab. 3.8. Its entries are, of course,

$$\bar{0} = 1, \qquad \bar{1} = 0 \tag{3.9}$$

Table 3.8 *Truth table for NOT A*

A	$\bar{A}$
0	1
1	0

Application of the NOT operation to $\bar{A}$, the second column of Tab. 3.8, yields the first column, A. Thus the complement of $\bar{A}$, which we denote $(\overline{\bar{A}})$ or $\bar{\bar{A}}$, is A. That is,

$$\bar{\bar{A}} = A \tag{3.10}$$

In the case of logic variables, if A represents a statement, then $\bar{A}$ is the complementary statement. If A is true ($A = 1$), then $\bar{A}$ is false ($\bar{A} = 0$), and vice versa. For example, if A is the statement "Allen was at the game," used in the previous section, then $\bar{A}$ is the statement "Allen was not at the game."

The AND, OR, and NOT operations are the basic Boolean operations, and they may be used to express any Boolean function or any logic statement. For example, consider the statement f: "You are right and he is wrong, or he is right and you are wrong." Letting A represent "You are right" and B represent "He is right," we may express f as

$$f = A\bar{B} + B\bar{A}$$

The truth table for f is Tab. 3.9, from which we see, for instance, that $f = 0$ (f is false) when $A = B = 1$; that is, when both A and B are true.

Table 3.9 *Truth table for $A\bar{B} + B\bar{A}$*

A	B	$\bar{A}$	$\bar{B}$	$A\bar{B}$	$B\bar{A}$	$A\bar{B} + B\bar{A}$
0	0	1	1	0	0	0
0	1	1	0	0	1	1
1	0	0	1	1	0	1
1	1	0	0	0	0	0

Now that we have defined the three basic operations of AND, OR, and NOT, let us summarize the basic rules of Boolean algebra, as shown in Tab. 3.10. These are sometimes called the *postulates* of Boolean algebra.

Table 3.10 *Postulates of Boolean algebra*

AND	*OR*	*NOT*
$0 \cdot 0 = 0$	$1 + 1 = 1$	$\bar{0} = 1$
$0 \cdot 1 = 1 \cdot 0 = 0$	$1 + 0 = 0 + 1 = 1$	$\bar{1} = 0$
$1 \cdot 1 = 1$	$0 + 0 = 0$	

It is evident from Tab. 3.10 that a certain *duality* exists between the AND and OR postulates and between the two NOT postulates. If we replace 0 by 1, 1 by 0, and $+$ by $\cdot$ or $\cdot$ by $+$, then the AND postulates become the OR postulates and vice versa. Similarly, interchanging 0 and 1 changes either of the NOT postulates to the other.

Since making the interchanges of 0 and 1 and of $+$ and $\cdot$ in the postulates leaves them unchanged, and therefore still valid, we may make the same interchanges in any Boolean statement without changing its validity. The result is called the *dual* of the statement. In the same manner we define a *dual* variable or a *dual* function as the variable or function with these interchanges made. For example, let us consider the statement

$$A + 1 = 1 \tag{3.11}$$

which the reader was asked to prove in Ex. 3.2.1(a). The dual statement, which is also true, is

$$A \cdot 0 = 0 \tag{3.12}$$

Here the $+$ is changed to $\cdot$ and the 1's to 0's. We may see by perfect induction that Eq. (3.12) holds, since $1 \cdot 0 = 0$ and $0 \cdot 0 = 0$.

As another example, consider the true statement from Ex. 3.2.1(c),

$$A(B + C) = AB + AC \tag{3.13}$$

Its dual is

$$A + BC = (A + B)(A + C) \tag{3.14}$$

which, therefore, is also true. In fact, this last statement was proved by perfect induction in Tab. 3.5 of Sec. 3.2. It may help, in seeing that Eq. (3.14) is the dual of Eq. (3.13), to note that the left member of Eq. (3.13) is A times the sum of B and C, and thus its dual, the left member of Eq. (3.14), is A plus the product of B and C, and so on.

Let us now consider the complementation of the functions AB and $A + B$. In other words, let us find f and g, where

$$f = (\overline{A \cdot B}) = \overline{AB} \tag{3.15}$$

and

$$g = (\overline{A + B}) = \overline{A + B} \tag{3.16}$$

To see if we can discover the answers, let us construct a truth table consisting of the sums and products $A + B$, AB, $\bar{A} + \bar{B}$, and $\bar{A} \cdot \bar{B}$. The result is Tab. 3.11.

Table 3.11 *Table of various combinations of A and B*

A	B	AB	$\overline{AB}$	$A + B$	$\overline{A + B}$	$\bar{A}$	$\bar{B}$	$\bar{A} + \bar{B}$	$\bar{A} \cdot \bar{B}$
0	0	0	1	0	1	1	1	1	1
0	1	0	1	1	0	1	0	1	0
1	0	0	1	1	0	0	1	1	0
1	1	1	0	1	0	0	0	0	0

The column labeled $\overline{AB}$ is identical to the column labeled $\bar{A} + \bar{B}$; therefore we must have

$$\overline{AB} = \bar{A} + \bar{B} \tag{3.17}$$

Thus we have found f. To find g, we note that the $\overline{A + B}$ column is the same as the $\bar{A} \cdot \bar{B}$ column. Therefore we have

$$\overline{A + B} = \bar{A} \cdot \bar{B} \tag{3.18}$$

Note that we could have derived Eq. (3.18) as the dual of Eq. (3.17). (Interchanging the addition and multiplication operations in one produces the other, there being no 0's or 1's present to interchange.)

The results in Eqs. (3.17) and (3.18) are known as *De Morgan's laws*, in honor of the English mathematician Augustus De Morgan (1806–1871), who was a founder, with George Boole, of symbolic logic. In words, the laws state that the complement of a product is the sum of the complements, and the complement of a sum is the product of the complements.

As an example, the complement of the function

$$f = A(1 + B) \tag{3.19}$$

is, by De Morgan's law,

$$\begin{aligned}\bar{f} &= \overline{A(1 + B)}\\ &= \bar{A} + (\overline{1 + B})\\ &= \bar{A} + 0 \cdot \bar{B}\end{aligned} \tag{3.20}$$

As the reader may show with a truth table, $\bar{f} = 0$ when $f = 1$ and $\bar{f} = 1$ when $f = 0$, as must be the case with complements.

In Eq. (3.20), as is the case in general, we may obtain the complement of a function by replacing $+$ by $\cdot$, $\cdot$ by $+$, and variables (A, B, etc.) and constants (0 and 1) by their complements. We may recall that these are the same steps that produce a dual expression, except that only the constants 0 and 1, and not the variables, are replaced by their complements. For example, the dual of f in Eq. (3.19) is

$$g = A + 0 \cdot B$$

whereas its complement is Eq. (3.20).

EXERCISES

3.4.1 Show by perfect induction that

$$A + \bar{A}B = A + B$$

3.4.2 Find the dual of the statement of Ex. 3.4.1 and show by perfect induction that it is also true.

Ans: $A(\bar{A} + B) = AB$

3.4.3 Evaluate the function

$$f = \overline{AB} + (\bar{B} + C)$$

when A, B, and C are

(a) 1, 0, 0

(b) 1, 1, 0

(c) 1, 1, 1

Ans: (a) 1; (b) 0; (c) 1

3.4.4 Find the complement of f in Ex. 3.4.3, and show that its values for the given cases are the complements of those of f.

Ans: $AB(B\bar{C})$

3.5 Basic Algebraic Properties

Many of the basic properties of ordinary algebra also apply to Boolean algebra. In this section we establish a number of these properties by means of the Boolean postulates of Tab. 3.10 and perfect induction. Also, since by the duality principle every true statement has a dual statement that is also true, we may state two valid properties for every one we derive.

Let us begin with the *commutative laws*:

$$\begin{aligned} &1(a) \quad AB = BA \\ &1(b) \quad A + B = B + A \end{aligned} \tag{3.21}$$

The first of these is easily established by perfect induction (Tab. 3.12), and the second is the dual of the first.

Table 3.12 *Truth table for $AB = BA$*

A	B	AB	BA
0	0	0	0
0	1	0	0
1	0	0	0
1	1	1	1

The *associative laws* are

$$\begin{aligned} &2(a) \quad A(BC) = (AB)C \\ &2(b) \quad A + (B + C) = (A + B) + C \end{aligned} \tag{3.22}$$

Table 3.13 establishes the first of these, and the second is the dual of the first.

Table 3.13 *Truth table for $A(BC) = (AB)C$*

A	B	C	AB	$(AB)C$	BC	$A(BC)$
0	0	0	0	0	0	0
0	0	1	0	0	0	0
0	1	0	0	0	0	0
0	1	1	0	0	1	0
1	0	0	0	0	0	0
1	0	1	0	0	0	0
1	1	0	1	0	0	0
1	1	1	1	1	1	1

Because the position of the parentheses is irrelevant, we may write the product and sum of three variables without ambiguity as ABC and $A + B + C$. This may be extended, of course, to any number of variables.

The *distributive laws* are

$$\begin{aligned} &3(\text{a}) \quad A(B + C) = AB + AC \\ &3(\text{b}) \quad A + BC = (A + B)(A + C) \end{aligned} \tag{3.23}$$

The first of these is familiar to us from ordinary algebra, but the second may seem somewhat strange. Nevertheless, it is the dual of the first and is also true. In this case we have already proved the second property in Sec. 3.2 by means of Tab. 3.5. The first follows from duality.

There are several properties, or theorems as we might call them, that hold for Boolean algebra but are invalid for ordinary algebra in most cases. Some of these are listed in dual pairs, as follows, and all of them may be proved readily by perfect induction:

$$\begin{aligned} &4(\text{a}) \quad A \cdot 0 = 0 \\ &4(\text{b}) \quad A + 1 = 1 \\ &5(\text{a}) \quad A \cdot 1 = A \\ &5(\text{b}) \quad A + 0 = A \\ &6(\text{a}) \quad A \cdot A = A \\ &6(\text{b}) \quad A + A = A \\ &7(\text{a}) \quad A \cdot \bar{A} = 0 \\ &7(\text{b}) \quad A + \bar{A} = 1 \end{aligned} \tag{3.24}$$

To illustrate the proof of some of these theorems, let us consider Tab. 3.14. Comparison of the A column with the $A \cdot A$ column establishes Theorem 6(a), and the $A \cdot \bar{A}$ column establishes Theorem 7(a). Theorems 6(b) and 7(b) are their duals.

Table 3.14 *Truth table for $A \cdot A$ and $A \cdot \bar{A}$*

A	$\bar{A}$	$A \cdot A$	$A \cdot \bar{A}$
0	1	0	0
1	0	1	0

To complete our list of basic properties and theorems, we repeat De Morgan's laws,

$$8(a) \quad \overline{AB} = \bar{A} + \bar{B}$$
$$8(b) \quad \overline{A + B} = \bar{A} \cdot \bar{B} \qquad (3.25)$$

which were derived in Sec. 3.4. As was pointed out there, these may be extended to any number of variables. In the general case for n variables, $A_1, A_2, \ldots, A_n$, we have

$$9(a) \quad \overline{A_1 A_2 \cdots A_n} = \bar{A}_1 + \bar{A}_2 + \cdots + \bar{A}_n$$
$$9(b) \quad \overline{A_1 + A_2 + \cdots + A_n} = \bar{A}_1 \bar{A}_2 \cdots \bar{A}_n \qquad (3.26)$$

We may establish the first of these by noting that there are two cases: (a) all the variables are 1's, or (b) at least one variable is 0. In case (a) we have

$$A_1 = A_2 = \cdots = A_n = 1$$

and therefore by the associative law

$$A_1 A_2 \cdots A_n = 1$$

or equivalently,

$$\overline{A_1 A_2 \cdots A_n} = 0 \qquad (3.27)$$

Also in case (a) we have

$$\bar{A}_1 = \bar{A}_2 = \cdots = \bar{A}_n = 0$$

and therefore by the associative law,

$$\bar{A}_1 + \bar{A}_2 + \cdots + \bar{A}_n = 0 \qquad (3.28)$$

By Eqs. (3.27) and (3.28) it is clear that Theorem 9(a) holds for case (a).

In case (b), since at least one of the variables is 0, we have by Theorem 4(a) and the associative law

$$A_1 A_2 \cdots A_n = 0$$

and therefore

$$\overline{A_1 A_2 \cdots A_n} = 1 \qquad (3.29)$$

For this case, the complement of a zero variable is 1; therefore, by Theorem 4(b) and the associative law

$$\bar{A}_1 + \bar{A}_2 + \cdots + \bar{A}_n = 1 \qquad (3.30)$$

From Eqs. (3.29) and (3.30) we see that in case (b) the theorem is also valid, which completes the proof.

We will consider other, more complicated, theorems in the next section.

EXERCISES

3.5.1 Show by perfect induction that the following statements are true:

(a) $\overline{A + \bar{A}B} = \overline{A + B}$

(b) $\overline{A\bar{B} + \bar{A}B} = AB + \bar{A}\,\bar{B}$

(c) $\overline{AB + \bar{C}} = \bar{A}C + A\bar{B}C$

(d) $AB + BC + C\bar{A} = AB + C\bar{A}$

3.5.2 Obtain four more true statements by taking the duals of the statements of Ex. 3.5.1.

Ans: (a) $\overline{A(\bar{A} + B)} = \overline{AB}$; (b) $\overline{(A + \bar{B})(\bar{A} + B)} = (A + B)(\bar{A} + \bar{B})$; (c) $\overline{(A + B)\bar{C}} = (\bar{A} + C)(A + \bar{B} + C)$; (d) $(A + B)(B + C)(C + \bar{A}) = (A + B)(C + \bar{A})$

3.5.3 Apply De Morgan's theorem to each case of Ex. 3.5.1 to obtain four more identities.

Ans: (a) $\bar{A}(A + \bar{B}) = \bar{A}\,\bar{B}$; (b) $A\bar{B} + \bar{A}B = (\bar{A} + \bar{B})(A + B)$; (c) $AB + \bar{C} = (A + \bar{C})(\bar{A} + B + \bar{C})$; (d) $(\bar{A} + \bar{B})(\bar{B} + \bar{C})(\bar{C} + A) = (\bar{A} + \bar{B})(\bar{C} + A)$

3.6 Other Boolean Theorems

In Sec. 3.5 we established a number of theorems in Boolean algebra by using perfect induction and duality. The method of perfect induction is direct and easy to apply as long as the number of cases under consideration is small. In this section we obtain more Boolean theorems that will be highly useful later, and in so doing we shall illustrate algebraic methods of proof that are, in many cases, easier to apply than perfect induction.

In every case the proof will be based on postulates and previously known theorems. For ready reference we numbered the theorems of the last section from 1(a) through 9(b). For example, if a step in the proof is not obvious but is justified by Theorem 1(a), we will write T1(a) after the step.

To illustrate the procedure, let us prove Theorem 6(b),

$$A + A = A$$

which was given in the previous section. We have

$$\begin{aligned} A + A &= A \cdot 1 + A \cdot 1 && \text{T5(a)} \\ &= A(1 + 1) && \text{T3(a)} \\ &= A \cdot 1 && \\ &= A && \text{T5(a)} \end{aligned}$$

Two very useful dual theorems are

$$\begin{aligned} &10\text{(a)} \quad A(A + B) = A \\ &10\text{(b)} \quad A + AB = A \end{aligned} \tag{3.31}$$

In these statements a smaller term (or factor) appears in a larger term (or factor), in which case the larger term (or factor) is redundant. For example, in Theorem 10(b), A appears in AB, so that AB is redundant.

To prove Theorem 10(a) we may write

$$\begin{aligned} A(A+B) &= AA + AB && \text{T3(a)} \\ &= A + AB && \text{T6(a)} \\ &= A(1+B) && \text{T3(a)} \\ &= A \cdot 1 && \text{T4(b)} \\ &= A && \text{T5(a)} \end{aligned}$$

As an example of the use of Eqs. (3.31) let us consider the function

$$f = X\bar{Y} + X\bar{Y}W + X\bar{Y}(\bar{W} + U)$$

The function $X\bar{Y}$ is common to all three terms and, like A in Theorem 10(b), may be used to absorb the other terms. That is,

$$f = X\bar{Y} + X\bar{Y}(W + \bar{W} + U) = X\bar{Y}$$

(Compare this result with T10(b) for $A = X\bar{Y}$ and $B = W + \bar{W} + U$.)

A theorem and its dual that are very similar to Eqs. (3.31) are

$$\begin{aligned} &11\text{(a)} \quad A(\bar{A} + B) = AB \\ &11\text{(b)} \quad A + \bar{A}B = A + B \end{aligned} \tag{3.32}$$

A proof of 11(a) is given by

$$\begin{aligned} A(\bar{A}+B) &= A\bar{A} + AB && \text{T3(a)} \\ &= 0 + AB && \text{T7(a)} \\ &= AB && \text{T5(b)} \end{aligned}$$

Theorem 11(b) follows as the dual of 11(a). However, an interesting proof is

$$\begin{aligned} A + \bar{A}B &= (A + \bar{A})(A + B) && \text{T3(b)} \\ &= 1 \cdot (A + B) && \text{T7(b)} \\ &= A + B && \text{T5(a)} \end{aligned}$$

Let us consider next the dual theorems

$$\begin{aligned} &12\text{(a)} \quad AB + A\bar{B} = A \\ &12\text{(b)} \quad (A + B)(A + \bar{B}) = A \end{aligned} \tag{3.33}$$

the first of which may be proved by

$$\begin{aligned} AB + A\bar{B} &= A(B + \bar{B}) \\ &= A \cdot 1 \\ &= A \end{aligned}$$

Theorem 12(a) illustrates the principle that if a sum or product of terms is present in which one term (such as A) appears fixed while the other terms are occurring in all possible combinations (such as the two possibilities B and $\bar{B}$), then the sum or product is simply equal to the fixed term.

As an example, let us consider the function

$$f = XYZ + X\bar{Y}Z + XY\bar{Z} + X\bar{Y}\bar{Z}$$

The variable X appears in every term while all combinations, YZ, $\bar{Y}Z$, $Y\bar{Z}$, and $\bar{Y}\bar{Z}$, of Y and Z are occurring. Therefore $f = X$.

Finally, let us consider the dual theorems

$$\begin{aligned} &\text{13(a)} \quad AB + BC + C\bar{A} = AB + C\bar{A} \\ &\text{13(b)} \quad (A + B)(B + C)(C + \bar{A}) = (A + B)(C + \bar{A}) \end{aligned} \tag{3.34}$$

In the case of 13(a), a proof is

$$\begin{aligned} AB + BC + C\bar{A} &= AB + BC \cdot 1 + C\bar{A} \\ &= AB + BC(A + \bar{A}) + C\bar{A} \\ &= (AB + ABC) + (C\bar{A} + C\bar{A}B) \\ &= AB + C\bar{A} \end{aligned}$$

The last step is a consequence of two applications of T10(b).

Theorems 13(a) and 13(b) are sometimes called the *included term* and *included factor* theorems, respectively. That is, if we write the letters A, B, and C in the cyclic order A, B; B, C; and C, A; with first or last (but not both) complemented, the included middle term is redundant.

As an example of the use of T13(a), suppose we wish to simplify the Boolean expression

$$f = AB + C\bar{A} + BCD \tag{3.35}$$

By T13(a) we may insert the term BC, obtaining

$$\begin{aligned} f &= AB + BC + C\bar{A} + BCD \\ &= AB + C\bar{A} + (BC + BCD) \end{aligned}$$

Since by T10(b), BCD is redundant in the terms in parentheses, we have

$$f = AB + BC + C\bar{A}$$

or by T13(a),

$$f = AB + C\bar{A} \tag{3.36}$$

The result in Eq. (3.36) could have been obtained by inspection of Eq. (3.35). Since by T13(a) the term BC may be inserted without changing f, imagine it there and use T10(b) to remove BCD.

Most of the theorems we have considered in this section are useful in simplifying Boolean expressions. The fewer letters involved in a Boolean function, the easier it is to interpret as a logic statement or to represent with electronic hardware. Thus it is important in a given situation not only to find the Boolean function but to simplify it as much as possible.

EXERCISE

3.6.1 Simplify the following Boolean expressions by applying algebraic methods:

(a) $A(\overline{AB})B$

(b) $A + \overline{A + B} + B$

(c) $A\bar{B} + \bar{A}B + AB$

(d) $\bar{A}\bar{B}\bar{C} + \bar{A}B\bar{C} + AB\bar{C} + A\bar{B}\bar{C}$

(e) $(A + B + \bar{C})(A + B + C)$

(f) $(A + B)(A + \bar{B} + C)$

(g) $(A + B)(A + C)(A + D)$

(h) $(A + C)(\bar{A} + B)$

Ans: (a) 0; (b) 1; (c) $A + B$; (d) $\bar{C}$; (e) $A + B$; (f) $A + BC$; g) $A + BCD$; (h) $AB + C\bar{A}$

3.7 Switching Circuits

Perhaps the best example of a physical device that can be represented by a Boolean variable is a switch in an electric circuit. In fact, it is possible to design circuits of switches, or *switching circuits*, to represent any Boolean function, as we shall see in this section. For this reason Boolean functions are sometimes called *switching functions*.

As we know, a switch is a two-state device that is either OPEN or CLOSED. We could use the terms FALSE or 0 for OPEN, and TRUE or 1 for CLOSED, and thus use switches to represent logic variables, as well as any other Boolean variable. For example, the switch shown in Fig. 3.1(a) is in the OPEN, or 0, state, and the one in Fig. 3.1(b) is in the CLOSED, or 1, state.

To illustrate the OR and AND operations with switches, let us consider Fig.

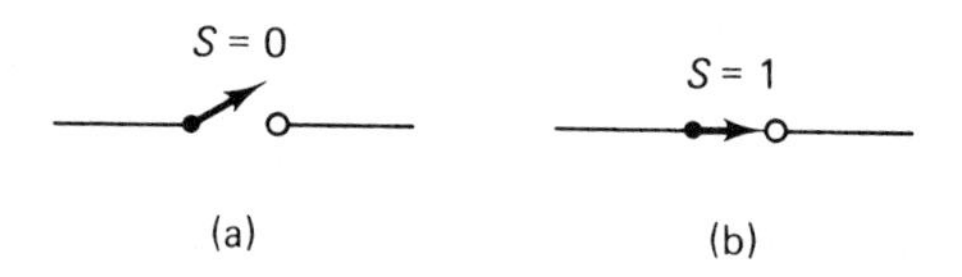

Figure 3.1 *Switch S representing (a) an OPEN or 0 state, and (b) a CLOSED or 1 state*

3.2. The switches A and B in (a) are said to be in *series*, and those in (b) are in *parallel*. Evidently the path (say, P) through the switches from x to y is closed if in (a) A AND B are closed, and in (b) A OR B is closed. In other words, for Fig. 3.2(a), we may write

$$P = AB \tag{3.37}$$

and for Fig. 3.2(b)

$$P = A + B \tag{3.38}$$

This is because in (a), $P = 1$ (the path is closed) only when $A = B = 1$ (both A and B are closed), and in (b), $P = 1$ when $A = 1$ or $B = 1$ (or both). Otherwise, $P = 0$ (open) in both figures. These statements are consistent with the truth tables for AB and $A + B$ (Tabs. 3.2 and 3.1, in Sec. 3.1).

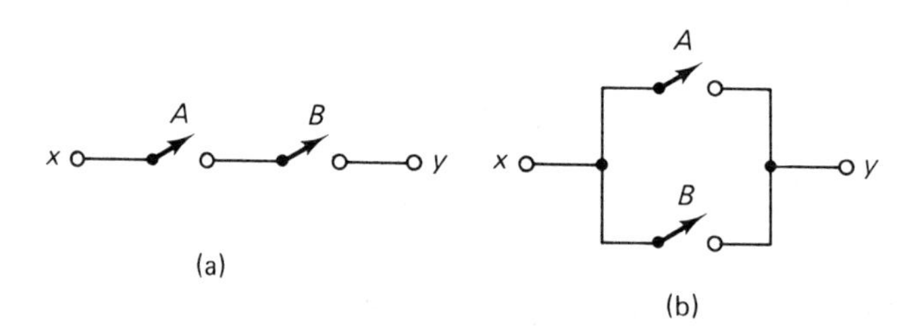

Figure 3.2 *Switches representing (a) AB and (b) A + B*

To interpret the NOT operation in connection with a switch, observe that a switch labeled A in a circuit is open when one labeled $\bar{A}$ is closed, and vice versa. To be more specific concerning $\bar{A}$, let us consider the switch in Fig. 3.3, which is shown with two possible positions A and $\bar{A}$. If the path P from x to y is to be closed ($P = 1$) when the switch is in the position $\bar{A}$, that is, $P = \bar{A}$, then we must have $A = 0$ and $\bar{A} = 1$.

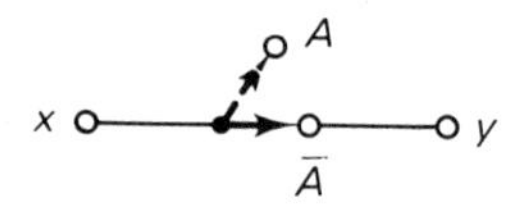

Figure 3.3 *Switch representing $\bar{A}$*

As we have seen, all three of the Boolean operations AND, OR, and NOT may be interpreted in terms of switches. Therefore we may represent any Boolean function by a switching circuit.

To illustrate the procedure, suppose we wish to represent the function

$$P = AB(C + D) \tag{3.39}$$

by a switching circuit. In terms of AND and OR operations we may write

$$P = (A \text{ AND } B) \text{ AND } (C \text{ OR } D)$$

Therefore the path P contains a series combination of A and B in series with a parallel combination of C and D. The result is shown in Fig. 3.4, where it may be seen that $P = 1$ (the path is closed) when $A = B = 1$ (A AND B closed) and either $C = 1$ or $D = 1$ (C OR D closed).

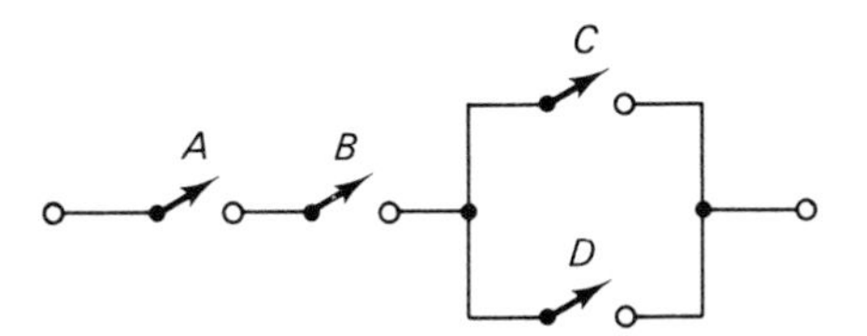

Figure 3.4 *Switching circuit representing Eq. (3.39)*

A switching circuit could connect a source of electrical power, such as a battery, with a light bulb so that the light will be ON when a path through the switches is closed and OFF when the path is open. Thus if the switching circuit represents a logic statement, the light would be on when the statement is true and off when it is not true.

As an example, suppose a red-haired, blue-eyed man insists that his daughter date only people like himself. Specifically, his rule is that her date must be a red-haired man or a blue-eyed man. Therefore, if M represents man, R represents red-haired, and B represents blue-eyed, then D representing the date must be

$$D = MR + MB \tag{3.40}$$

That is, D must be (M AND R) OR (M AND B). A switching circuit that will turn on a light when a prospective suitor is acceptable will have a series combination of switches M and R in parallel with a series combination of M and B. The result is shown in Fig. 3.5, together with the source and light. The reader may check the result by considering the circuit for all possible states of the switches or by constructing the truth table for D.

The father could have stated his rule in an equivalent manner by requiring that the date be a man with either red hair or blue eyes. In this case the Boolean function is

$$D = M(R + B) \tag{3.41}$$

which is evidently equivalent to Eq. (3.40). The corresponding switching circuit

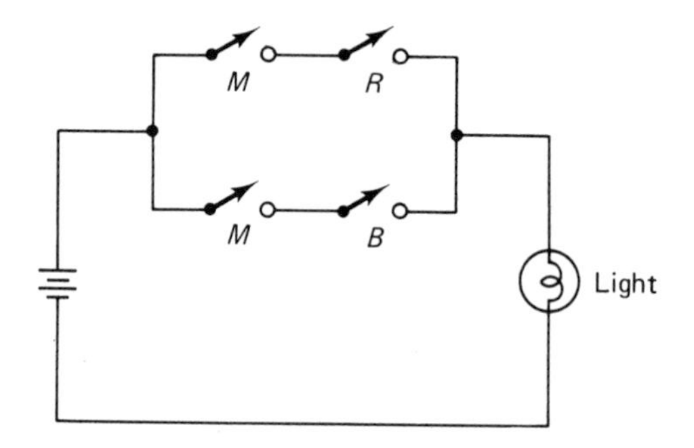

Figure 3.5 *Circuit representing Eq. (3.40)*

contains M in series with the parallel combination of R and B, as shown in Fig. 3.6.

This last example illustrates the importance of minimizing the number of letter appearances in a Boolean expression. In the case of Fig. 3.5 four single-pole switches are required, or else two single-pole switches and one double-pole switch to open or close both M contacts at once. In Fig. 3.6 three single-pole switches perform the same function. Regardless of the type of switches used, there are four contacts to open and close in Fig. 3.5 and only three in Fig. 3.6.

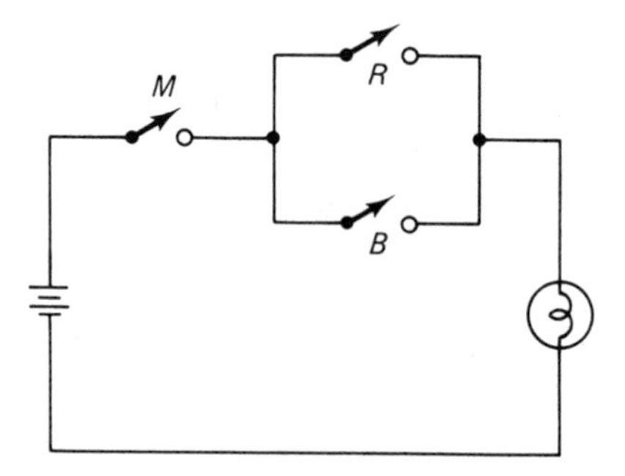

Figure 3.6 *Circuit representing Eq. (3.41)*

As a final example, suppose we wish to install two switches A and B at opposite ends of a hall so that the hall light can be turned on or off by either switch. Let L represent the light, so that $L = 1$ when the light is on and $L = 0$ when it is off. If the light is on for some state of the switches, say, for both switches on, then we have

$$L = 1 \quad \text{when} \quad A = B = 1$$

If we change either switch to off, we wish to have the light off; that is,

$$L = 0 \quad \text{when} \quad A = 0, B = 1, \quad \text{or} \quad A = 1, B = 0$$

Changing either switch at this point must return the light to on, which requires

$$L = 1 \quad \text{when} \quad A = 1, B = 1, \quad \text{or} \quad A = 0, B = 0$$

All the possible states of A and B have been enumerated, and we see that the light is ON if A AND B are both on or if A AND B are both off. Otherwise, the light is off. Therefore we have

$$L = AB + \bar{A}\bar{B} \tag{3.42}$$

We may show with a truth table that this function is consistent with the states enumerated earlier.

The switching arrangement showing the contacts is given in Fig. 3.7(a), and in Fig. 3.7(b) double-throw switches are used. It should be clear, particularly from the latter figure, that we have the two-way switch, as required.

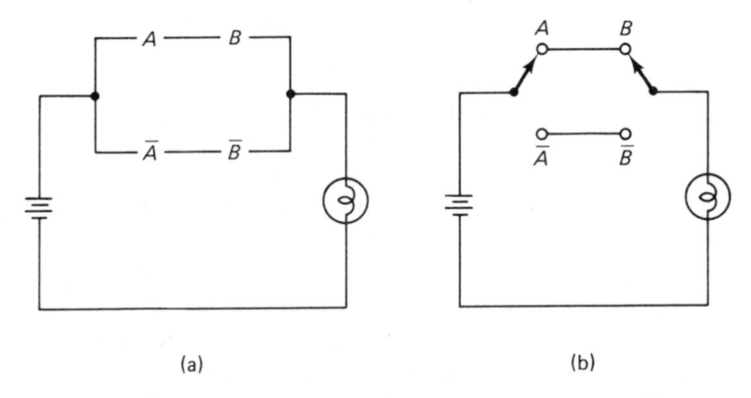

Figure 3.7 *Switching circuit for Eq. (3.42) showing (a) contacts and (b) double-throw switches*

Switching circuits are important because they may be used to perform the logic operations AND, OR, and NOT, which are the basic operations performed by the digital computer. In all modern computers, however, the switching operations are done by electronic circuits known as *logic gates*. These devices will be considered in the next chapter.

EXERCISES

3.7.1 Solve the two-way switch problem of this section by another method.

Ans: $L = A\bar{B} + \bar{A}B$

3.7.2 Find a Boolean function f that describes the switching circuit shown. We require that $f = 1$ when the path xy is closed.

Ans: $A(BC + D) + AC$

3.7.3 Note that $f = 1$ in Ex. 3.7.2 only when A and C are both 1 or when A and D are both 1. Therefore it does not matter whether $B = 0$ or $B = 1$ (open or closed). Show this mathematically by simplifying f.

Ans: $f = A(C + D)$

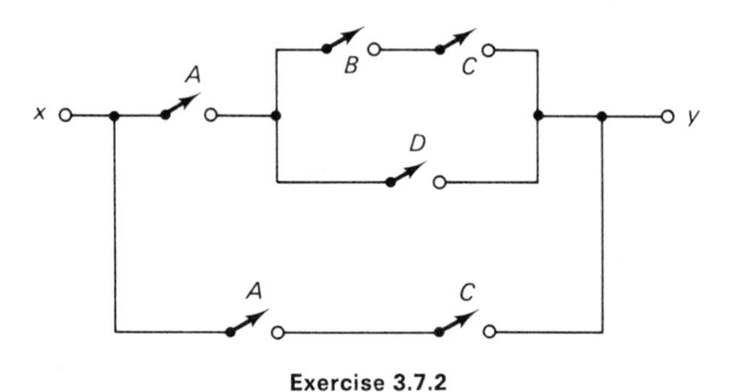

Exercise 3.7.2

3.8 Standard Forms of Boolean Functions

In many cases a Boolean function is not given as an explicit expression containing Boolean variables but is defined solely by a truth table. For example, in designing the two-way switch of the previous section we observed that if the state of exactly one of the switches A and B is changed, the state of the light must change. Writing the states of the switches in the order 00, 01, 11, and 10, we see that only one number, or the state of one switch, changes as we progress from one state to the next. Thus we need only to require that the state of the light L changes with each state. That is, we could choose for the given ordering of the states the light states 1, 0, 1, 0. This results in L being defined by Tab. 3.15, where the states have been reordered.

Table 3.15 *Truth table for a two-way light*

A	B	L
0	0	1
0	1	0
1	0	0
1	1	1

At this point we have L defined by a truth table, but to construct the two-way switch we need a Boolean expression for L. Toward this end let us consider the expanded truth table of Tab. 3.16, in which two additional columns F_1 and F_2 have been added. Column F_1 contains one of the 1's of L, and all its other entries are

Table 3.16 *Truth table for L, F_1, and F_2*

A	B	L	F_1	F_2
0	0	1	1	0
0	1	0	0	0
1	0	0	0	0
1	1	1	0	1

0's. Column F_2 contains the other 1 of L and is otherwise all 0's. Therefore $L = 1$ if $F_1 = 1$ OR if $F_2 = 1$, and $L = 0$ otherwise. Thus L may be written

$$L = F_1 + F_2 \tag{3.43}$$

Also from Tab. 3.16 we see that $F_1 = 1$ if either $A = B = 0$ or $\bar{A} = \bar{B} = 1$, and in every other case $F_1 = 0$. Therefore we have

$$F_1 = \bar{A}\bar{B}$$

Similarly, $F_2 = 1$ if and only if $A = B = 1$, so that

$$F_2 = AB$$

By Eq. (3.43) we have

$$L = \bar{A}\bar{B} + AB \tag{3.44}$$

which agrees with the result of the previous section.

The function L in Eq. (3.44) is expressed as a sum of products, with each product term containing all the inputs in complemented or uncomplemented form. Such an expression is a *standard sum of products* (SOP) form, and the terms, which in this case are $\bar{A}\bar{B}$ and AB, are sometimes called *minterms*. In general, there is a minterm corresponding to each input in the truth table of a function f, but only those corresponding to $f = 1$ appear in the SOP form of f. Minterms are formed by complementing variables whose values are 0 and leaving uncomplemented those whose values are 1. For example, in Tab. 3.16 minterm $\bar{A}\bar{B}$ corresponds to input 00, $\bar{A}B$ to 01, $A\bar{B}$ to 10, and AB to 11.

As an example of a function, or *output*, of three input variables, let us consider the function f defined by Tab. 3.17. The minterms are shown in the last column, and only the second, third, and seventh minterms correspond to $f = 1$. Therefore f may be written in SOP form, consisting of these minterms, as

$$f = \bar{A}\bar{B}C + \bar{A}B\bar{C} + AB\bar{C} \tag{3.45}$$

Since the complement $\bar{f}$ of a function f is 1 when $f = 0$, its SOP form will consist of those minterms corresponding to $f = 0$ in the truth table. Taking the comple-

Table 3.17 *Truth table with minterms*

A	B	C	Output f	Minterm
0	0	0	0	$\bar{A}\bar{B}\bar{C}$
0	0	1	1	$\bar{A}\bar{B}C$
0	1	0	1	$\bar{A}B\bar{C}$
0	1	1	0	$\bar{A}BC$
1	0	0	0	$A\bar{B}\bar{C}$
1	0	1	0	$A\bar{B}C$
1	1	0	1	$AB\bar{C}$
1	1	1	0	ABC

ment of $\bar{f}$ then yields f, which is the complement of the SOP. By De Morgan's law the resulting expression for f is a *product of sums* (POS), which is another standard form. The terms in the POS are sometimes called *maxterms*, which, like minterms, contain all the input variables.

As an example, from Tab. 3.17 we see that the minterms corresponding to $f = 0\ (\bar{f} = 1)$ are $\bar{A}\bar{B}\bar{C}$, $\bar{A}BC$, $A\bar{B}\bar{C}$, $A\bar{B}C$, and ABC. Therefore $\bar{f}$ is given by

$$\bar{f} = \bar{A}\bar{B}\bar{C} + \bar{A}BC + A\bar{B}\bar{C} + A\bar{B}C + ABC$$

and by De Morgan's law we have $f = \bar{\bar{f}}$, or

$$f = (A + B + C)(A + \bar{B} + \bar{C})(\bar{A} + B + C)(\bar{A} + B + \bar{C})(\bar{A} + \bar{B} + \bar{C}) \tag{3.46}$$

This is the POS form of Eq. (3.45).

There are cases where it is preferable to express a Boolean function in a standard form such as the SOP form, even though an equivalent expression may be much simpler. For example, a *read only memory* (ROM), which we will consider in Chap. 6, is a device in which a truth table may be stored for future use. Thus if we have a function such as

$$f = AB\bar{C} + ABC \tag{3.47}$$

to be stored, we must place 1's at positions 110 and 111 and 0's elsewhere. If f is given in the simpler form

$$f = AB$$

then we cannot store it without converting it to the SOP form of Eq. (3.47). The conversion may be done in this case by writing

$$\begin{aligned} AB &= AB \cdot 1 = AB(\bar{C} + C) \\ &= AB\bar{C} + ABC \end{aligned}$$

EXERCISES

3.8.1 **(a)** How many minterms will a five-variable system have?

(b) Find the minterms corresponding to inputs 01000, 01011, and 10001, if the variables are A, B, C, D, and E.

Ans: (a) $2^5 = 32$; (b) $\bar{A}B\bar{C}\bar{D}\bar{E}$, $\bar{A}B\bar{C}DE$, $A\bar{B}\bar{C}\bar{D}E$

3.8.2 Find from Tab. 3.16 the POS form of L. Show by multiplying the factors together that the POS form is equivalent to the SOP form of Eq. (3.44).

Ans: $(A + \bar{B})(\bar{A} + B)$

3.8.3 Find (a) the SOP and (b) the POS forms of $f = A + BC$ given by Tab. 3.5. Simplify each of these to the given form of f.

Ans: (a) $\bar{A}BC + A\bar{B}\bar{C} + A\bar{B}C + AB\bar{C} + ABC$;
(b) $(A + B + C)(A + B + \bar{C})(A + \bar{B} + C)$

3.8.4 Show by finding f and $\bar{f}$ from Tab. 3.17 as a SOP form of minterms that $f + \bar{f} = 1$. That is, the sum of all the minterms is 1. This is also true of the maxterms.

PROBLEMS

3.1 Prove the following Boolean statements by perfect induction:

(a) $(A + B)(A + AB) = A$

(b) $AB + A(A + B) = A$

(c) $(A + B)(A + BC) = A + BC$

(d) $A(A + B)(B + C) = A(B + C)$

(e) $(A + B)(B + C)(C + A) = AB + BC + CA$

3.2 Show by a truth table that if

$$AB = AC$$

then it is not necessarily true that $B = C$; however, if in addition

$$A + B = A + C$$

then $B = C$.

3.3 A father offers to reward his sons A, B, and C if in a series of forthcoming games, either A or B wins, providing that either A or C also wins. Write a Boolean statement of R, the condition that a reward is made, and determine under what conditions there is no reward.

3.4 Prove the following statements by perfect induction:

(a) $(A + C)(A + D)(B + C)(B + D) = AB + CD$

(b) $(A + B)(A + C)(A + D) = A + BCD$

3.5 Prove the following statements by perfect induction:

(a) $X + \overline{XY} + Y = 1$

(b) $(A + B)(\bar{A} + B)(\bar{A} + \bar{B}) = \bar{A}B$

(c) $(A + B + C)(A + \bar{B} + C) = A + C$

(d) $\bar{A}BC + \bar{A}\bar{B}C + AB + A\bar{B} + \bar{A}\bar{B}\bar{C} = A + \bar{B} + C$

(e) $A\bar{B} + AB\bar{C} + ABCD + ABC\bar{D} = A$

(f) $AB + \bar{C}\bar{D} + \bar{A}BC\bar{D} + A\bar{B}\bar{C}D = (A + \bar{D})(B + \bar{C})$

3.6 Solve Prob. 3.1 by algebraic methods.

3.7 Solve Prob. 3.4 by algebraic methods.

3.8 Solve Prob. 3.5 by algebraic methods.

3.9 Find the complement $\bar{f}$ of the function f in the following cases and show that $f\bar{f} = 0$ and $f + \bar{f} = 1$:

(a) $AB + C(\bar{A} + B)$

(b) $AB + \bar{B}C + \bar{C}A$

(c) $\overline{A + B} + BC$

(d) $\overline{(A\bar{B} + C)D}$

3.10 Show by demonstrating the equivalence of the two corresponding Boolean functions that the statement, "Either the union will strike or the union is weak, and the union is not weak," is equivalent to "The union is not weak, and it will strike."

3.11 Repeat Prob. 3.10 for the equivalence of the following two statements: "A coin will fall heads or it will fall tails, and it cannot fall both heads and tails. It falls heads," and "A coin falls heads and not tails."

3.12 Consider the Dilemma of Epicurus: "God is either all-powerful (A) or He is not ($\bar{A}$), and He either wishes to prevent evil (P) or He does not ($\bar{P}$). There are four situations. (a) He cannot prevent evil but He wishes to, in which case He is impotent. (b) He can prevent evil but He does not wish to, in which case He is perverse. (c) He cannot prevent evil and He wishes not to, in which case He is both impotent and perverse. This leaves the last case, (d) He wishes to prevent evil and He can. But then why does He not?" Show by the use of Boolean algebra that if Epicurus had the correct view of God, there is indeed a dilemma. [*Suggestion:* Start with $(A + \bar{A})(P + \bar{P}) = 1$, which is the equivalent of the first statement above.]

3.13 To be eligible to serve as president of a company a person must (1) be a member of the majority stockholder's family or have been with the company at least ten years, and (2) be a nonmember of the family or have at least ten years with the company or own at least 5% of the stock, or else (3) be an outsider as far as the family is concerned but be with the company at least ten years and own at least 5% of the stock. Use Boolean algebra to simplify these regulations.

3.14 Prove the following statements by algebraic methods:

(a) $\overline{AB + \bar{A}\bar{B}} = A\bar{B} + \bar{A}B$

(b) $X\bar{Y} + \bar{X}YZ + XZ = X\bar{Y} + YZ$

(c) $AB + \bar{A}CD + BCD = AB + \bar{A}CD$

(d) $AB + \bar{A}C + A\bar{B}C + A\bar{B}\bar{C} + \bar{A}B = A + B + C$

(e) $[\overline{\overline{(ABC)}\,\overline{(BD)}\bar{B}}]E = BE$

(f) $\bar{A}\bar{B}C + A\bar{B}C + \bar{A}\bar{B}\bar{C} + BD + A\bar{B}\bar{C} = \bar{B} + D$

(g) $(A + BCD)(A + \overline{BC} + D) = A + BCD$

(h) $(A + B)(A + \bar{B} + C)(A + BC) = A + BC$

(i) $(X + YZ)(\bar{W} + YZ)(\bar{W} + \bar{X}) = \bar{W}X + \bar{X}YZ$

(j) $ABC + BCD + A\bar{B}\bar{D} + \bar{A}BC\bar{D} + \bar{A}\bar{B}\bar{D} = BC + \bar{B}\bar{D}$

(k) $AB + \bar{A}C + \bar{B}D + CD = AB + \bar{A}C + \bar{B}D$

3.15 Draw a switching circuit representing the Boolean function f of Probs. 3.9(a) and (b).

3.16 Three people, A, B, and C, form a committee which is to vote on several issues. Design a switching circuit that will turn on a light when a majority vote for a given proposition. Use no more than five contacts.

3.17 Design a switching circuit that will turn on a light when a proposition fails to pass in the committee of Prob. 3.16. (*Suggestion:* Use $\bar{f}$, the complement of f in Prob. 3.16.)

3.18 Combine the circuits of Probs. 3.16 and 3.17 so that a green light indicates passage and a red light indicates failure of a proposition.

3.19 Design a switching circuit in which any one of three switches may be used to control a light independently of the other two switches. (*Suggestion:* Note that in the ordering 000, 010, 011, 001, 101, 100, 110, 111, of the eight inputs, only the state of one switch at a time changes as we progress from one input to the next.)

3.20 **(a)** Obtain in SOP form the function $f(A, B, C)$ defined by $f = 1$ when A, B, C are given by 001, 011, and 101, and $f = 0$ otherwise.

(b) Simplify f to contain only three letter appearances.

3.21 **(a)** Obtain the function f of Prob. 3.20 in POS form.

(b) Simplify the result to that of Prob. 3.20(b).

3.22 Given the three-variable Boolean function

$$f = AB + C$$

find

(a) the SOP form

(b) the POS form

(*Suggestion:* Note that $AB = AB(C + \bar{C})$, etc., and the SOP form of $\bar{f}$ is the SOP terms missing in f.)

4

LOGIC GATES

A *logic gate*, or simply a *gate*, is an electronic device that performs a Boolean operation on one or more inputs to produce an output. The inputs and output are Boolean variables whose two states may be LOW and HIGH voltages, such as 0 volts and 5 volts, or LOW and HIGH currents, such as 0 milliamperes and 20 milliamperes. However, as in the previous chapter, we will use the binary numbers 0 and 1 to represent the states.

A *logic network*, or *digital network*, is a collection of gates that are interconnected so as to perform some specified function. Other terms that are sometimes used are *binary networks* and *switching networks*, and, of course, the word *circuit* often is used interchangeably with *network*. There are many examples of logic circuits, but the most common example is the digital computer.

We will consider digital circuits in detail in later chapters, but in this chapter we are concerned primarily with the characteristics of the logic gates that are used in the circuits. As we will see, the logic gates perform the switching functions done by the switches of Sec. 3.7. Since these switching operations correspond to the Boolean operations described in Chap. 3, the gates may also be used to implement Boolean functions.

Our primary interest here is in the functions the gates perform; however, for the reader who is interested in some of the more practical aspects of gates, a section on this subject is included at the end of the chapter.

4.1 The OR, AND, and NOT Gates

A logic circuit or gate that performs the Boolean operation OR on two or more input variables is called an OR gate. Its symbol, for two input variables A and B, is shown in Fig. 4.1. Its output in this case is A OR B, or $A + B$, as shown. The truth table for the OR gate is, of course, that for the OR operation given earlier in Tab. 3.1.

Figure 4.1 *Two-input OR gate*

In the case of three or more inputs, $A, B, C, \ldots$, the output of the OR gate is

$$f = A + B + C + \cdots \tag{4.1}$$

Evidently, by the definition of logic addition, $f = 1$ if at least one of the input variables is 1. Or, to put it another way, $f = 0$ if and only if each input variable is 0, that is, $A = B = C = \cdots = 0$. A three-input OR gate is shown in Fig. 4.2,

A
B
C
A + B + C

Figure 4.2 *Three-input OR gate*

and its associated truth table is given in Tab. 4.1. Note that because of the asso-

Table 4.1 *Truth table for three-input OR gate*

A	B	C	$A + B + C$
0	0	0	0
0	0	1	1
0	1	0	1
0	1	1	1
1	0	0	1
1	0	1	1
1	1	0	1
1	1	1	1

ciative law of addition, the expression for the output, which is

$$f = A + B + C$$

does not need parentheses to indicate the order in which the additions are performed.

An AND gate is a device that performs the AND operation on two or more input variables. The symbol for an AND gate with two input variables A and B is shown in Fig. 4.3. The output in this case is A AND B, or AB, as shown. In the

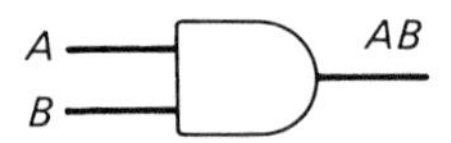

Figure 4.3 *Two-input AND gate*

case of three or more input variables $A, B, C, \cdots$, the output is

$$f = ABC \cdots \tag{4.2}$$

The truth table for the two-input AND gate is that of AB given earlier in Tab. 3.2. In the general case we see from Eq. (4.2) that the output f is 1 if all of the input variables are 1.

As an example, a three-input AND gate is shown in Fig. 4.4, and the truth table

Figure 4.4 *Three-input AND gate*

for this case is Tab. 4.2. Because of the associative law of logic multiplication we may write the output ABC without ambiguity.

Table 4.2 *Truth table for three-input AND gate*

A	B	C	ABC
0	0	0	0
0	0	1	0
0	1	0	0
0	1	1	0
1	0	0	0
1	0	1	0
1	1	0	0
1	1	1	1

A NOT gate, or *inverter*, is a device, symbolized by Fig. 4.5, that performs the NOT operation. There is a single input, shown as A, and the output is the complement of the input, which in this case is $\bar{A}$. The truth table for the inverter is, of

Figure 4.5 *Inverter*

course, that for the NOT operation given earlier in Tab. 3.8. The small circle, or *node*, at the output of the NOT gate is a standard notation for inversion or complementation.

As an example illustrating the use of logic gates, suppose we wish to find the output F of the network of gates shown in Fig. 4.6. The inputs are the Boolean variables A, B, and C.

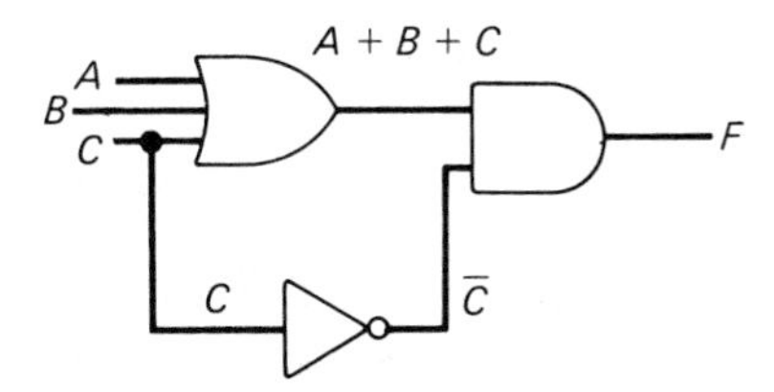

Figure 4.6 *Network of gates*

Note that C is an input to the three-input OR gate as well as the input to the inverter. The dot connecting the two input lines labeled C is a standard way of indicating that C "flows" both into the OR gate and into the inverter. The output $A + B + C$ of the OR gate and $\bar{C}$ of the inverter are the inputs to the AND gate. Therefore the output F is given by

$$F = (A + B + C)\bar{C}$$

which may be simplified to

$$F = (A + B)\bar{C} \tag{4.3}$$

EXERCISES

4.1.1 Find the output F of the given network.

Ans: $AB + C$

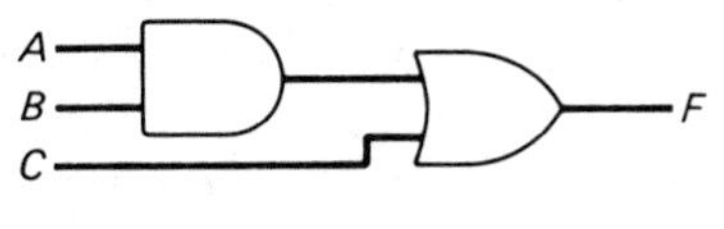

Exercise 4.1.1

4.1.2 Find F.

Ans: $AB + \bar{A}B = B$

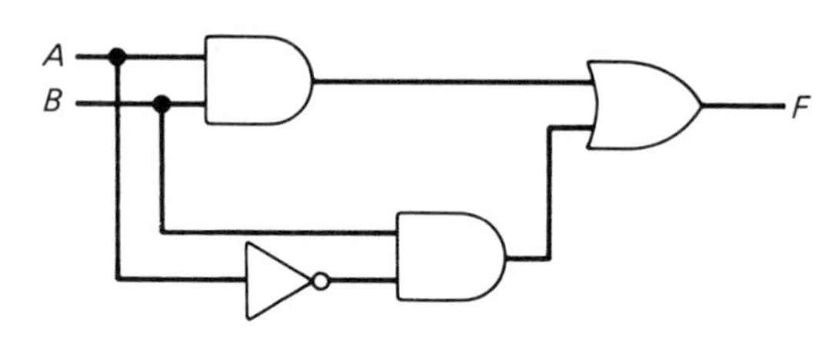

Exercise 4.1.2

4.2 Complete Set of Operations

A set of logic operations that may be used to represent completely any Boolean function is said to be a *complete set*. Since any Boolean expression may be written in terms of the OR, AND, and NOT operations, it is clear that these operations form a complete set. We may also think of the OR, AND, and NOT gates as a complete set. That is, any Boolean expression may be implemented by means of OR, AND, and NOT gates alone. No other gates are required; in fact any other gates, such as those we shall discuss in the remainder of this chapter, may be constructed using OR, AND, and NOT gates.

Conversely, we may say that any logic circuit made up of gates belonging to a complete set represents one or more Boolean functions. The process of finding the Boolean function represented by the circuit is called *analyzing* the circuit. We considered an example of logic circuit analysis at the end of the previous section. The opposite problem, that of finding the logic circuit, given the inputs and the output, is called *synthesizing* the circuit. If we have gates representing a complete set of logic operations, it is always possible to perform circuit synthesis for a given Boolean function. We will consider logic circuits in detail in Chaps. 5, 6, and 7.

As an illustration let us consider the function

$$F = (A + B)\bar{C} \tag{4.4}$$

which was shown in Sec. 4.1 to be the output of the circuit of Fig. 4.6. To construct a circuit *realizing* F (that is, having F as its output), we note that F may be the output of a two-input AND gate having $A + B$ and $\bar{C}$ as its inputs. Evidently $\bar{C}$ may be obtained as the output of an inverter with input C, and $A + B$ is the output of an OR gate with inputs A and B. The resulting circuit is shown in Fig. 4.7.

The circuit in Fig. 4.7 uses a two-input OR gate and thus is somewhat simpler than the one in Fig. 4.6, which used a three-input OR gate. Both circuits implement the same function F, but the version of F used to derive Fig. 4.7 is less complex. Generally, the simpler the form of the Boolean function, the less hardware it takes to implement it.

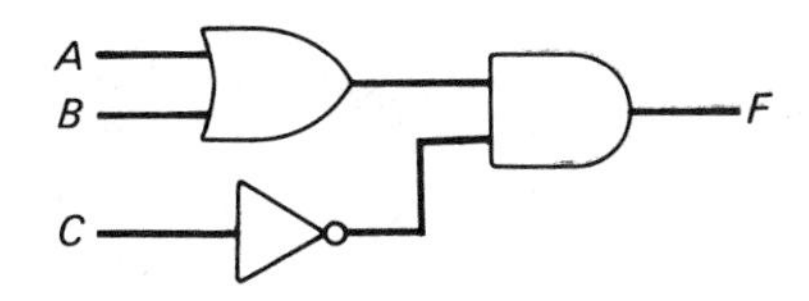

Figure 4.7 *Logic circuit implementation of Eq. (4.4)*

EXERCISE

4.2.1 Using OR, AND, and NOT gates, implement the functions

(a) $F = A + \bar{A}B$

(b) $F = \overline{AB + C}$

Ans:

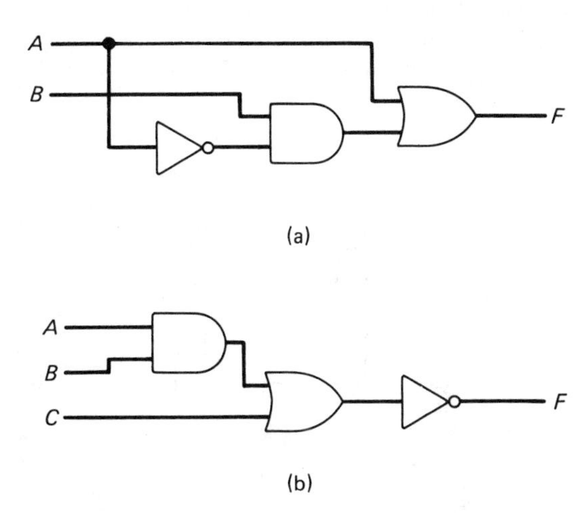

Exercise 4.2.1

4.3 The EXCLUSIVE-OR Gate

The normal OR operation applied to two Boolean variables A and B results in 1 if either $A = 1$ or $B = 1$ or if *both* $A = 1$ and $B = 1$. A very useful Boolean operation known as the EXCLUSIVE-OR, or XOR, operation *excludes* the last case $A = B = 1$. That is, A XOR $B = 1$ if $A = 1$ or $B = 1$ *but not both.* The truth table for the EXCLUSIVE-OR operation, in the two-variable case, is given in Tab. 4.3.

Table 4.3 *Truth table for EXCLUSIVE-OR operation*

A	B	A XOR B
0	0	0
0	1	1
1	0	1
1	1	0

An alternative symbol for the EXCLUSIVE-OR operation is $\oplus$, called the *ring sum*. That is,

$$A \text{ XOR } B = A \oplus B \tag{4.5}$$

Evidently, from the truth table we may write the SOP form in the two-variable case as

$$A \oplus B = \bar{A}B + A\bar{B} \tag{4.6}$$

Since the OR, AND, and NOT gates represent a complete set of operations, we may implement the function $A \oplus B$ with these gates. In fact, it is clear from Eq. (4.6) that the implementation may be carried out by two AND gates, two inverters, and an OR gate. The result, as the reader may verify, is the circuit shown in Fig. 4.8.

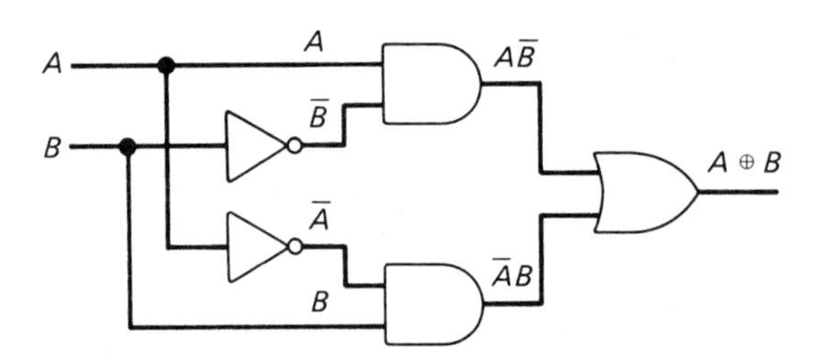

Figure 4.8 *Implementation of the XOR operation*

The XOR operation is of sufficient importance in logic circuits to justify the construction of a single gate to implement it. This gate, called the XOR, or EXCLUSIVE-OR, gate is symbolized for two inputs in Fig. 4.9.

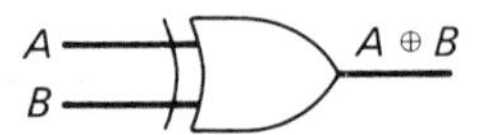

Figure 4.9 *EXCLUSIVE-OR gate*

Since $A \oplus B = 1$ if and only if either A or B, but not both, is 1, it is easy to show by means of a truth table that

$$(A \oplus B) \oplus C = A \oplus (B \oplus C) \tag{4.7}$$

In other words, the associative law holds for the EXCLUSIVE-OR operation, and we may write $A \oplus B \oplus C$ to represent either member of Eq. (4.7). The truth table for $A \oplus B \oplus C$, shown in Tab. 4.4, is the output of the three-input XOR gate of Fig. 4.10.

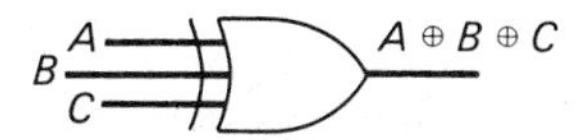

Figure 4.10 *Three-input EXCLUSIVE-OR gate*

Table 4.4 *Truth table for $A \oplus B \oplus C$*

A	B	C	$A \oplus B \oplus C$
0	0	0	0
0	0	1	1
0	1	0	1
0	1	1	0
1	0	0	1
1	0	1	0
1	1	0	0
1	1	1	1

In the general case, the EXCLUSIVE-OR operation on two or more variables yields an output of 1 if and only if an *odd number* of the variables are 1. The reader may see this from the definition of the EXCLUSIVE-OR operation and may readily verify it for the two- and three-input cases of Tabs. 4.3 and 4.4.

EXERCISES

4.3.1 Show that the following statements are true:

(a) $A \oplus A = 0$

(b) $A \oplus \bar{A} = 1$

(c) $A \oplus 0 = A$

(d) $A \oplus 1 = \bar{A}$

(e) $A \oplus B = B \oplus A$

4.3.2 Find the POS form of $A \oplus B$ and verify that the implementation is as shown.

Ans: $(A + B)(\bar{A} + \bar{B})$

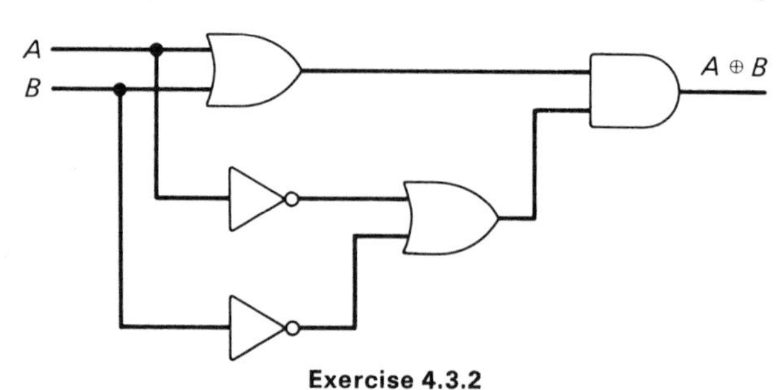

Exercise 4.3.2

4.3.3 Show that $A \oplus (B \oplus C)$ and $(A \oplus B) \oplus C$ are equal by finding their common equivalent form using AND, OR, and NOT operations. Thus the EXCLUSIVE-OR operation satisfies the associative law.

Ans: $ABC + \bar{A}\bar{B}C + \bar{A}B\bar{C} + A\bar{B}\bar{C}$

4.4 The NAND Gate

Another very important Boolean operation is the NOT AND, or NAND, operation, defined by

$$A \text{ NAND } B = \overline{AB} \tag{4.8}$$

The truth table, given in Tab. 4.5, shows that A NAND $B = 0$ if and only if both $A = 1$ and $B = 1$. Another form of Eq. (4.8), by De Morgan's law, is

$$A \text{ NAND } B = \bar{A} + \bar{B} \tag{4.9}$$

Table 4.5 *Truth table for A NAND B*

A	B	A NAND $B = \overline{AB}$
0	0	1
0	1	1
1	0	1
1	1	0

From Eq. (4.8) we see that the NAND operation can be implemented with an AND gate and an inverter, as shown in Fig. 4.11. However, because of the importance of the NAND operation, a single gate, called a NAND gate, is available for performing the operation. The NAND gate in the two-input case is symbolized as shown in Fig. 4.12. Note that the symbol is that of an AND gate followed by a small circle, or node, indicating complementation.

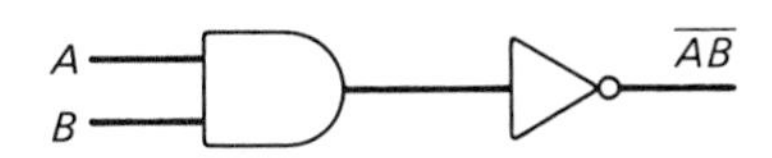

Figure 4.11 *Implementation of the NAND operation*

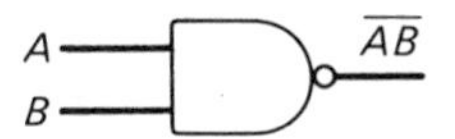

Figure 4.12 *Two-input NAND gate*

In general, NAND gates may have more than two inputs, such as $A, B, C, \ldots$. In this case the output is the complement of the product of the inputs $\overline{ABC \cdots}$. The NAND operation does not obey the associative law, as the reader is asked to show in Ex. 4.4.3, and therefore we will avoid the use of expressions such as

$$F = A \text{ NAND } B \text{ NAND } C$$

for the output of the three-input NAND gate shown in Fig. 4.13. Instead we will

Figure 4.13 *Three-input NAND gate*

use the function

$$F = \overline{ABC}$$

as indicated.

In summary, the NAND gate is a logic gate whose output is 1 when any or all inputs are 0, and its output is 0 only when all inputs are 1.

The NAND operation has the interesting property of being a complete set in itself. To show this we need only to demonstrate that any Boolean function can be expressed in terms of the NAND operation. Since any Boolean function can be expressed in terms of OR, AND, and NOT operations, we need only show that these three operations may be expressed in terms of the NAND operation.

Let us begin by noting from Eq. (4.8) that

$$A \text{ NAND } A = \overline{AA} = \bar{A} \tag{4.10}$$

Therefore, when both its inputs are the same, a NAND operation acts as a NOT operation.

We also may write

$$\begin{aligned} AB &= \overline{\overline{AB}} \\ &= \overline{A \text{ NAND } B} \end{aligned}$$

which by Eq. (4.10) yields

$$AB = (A \text{ NAND } B) \text{ NAND } (A \text{ NAND } B) \tag{4.11}$$

Thus the AND operation may be performed with only NAND operations.

Finally, we may write

$$\begin{aligned} A + B &= \overline{\bar{A}\bar{B}} \\ &= \bar{A} \text{ NAND } \bar{B} \end{aligned}$$

Again by Eq. (4.10), we have

$$A + B = (A \text{ NAND } A) \text{ NAND } (B \text{ NAND } B) \tag{4.12}$$

which shows that the OR operation may be performed with only NAND operations. Therefore the NAND operation is a complete set. The results given in Eqs. (4.10), (4.11), and (4.12) are used in Fig. 4.14 to demonstrate that AND, OR, and NOT gates may be obtained using only NAND gates.

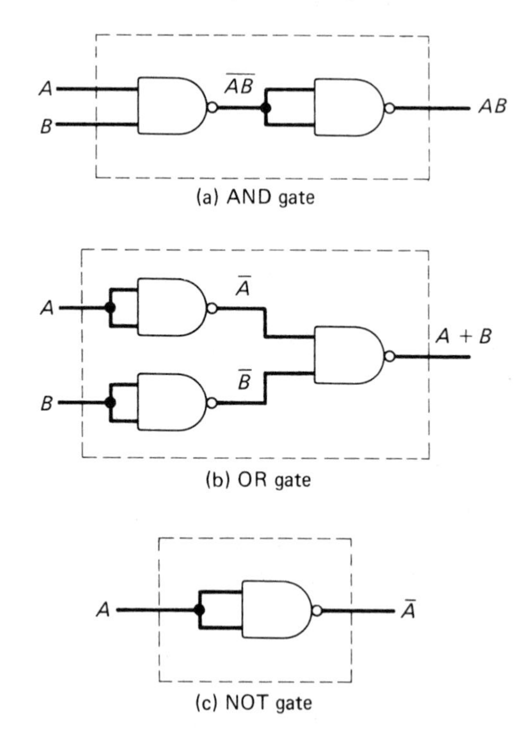

Figure 4.14 *Implementation of (a) AND, (b) OR, and (c) NOT gates using NAND gates*

As the reader is asked to show in Ex. 4.4.1, we may write

$$1 \text{ NAND } A = \bar{A}$$

Therefore we also may obtain the complement of A with a NAND gate whose inputs are A and 1. This is sometimes preferable to the arrangement in Fig. 4.14(c).

To illustrate the use of NAND gates, let us synthesize the function

$$F = AB + \bar{C} \tag{4.13}$$

Since the output of a NAND gate is a complement of a product, the implementation is more direct if F is expressed in that form. This form is especially easy to obtain when F is a sum of products, as it is in this case. We simply write F as the *double complement* of F and apply De Morgan's law, which removes one complementation. In the case of Eq. (4.13) we have

$$\begin{aligned} F = \bar{\bar{F}} &= \overline{\overline{AB + \bar{C}}} \\ &= \overline{(\overline{AB})C} \\ &= (\overline{AB}) \text{ NAND } C \end{aligned}$$

Therefore F is the output of a NAND gate having $\overline{AB}$ and C as its inputs. Furthermore, $\overline{AB}$ is the output of a NAND gate with inputs A and B. From this information we may draw the circuit, which is given in Fig. 4.15.

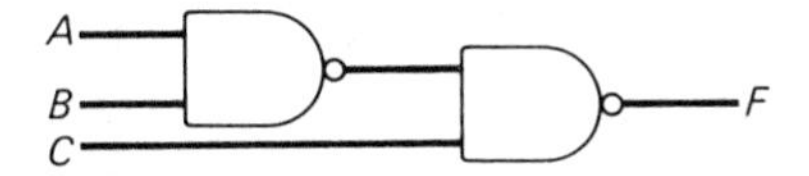

Figure 4.15 *Implementation of Eq. (4.13) using only NAND gates*

The small circle placed at the output of the NOT gate and the NAND gate indicates complementation, as we have noted. Equivalent symbols often are used in which the circle is placed on one or more inputs, indicating that these inputs are inverted. For example, the output of a NAND gate having inputs A and B is given by $\bar{A} + \bar{B}$, which may be interpreted as the output of an OR gate if both inputs are first inverted. Thus an equivalent symbol of the NAND gate is an OR gate with circles placed at the inputs, as shown in the two-input case in Fig. 4.16. Because $\bar{A} + \bar{B}$ is obtained from the output $\overline{AB}$ by De Morgan's law, this equivalent symbol is sometimes called the *De Morgan equivalent* NAND gate. It will be useful later when we synthesize functions using NAND gates.

A
B
$\overline{AB} = \bar{A} + \bar{B}$

Figure 4.16 *De Morgan equivalent NAND gate*

EXERCISES

4.4.1 Verify the following statements:

(a) $0 \text{ NAND } A = 1$

(b) $1 \text{ NAND } A = \bar{A}$

(c) $\bar{A} \text{ NAND } A = 1$

(d) $A \text{ NAND } (\bar{A} + B) = A \text{ NAND } B$

(e) $AB + \bar{A}C + D = [(\bar{A} \text{ NAND } \bar{C}) \text{ NAND } (A \text{ NAND } \bar{B})] \text{ NAND } \bar{D}$

4.4.2 Express $A \oplus B$ in an equivalent form using only NAND operations.

Ans: $[A \text{ NAND } (B \text{ NAND } B)] \text{ NAND } [(A \text{ NAND } A) \text{ NAND } B]$

4.4.3 Show that the associative law does not hold for the NAND operation by expressing the following in terms of AND, OR, and NOT operations:

(a) $(A \text{ NAND } B) \text{ NAND } C$

(b) $A \text{ NAND } (B \text{ NAND } C)$

Ans: (a) $AB + \bar{C}$; (b) $\bar{A} + BC$

4.5 The NOR Gate

A Boolean operation which, like the NAND operation, is also complete in itself is the NOT OR, or NOR, operation, defined by

$$A \text{ NOR } B = \overline{A + B} \tag{4.14}$$

The truth table for the NOR operation is Tab. 4.6, from which we see that $A \text{ NOR } B = 1$ if and only if $A = B = 0$. Applying De Morgan's law, we have another form of Eq. (4.14), given by

$$A \text{ NOR } B = \bar{A}\bar{B} \tag{4.15}$$

Table 4.6 *Truth table for A NOR B*

A	B	$A \text{ NOR } B = \overline{A + B}$
0	0	1
0	1	0
1	0	0
1	1	0

The NOR operation can be performed with an OR gate and an inverter, as shown in Fig. 4.17. However, as in the case of the NAND operation, there is also

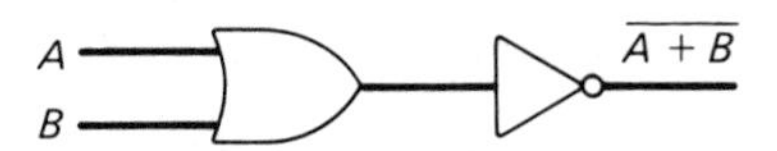

Figure 4.17 *Implementation of the NOR operation*

a single gate, called a NOR gate, that performs the NOR operation. The symbol for a two-input NOR gate is shown in Fig. 4.18, and it is simply an OR gate followed by the small circle indicating complementation.

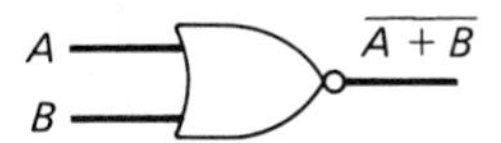

Figure 4.18 *Two-input NOR gate*

Like NAND gates, NOR gates also may have more than two inputs. The output of a NOR gate with inputs $A, B, C, \ldots$ is the complement of the sum of the inputs $\overline{A + B + C + \cdots}$. Also like the NAND operation, the NOR operation does not obey the associative law (see Ex. 4.5.3). Therefore, rather than use expressions such as

$$F = A \text{ NOR } B \text{ NOR } C$$

for the output of the three-input NOR gate in Fig. 4.19, we will write

Figure 4.19 *Three-input NOR gate*

$$F = \overline{A + B + C}$$

as shown.

In summary, the NOR gate has an output of 0 when any or all inputs are 1, and its output is 1 only when all inputs are 0.

To show that the NOR operation is complete in itself we need to demonstrate that the OR, AND, and NOT operations may be performed by means of the NOR operation. We begin by noting from Eq. (4.14) that

$$A \text{ NOR } A = \overline{A + A} = \bar{A} \tag{4.16}$$

and thus the NOR operation can be used to perform the NOT operation.

We also may write

$$\begin{aligned} A + B &= \overline{\overline{A + B}} \\ &= \overline{A \text{ NOR } B} \end{aligned}$$

which by Eq. (4.16) yields

$$A + B = (A \text{ NOR } B) \text{ NOR } (A \text{ NOR } B) \tag{4.17}$$

Therefore the OR operation may be performed with NOR operations.

By De Morgan's law we may write

$$AB = \overline{\bar{A} + \bar{B}}$$

which by definition is equivalent to

$$AB = \bar{A} \text{ NOR } \bar{B}$$

Using Eq. (4.16) again, we may write this result as

$$AB = (A \text{ NOR } A) \text{ NOR } (B \text{ NOR } B) \tag{4.18}$$

Therefore the NOR operation is complete.

The results given in Eqs. (4.16), (4.17), and (4.18) are summarized in Fig. 4.20.

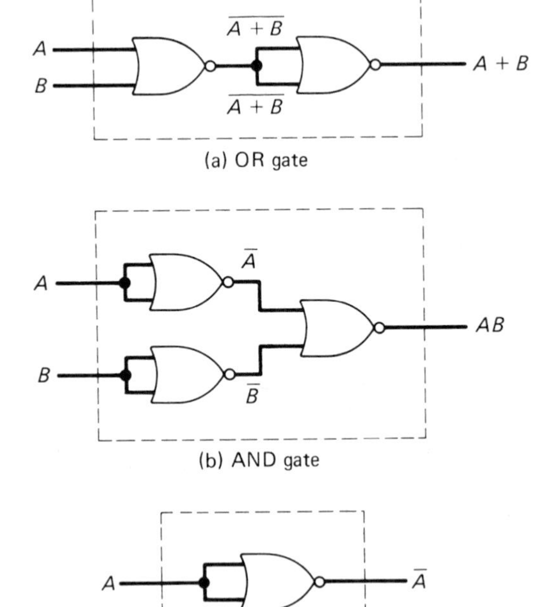

Figure 4.20 *Implementation of (a) OR, (b) AND, and (c) NOT gates using NOR gates*

Since OR, AND, and NOT gates may be constructed with NOR gates, any Boolean function may be implemented with NOR gates.

Since the output of a NOR gate is a complement of a sum, a function F may be directly synthesized with NOR gates if F is a complement of a sum. This form is easily obtained if F is a product of sums, such as

$$F = (A + B)(B + C) \tag{4.19}$$

We simply write F as the double complement of F, as we did in the previous section for NAND gates. Applying De Morgan's law removes one complementation and leaves F in the desired form.

In the case of Eq. (4.19) we have

$$\begin{aligned} F = \bar{\bar{F}} &= \overline{\overline{(A + B)(B + C)}} \\ &= \overline{\bar{A}\bar{B} + \bar{B}\bar{C}} \\ &= (\bar{A}\bar{B}) \text{ NOR } (\bar{B}\bar{C}) \end{aligned}$$

Therefore F is the output of a NOR gate having inputs $\bar{A}\bar{B}$ and $\bar{B}\bar{C}$. Furthermore, by Eq. (4.15), $\bar{A}\bar{B}$ is the output of a NOR gate with inputs A and B, and $\bar{B}\bar{C}$ is the output of a NOR gate with inputs B and C. Using these facts we may draw the circuit, which is shown in Fig. 4.21.

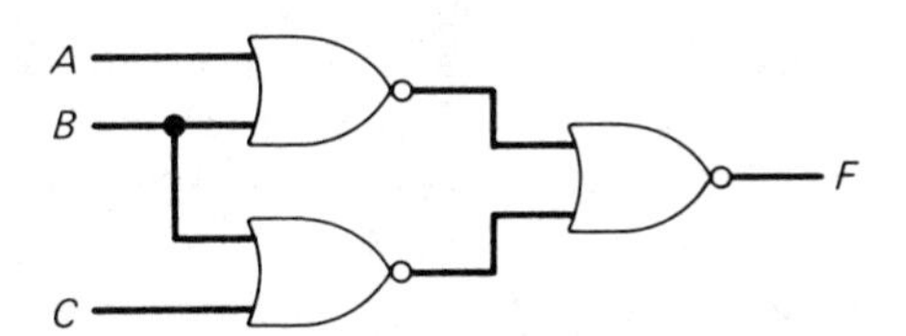

Figure 4.21 *Implementation of Eq. (4.19) using only NOR gates*

The NOR gate, like the NAND gate, has a De Morgan equivalent symbol, which is an AND gate with small circles at its inputs. The two-input case is shown in Fig. 4.22, where it is seen that the output is $\bar{A}$ AND $\bar{B}$, which by De Morgan's law is $\overline{A + B}$, as required.

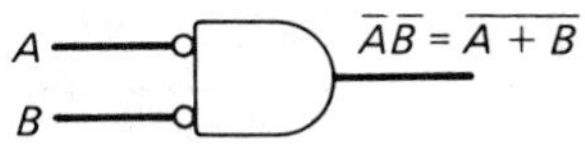

Figure 4.22 *De Morgan equivalent NOR gate*

EXERCISES

4.5.1 Verify the following statements:

(a) 0 NOR $A = \bar{A}$

(b) 1 NOR $A = 0$

(c) $\bar{A}$ NOR $A = 0$

(d) $\overline{(\bar{A} \text{ NAND } \bar{B})} = A$ NOR B

(e) A NOR $(\bar{A}B) = A$ NOR B

(f) $\bar{A}(B + C) = [(A \text{ NOR } \bar{B}) \text{ NOR } (A \text{ NOR } \bar{C})] \text{ NOR } A$

4.5.2 Express $AB + \bar{A}\bar{B}$ in an equivalent form using only NOR operations.

Ans: $[A \text{ NOR } (B \text{ NOR } B)] \text{ NOR } [(A \text{ NOR } A) \text{ NOR } B]$

4.5.3 Show that the associative law does not hold for the NOR operation by expressing the following in terms of AND, OR, and NOT operations:

(a) $(A$ NOR $B)$ NOR C

(b) A NOR $(B$ NOR $C)$

Ans: (a) $(A + B)\bar{C}$; (b) $\bar{A}(B + C)$

4.6 Applications of Gates

In this section we consider some common applications of logic gates. The examples chosen are relatively simple, but the extension to much more complex cases will be easily visualized. The major applications of gates will be deferred until the chapters on digital circuits, where systematic uses of gates will be discussed.

Let us consider first the design of a one-bit *digital comparator*, which is a device that will compare two one-bit numbers A and B and signal when $A > B$, when $A = B$, and when $A < B$. This is a special case of a comparator that compares N-bit binary numbers.

Let F_1 be the output that actuates a light when $A = B$. That is, $F_1 = 1$ when $A = B = 0$ and when $A = B = 1$. Also, if we let F_2 be the output that is 1 when $A < B$ $(A = 0, B = 1)$, and F_3 the output that is 1 when $A > B$ $(A = 1, B = 0)$, then we have the truth table of Tab. 4.7. There are, of course, only four possibilities in the one-bit case.

Table 4.7 *Truth table for one-bit digital comparator*

A	B	F_1	F_2	F_3
0	0	1	0	0
0	1	0	1	0
1	0	0	0	1
1	1	1	0	0

From Table 4.7, using minterms, we see that

$$F_1 = \bar{A}\bar{B} + AB$$
$$F_2 = \bar{A}B$$
$$F_3 = A\bar{B}$$

From these expressions we may obtain the digital circuit, using AND, OR, and NOT gates. The result is shown in Fig. 4.23. The outputs F_1, F_2, and F_3 may be used to actuate a light source of some type, such as a *light-emitting diode* (LED), to signal when $A = B$, $A < B$, or $A > B$.

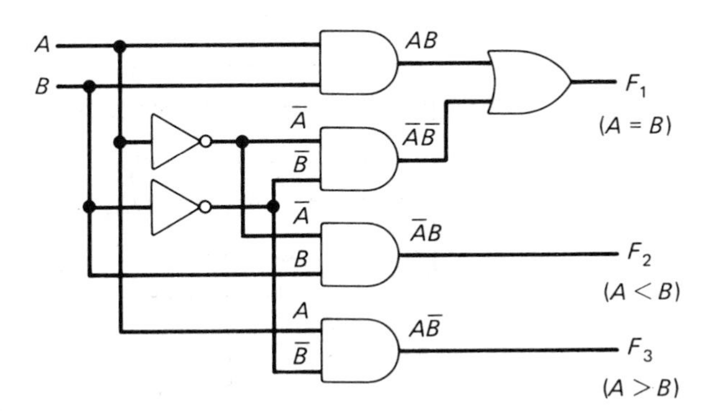

Figure 4.23 *One-digit comparator*

Codes are used to transmit numbers or characters over telephone lines, put them into a computer, and so on. For example, the number or character may be represented by a binary number equivalent. An *encoder* is a device that translates the character into its coded equivalent, and a *decoder* translates the code back into the character it represents. As our next example, let us design a *two-bit binary decoder*, which is a digital circuit that will display the decimal equivalent of inputs that are two-bit binary numbers. This is a special case of a more general circuit that will decode *N*-bit binary integers.

Let F_0, F_1, F_2, and F_3 be outputs that actuate a light to display the decimal integers 0, 1, 2, and 3 (the only possible cases corresponding to two-bit binary digits). The truth table is then the one shown in Tab. 4.8, where the inputs A and B are the two bits in the binary number $(AB)_2$.

From the truth table we see that

$$F_0 = \bar{A}\bar{B} = \overline{A + B} = A \text{ NOR } B$$
$$F_1 = \bar{A}B = \overline{A + \bar{B}} = A \text{ NOR } \bar{B}$$
$$F_2 = A\bar{B} = \overline{\bar{A} + B} = \bar{A} \text{ NOR } B$$
$$F_3 = AB = \overline{\bar{A} + \bar{B}} = \bar{A} \text{ NOR } \bar{B}$$

Table 4.8 *Truth table for two-bit binary decoder*

A	B	F_0	F_1	F_2	F_3
0	0	1	0	0	0
0	1	0	1	0	0
1	0	0	0	1	0
1	1	0	0	0	1

Therefore we may construct the circuit with four NOR gates and two inverters. (The latter are needed to make $\bar{A}$ and $\bar{B}$ available.) The result is given in Fig. 4.24.

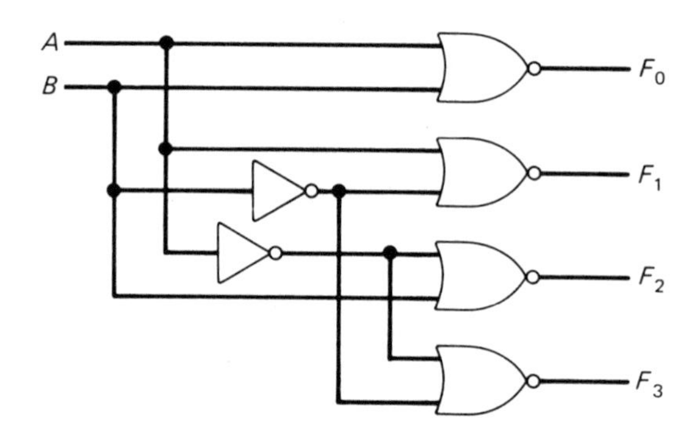

Figure 4.24 *Two-bit binary decoder*

For example, if the binary number 10 (decimal 2) is the input, then $F_2 = 1$, which turns on a light with the digit 2 printed over it.

Since each output function in Fig. 4.24 is 1 for only one of the states A, B, the circuit decodes the inputs into one of four decimal outputs. For this reason it is called a 1-of-4 *decoder*. In general, a 1-of-N decoder decodes the inputs into one of N possibilities.

As a last example, suppose we wish to have the binary decoder in Fig. 4.24 display the decimal equivalents of the binary inputs in a single position rather than at the different locations of the light sources. This is the type of display used extensively in hand calculators, which can show any decimal digit from 0 to 9 in each of its display positions. (The extension to the case of ten digits, rather than four as in our example, merely requires a larger truth table and more gates. The principle is exactly the same as in our simpler case.)

A decimal number can be displayed by lighting the appropriately numbered segments shown in Fig. 4.25(a). As may be seen in Fig. 4.25(b), the digit 0 is displayed when segments 0, 1, 2, 3, 4, and 5 are lit, 1 is displayed when 1 and 2 are lit, and so on. Any decimal integer can be displayed by lighting suitable segments, as the reader is asked to verify in Prob. 4.19.

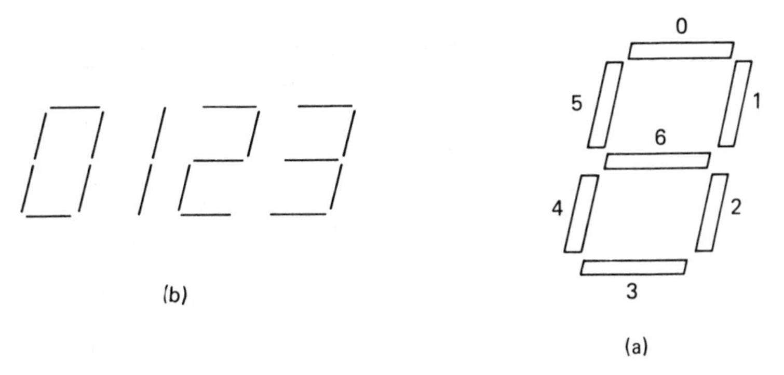

Figure 4.25 *(a) Sample numbers and (b) number segments*

In our example, however, we are interested only in the digits 0, 1, 2, and 3. Table 4.9 is the truth table showing which of the segments S_0, S_1, S_2, S_3, S_4, S_5, and S_6 are required to produce each of the decimal integers. From the truth table, using minterms or maxterms, we may find the various segment functions. For example, we have

$$\bar{S}_0 = \bar{A}B$$

and therefore

$$S_0 = A + \bar{B}$$

Similarly, we have

$$S_1 = 1 = A + \bar{A}$$
$$S_2 = \bar{A} + B$$
$$S_3 = A + \bar{B}$$
$$S_4 = \bar{A}\bar{B} + A\bar{B} = \bar{B}$$
$$S_5 = \bar{A}\bar{B} = A \text{ NOR } B$$
$$S_6 = A\bar{B} + AB = A$$

Table 4.9 *Truth table for segments displaying decimal numbers*

	Binary								
Decimal	*A*	*B*	S_0	S_1	S_2	S_3	S_4	S_5	S_6
0	0	0	1	1	1	1	1	1	0
1	0	1	0	1	1	0	0	0	0
2	1	0	1	1	0	1	1	0	1
3	1	1	1	1	1	1	0	0	1

The circuit is obtained with the Boolean functions $S_0, S_1, \ldots, S_6$ as outputs and the bits A and B as inputs. The result is shown in Fig. 4.26.

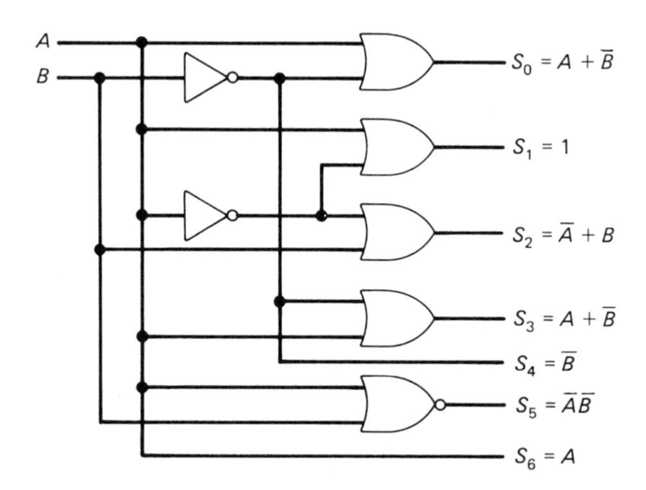

Figure 4.26 *Circuit for displaying decimal numbers*

As an example, suppose the binary number to be decoded and displayed is 11 (decimal 3). That is, $A = B = 1$. Then tracing the signals through Fig. 4.26 we see that $S_0 = 1$, $S_1 = 1$, $S_2 = 1$, $S_3 = 1$, $S_4 = 0$, $S_5 = 0$, and $S_6 = 1$. From Fig. 4.25 we see that this combination of signals displays the decimal number 3.

Note that $S_0 = S_3$ in this simple case (this will not be true if the decoder is designed for any decimal digit from 0 to 9), so that we could have used one signal to actuate both segments. Moreover, $S_1 = 1$ represents a light that is always on (again, this is not the case in general), but for illustrative purposes we have chosen to synthesize it by $A + \bar{A} = 1$. Finally, $S_6 = A$ is *driven* directly by the input A, which may or may not be done in practice, depending on the source and strength of A.

EXERCISES

4.6.1 Design a one-bit *equality comparator*, or *coincidence* circuit, which has an output $F = 1$ when two one-bit integers A and B are equal, and $F = 0$ otherwise.

Ans: $F = AB + \bar{A}\bar{B}$

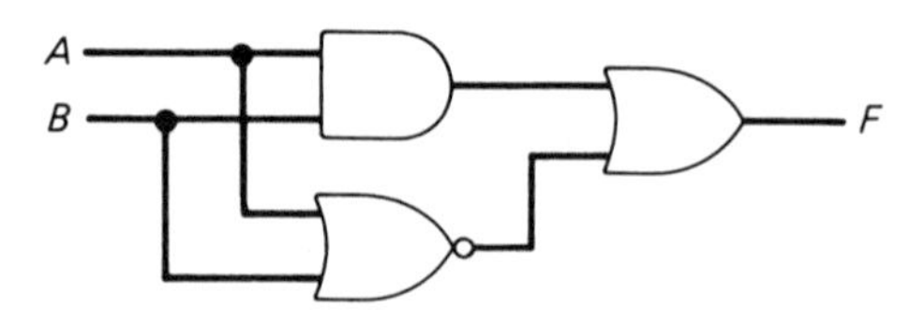

Exercise 4.6.1

4.6.2 The *coincidence* function $F = AB + \bar{A}\bar{B}$ of Ex. 4.6.1 is sometimes written, using a *ring product* operation $\odot$, as

$$F = A \odot B$$

and a single gate, called the COINCIDENCE gate, is used to perform the function. The symbol of the COINCIDENCE gate is as shown. Prove that

$$A \odot B = \overline{A \oplus B}$$

and therefore the COINCIDENCE gate may be constructed with an EXCLUSIVE-OR gate and an inverter.

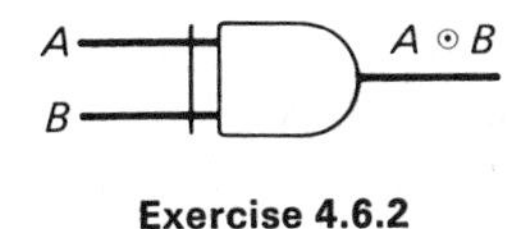

Exercise 4.6.2

4.6.3 Show that the ring product operation of Ex. 4.6.2 obeys the associative law by finding equivalent AND, OR, NOT expressions for both members of

$$A \odot (B \odot C) = (A \odot B) \odot C$$

Ans: $ABC + A\bar{B}\bar{C} + \bar{A}B\bar{C} + \bar{A}\bar{B}C$

4.7 Some Practical Aspects of Gates

In this chapter we have said that a logic gate is an electronic device, and we have considered in some detail the mathematical properties of the various gates that are used in digital circuits. We are not interested here in discussing the details of how the gates are made, because the nature of the technology normally makes little difference to the user of the device. Moreover the technology is continually changing; the most popular technology of today may be completely obsolete by tomorrow.

It seems appropriate, however, to make some general remarks about the evolution of logic gates and consider some of their practical aspects. For those readers who are interested in more details, there is a wealth of material available in the literature.

In the 1940s the electronic device used in manufacturing gates was the vacuum tube, and consequently the gates were relatively large, as were the power requirements. In the late 1940s and early 1950s semiconductor devices, such as transistors and diodes, replaced the vacuum tube, with the result that the gates were much smaller and much more reliable and consumed much less power. These separate, or *discrete*, components were superseded in the late 1950s by the *integrated circuit* (IC), which is a single monolithic *chip* of semiconductor in which the electronic circuit elements are fabricated.

A chip is about 0.01 inch thick and may vary in size from about 0.03 × 0.03 inch to 0.3 × 0.3 inch or 0.4 × 0.4 inch, with a typical size of about 0.05 × 0.05 inch. The number of gates on one chip may vary from 1 or 2, and usually a minimum of 4, to many thousands. The limit to the number of gates that can be put on a single chip is basically a function of manufacturing techniques, and at the present time the limit is nowhere in sight.

IC's are classified according to the number of gates they contain. *Small-scale integration* (SSI) refers to circuits containing 12 or fewer gates, *medium-scale integration* (MSI) to circuits with more than 12 but fewer than 100 gates, and *large-scale integration* to those having 100 or more gates. With chips in existence containing 10,000 gates, these figures seem meaningless, and indeed, there are some who think they should be multiplied by 10.

Connecting pins are attached to the terminals of the gates and the power supply terminals (of which there are two or more—one to ground and the others to dc voltage sources), and the chip is then enclosed in a protective *package.* In one type of package, called a *dual in-line package* (DIP), the connecting pins are arranged in two (dual) lines for easy plug-in operation using printed circuit boards. A top schematic view of a DIP containing four two-input NAND gates is shown in Fig. 4.27. There are 14 pin connections in this case, including the dc voltage source (V_{cc}) and ground (Gnd). A side view showing the pins is given in Fig. 4.28.

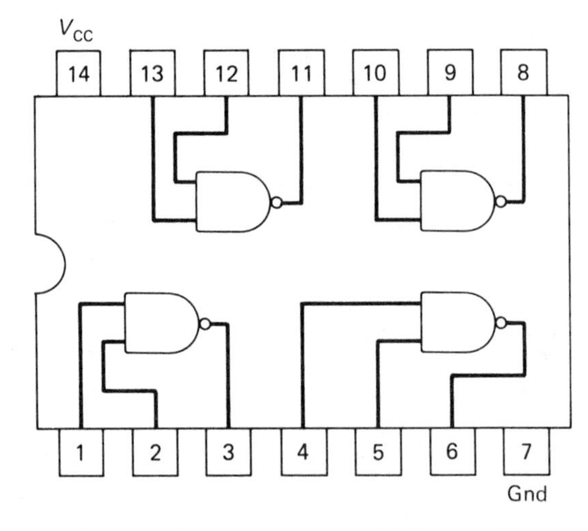

Figure 4.27 *Quad two-input NAND gate DIP*

The user should be aware of certain practical limitations on the logic gate. Two of these are the *fan-in* and *fan-out* of the gate, which are specified by the manufacturer. Fan-in may be defined as the number of inputs of the gate, and fan-out is the number of other gates that the output of the given gate can *drive.* Too many gates connected to the output of a given gate can cause the output voltage of the

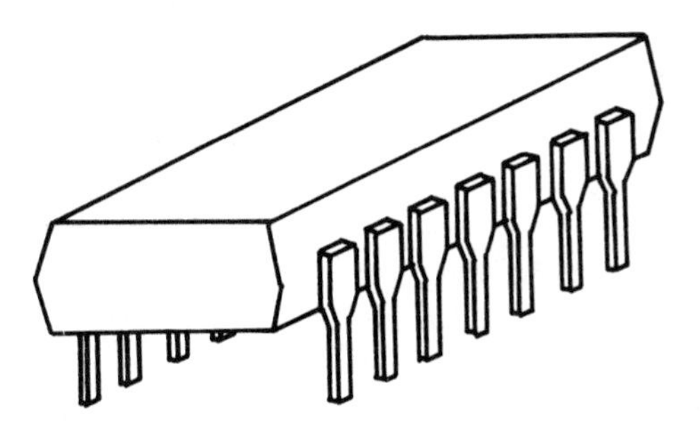

Figure 4.28 *Fourteen-pin DIP*

given gate to be too low in its high state or too high in its low state for the connected gates to function properly. As an example, in Fig. 4.8 the fan-in of each of the AND and OR gates is 2, but they need have only a fan-out of 1.

Another limitation is the *propagation delay time*, which is the time required for the output of a gate to change to its appropriate value after receiving an input signal. When a number of gates are operated in *cascade* (the output of one gate being an input of the next gate), propagation delay time can be a very important factor.

There are a number of types, or logic families, of gates, whose names indicate the technology used in their fabrication. Some of these are *transistor–transistor logic* (TTL or T^2L), *emitter-coupled logic* (ECL), *complementary metal oxide semiconductor* (CMOS), and *integrated injection logic* (IIL or I^2L). These different logic families have widely varying characteristics and, therefore, advantages and disadvantages. For example, the basic gate in TTL is the NAND gate with a typical fan-out of 10 and a propagation delay time of 9 nanoseconds. In the ECL family the basic gates are OR or NOR, and they have a higher fan-out (16 or more) and a shorter delay time (2 nanoseconds) than the TTL gates, but they dissipate more power (25 milliwatts as compared to 10 milliwatts per gate). The basic gate in CMOS logic is a NAND or a NOR. In this case the power dissipation per gate is small (10 nanowatts per gate), but the delay time is larger (25 nanoseconds). There are many other physical characteristics of the various gates, which one may find by consulting the literature or the manufacturers' specifications.

As we know, 0 and 1 may represent, respectively, LOW and HIGH, say 0 volts and 5 volts. A NAND gate truth table using these voltages is given in Tab. 4.10. Now suppose we choose to represent 0 volts by 1 and 5 volts by 0. Table 4.10 now becomes Tab. 4.11, which is the truth table of a NOR gate. Thus we see that if the *more positive* value (5 volts) is the TRUE or 1 condition, the gate is a NAND gate, and if the less positive, zero, or even *negative* value (0 volts in this case) is the TRUE or 1 condition, the gate is a NOR gate. This example illustrates the ideas of *positive logic* (or *positive*-TRUE condition) and *negative* logic (or *negative*-TRUE condition). To apply a logic gate properly, the user must know, of course, which logic is assumed in the truth table of the gate.

Table 4.10 *Truth table of a NAND gate using voltages*

A	B	F
0	0	5
0	5	5
5	0	5
5	5	0

Table 4.11 *Table 4.10 with the negative logic convention*

A	B	F
1	1	0
1	0	0
0	1	0
0	0	1

Another subject of practical interest is that of unused input leads. For example, if we have two inputs A and B and have three-input gates available, does it matter if the unused input is left open? It would if, for example, the open input acted as a 0 and we were performing an AND operation. In this case, since

$$A \cdot B \cdot 0 = 0$$

the output of the gate would always be 0. Similarly, if an open input of an OR gate acted like a 1, we would always have an output of 1. In cases where open inputs affect the output, we may *strap* unused inputs to a fixed voltage, represented by 0 in OR operations and 1 in AND operations. In these cases we have

$$A + B + 0 = A + B$$

and

$$A \cdot B \cdot 1 = A \cdot B$$

As a final note in this section we observe that in some cases an output signal of a gate may be too weak to drive a connected *load*. For example, the segments S_0 through S_6 shown in Fig. 4.25(a) may require more current than can be delivered by the gates in Fig. 4.26 or, in the case of S_6, by signal A directly. In this case a *buffer* gate may be required at each output to increase the output current. Buffer gates are standard components designed for the specific purpose of strengthening the output signals.

EXERCISES

4.7.1 Find

(a) the fan-in of gate X

(b) the required fan-out of gate X

(c) the required fan-out of gate Y

Note the symbol used when there are many inputs.

Ans: (a) 9; (b) 1; (c) 2

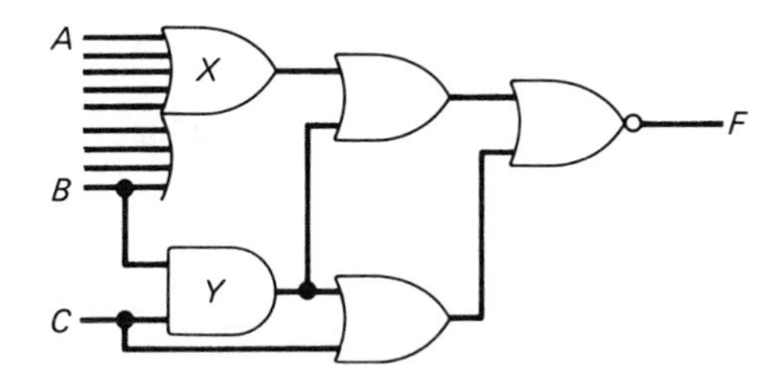

Exercise 4.7.1

4.7.2 Show that if the 0's and 1's are interchanged in the truth table of an AND gate it becomes the truth table of an OR gate, and vice versa.

4.7.3 **(a)** Make all the unlabeled inputs of gate X in Ex. 4.7.1 equal to 0 and find F

(b) Find F if all the unlabeled inputs are made 1.

Ans: (a) $\overline{A + B + C}$; (b) 0

PROBLEMS

4.1 Find logic circuits using OR, AND, or NOT gates implementing the following functions:

(a) $A + B\bar{C} + D\overline{EF}$

(b) $AB\bar{C} + A\bar{B}C + \bar{A}BC$

(c) $A(\bar{B} + C) + \bar{A}BC$

(d) $(A + B + C)(A + \bar{B} + \bar{C})(\bar{A} + B + C)$

4.2 Find F.

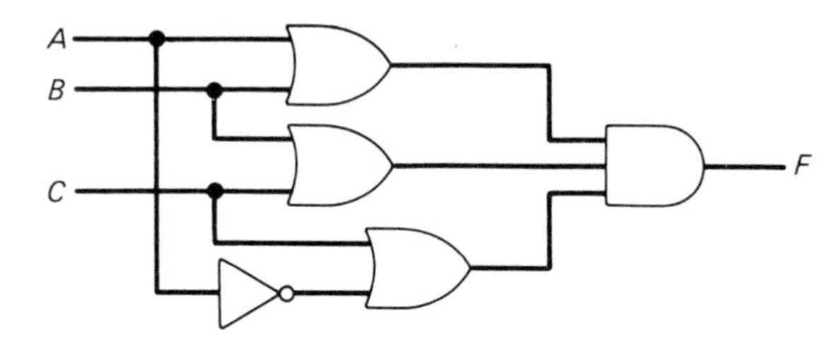

Problem 4.2

4.3 Obtain an implementation of F in Prob. 4.2 using four gates (OR, AND, or NOT).

4.4 Show that

$$(A \oplus B) \oplus (B \oplus C) = A \oplus C$$

Implement the left member with XOR gates.

4.5 Find F.

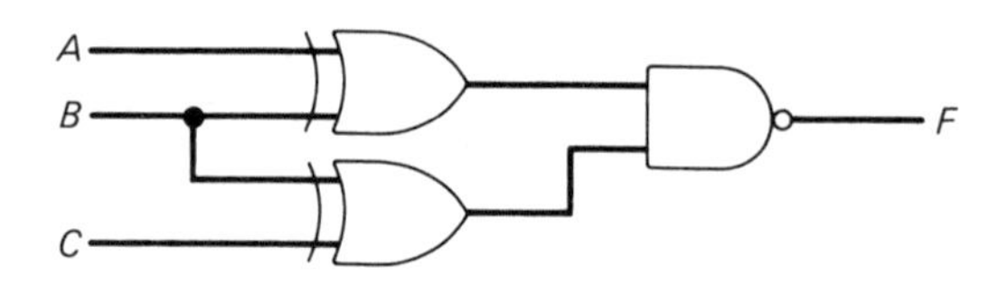

Problem 4.5

4.6 Show that the given circuit is equivalent to an XOR gate; that is, $F = A \oplus B$.

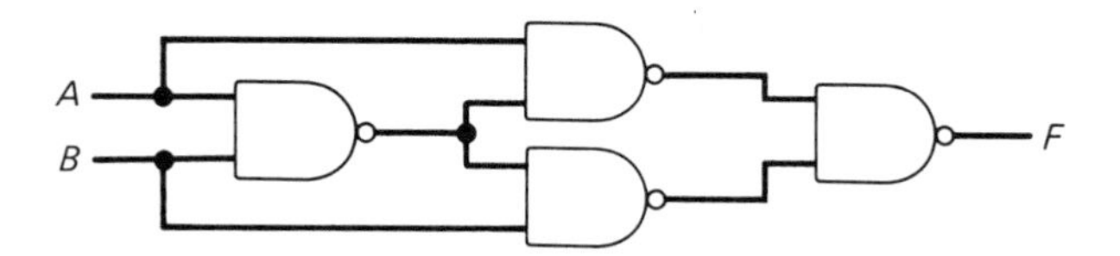

Problem 4.6

4.7 Implement the following functions using only NAND gates:

(a) $AB + BC$

(b) $\bar{A} + \bar{B} + C$

(c) $AB + CD$

(d) ABC

4.8 Design a logic circuit to energize an automobile starter S if and only if the key K is turned to start, the driver's seat belt B_d is fastened, and either there is no front seat passenger $\bar{P}$ or there is a front seat passenger P with seat belt B_p fastened.

4.9 Show that the two-way l'ght problem of Sec. 3.8 may a so be solved by taking the state of the light as

$$L = A \oplus B$$

where A and B are the two switches. Thus the problem may be solved by means of an XOR gate.

4.10 Solve Prob. 3.16 using four or fewer gates of the OR, AND, or NOT types.

4.11 Solve Prob. 3.16 using four or fewer NAND gates.

4.12 A plastic card is to be designed to open a certain door. The card is to have holes punched in appropriate places so that when it is inserted in a slot in the door a light shines through the holes actuating a voltage (denoted by 1). The absence of a hole blocks the light (no vol age, denoted by 0). Let the possible locations of the holes be denoted by A, B, C, D, E, and F, so that there are $2^6 = 64$ possible lock combinations. Design a logic circuit that will cause the door X to open for the combination $A = B = E = F = 1$ and $C = D = 0$ and will actuate an alarm Y if any other combination is tried.

4.13 Solve Prob. 3.19 using logic ga es. For the given ordering of the states let the respective ordering of the light states be 0, 1, 0, 1, 0, 1, 0, 1.

4.14 Construct an equivalent two-input NAND gate using only NOR gates.

4.15 Using only NOR gates, synthesize the function

$$F = (A + B + C)(C + D)$$

4.16 Find F.

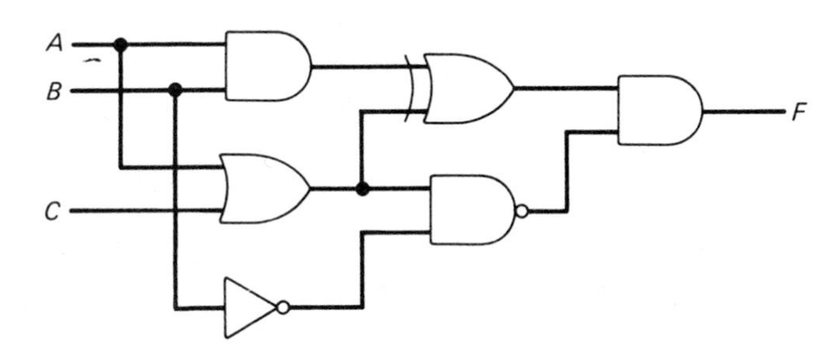

Problem 4.16

4.17 Find F. Note from this result that

$$(A \odot B) \odot (B \odot C) = A \odot C$$

Problem 4.17

4.18 Find F.

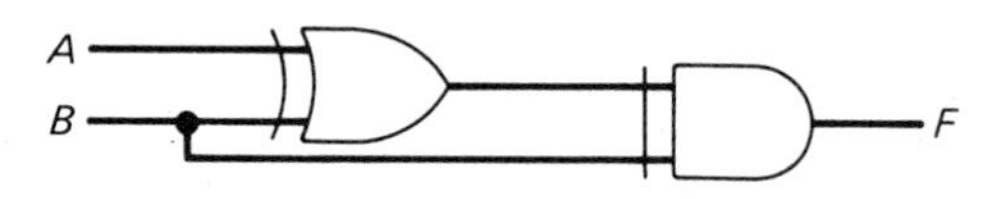

Problem 4.18

4.19 Construct a truth table similar to Table 4.9 that may be used for displaying any decimal digit from 0 through 9 by means of the number segments of Fig. 4.25(a). Note that four-bit binary numbers $(ABCD)_2$ are required.

4.20 Obtain a digital circuit for implementing S_0 and S_1 in the truth table of Prob. 4.19.

4.21 Show that we may *expand the inputs* available, in the case of OR gates by connecting three gates in cascade as shown. That is, show that

$$F = A + B + C + D$$

so that, in effect, we have an OR gate with four inputs constructed from three two-input gates.

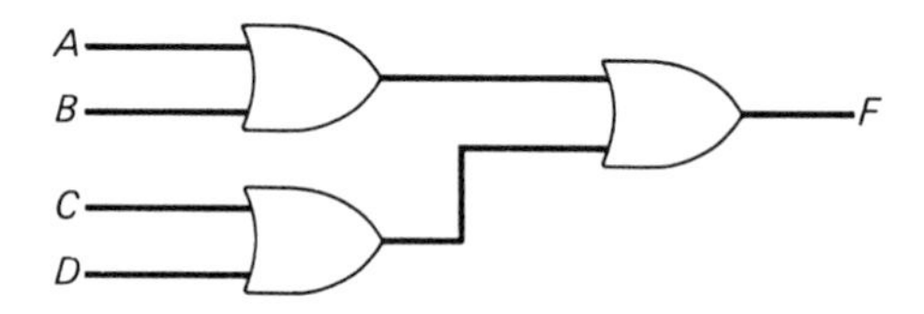

Problem 4.21

4.22 Show that

$$G = A + B + C + D + E + F$$

so that, in effect, the cascaded connection is a six-input OR gate. Compare this result with that of Prob. 4.21.

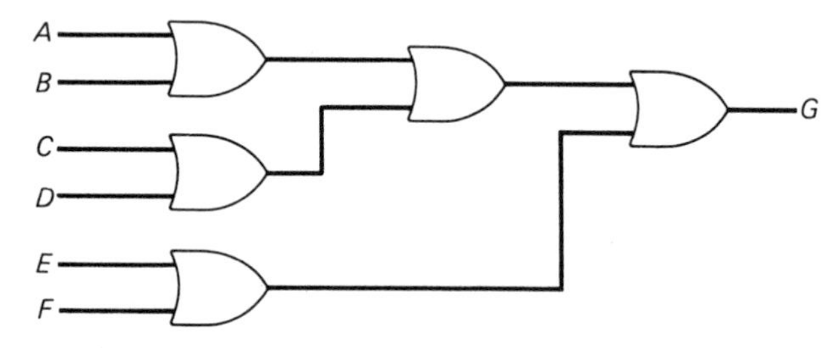

Problem 4.22

4.23 Replace the OR gates in Probs. 4.21 and 4.22 by AND gates and show that

$$F = ABCD$$

and

$$G = ABCDEF$$

Therefore cascaded AND gates may also be used to *expand input*.

4.24 The Boolean function F is given by the truth table of voltages, as shown. Find F as a function of A and B if

(a) negative logic is used

(b) negative logic is used on the inputs and positive logic is used on the output

(c) positive logic is used on the inputs and negative logic is used on the output

Problem 4.24

A (*volts*)	B (*volts*)	F (*volts*)
0	0	0
0	5	5
5	0	5
5	5	5

4.25 Show that

$$\begin{aligned} f &= A \oplus B \\ &= (\bar{A} + \bar{B})(A + B) \\ &= \overline{AB}\ A + B) \\ &= \overline{\overline{(\overline{AB})A}} + \overline{\overline{(\overline{AB})B}} \\ &= (\overline{AB} \text{ NAND } A) \text{ NAND } (\overline{AB} \text{ NAND } B) \end{aligned}$$

and that this result may be used to find the circuit of Prob. 4.6.

5

COMBINATIONAL NETWORKS

There are two basic types of digital networks, both of which, of course, are interconnections of logic gates. The first type, which we consider in this chapter, is called a *combinational* network and is characterized by the fact that the value of the output variable or variables at each instant is determined by the values of the input variables at that instant.

As an example, the network shown in Fig. 5.1 is a combinational network

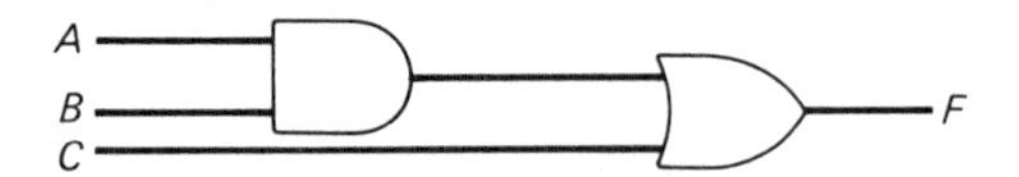

Figure 5.1 *Example of a combinational network*

because the output F, given by

$$F = AB + C \tag{5.1}$$

clearly depends at any instant on the values of the inputs at that instant. This is also seen from the truth table for F, given earlier in Table 3.3. In general, any circuit that can be represented by a truth table, as previously defined, or by a function such as Eq. (5.1) is a combinational circuit.

The other type of digital circuit is the *sequential* circuit, in which the outputs depend on previous inputs as well as present inputs. In other words, a sequential circuit "remembers" the sequence of inputs in its immediate past history. Sequential circuits will be discussed in Chaps. 7 and 9.

Let us consider, as an example of a sequential circuit, the network shown in Fig. 5.2. Suppose at some particular instant we have the inputs $A = B = 0$. This

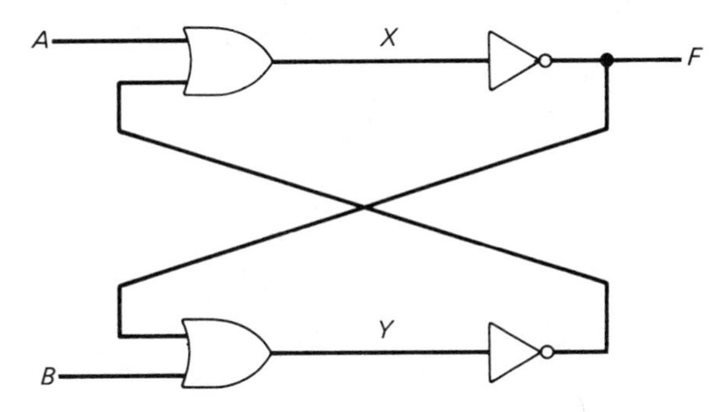

Figure 5.2 *Example of a sequential network*

information is not enough to determine F, since for these inputs F may be *either* 0 or 1. To see this, let us suppose first that $F = 0$. Then we have

$$Y = B + F = 0$$

and $\bar{Y} = 1$. Therefore we have

$$X = A + \bar{Y} = 1$$

and, of course, $F = \bar{X} = 0$. On the other hand, suppose $F = 1$. In this case we have $Y = 1$, $\bar{Y} = 0$, $X = 0$, and, of course, $F = 1$. Thus the inputs A and B do not, by themselves, determine whether F is 0 or 1.

As we pointed out earlier, in Sec. 4.2, there are two basic problems in digital circuit theory, known as *analysis* and *synthesis*. The analysis of a circuit consists in finding its output (or outputs) given the circuit and its inputs. The synthesis of a circuit consists in finding the circuit itself, given the inputs and outputs. In Chap. 4 we considered many relatively simple examples of both analysis and synthesis. In this chapter we consider more systematic methods of analyzing and synthesizing circuits. We will also consider a very powerful method of simplifying a Boolean expression, called the *Karnaugh map* method. This procedure will allow us to obtain the simplest Boolean expression for any function that is to be synthesized.

5.1 Network Analysis

The analysis of a logic circuit can be performed in a straightforward way by starting at the inputs and tracing forward through the circuit to the required output. As we proceed through each gate we must note the function appearing as its output, so that when we arrive at the output we may write its Boolean expression.

To illustrate, let us analyze the logic circuit of Fig. 5.3; that is, let us find the output function F. The vertical dashed lines shown represent *gate levels* or *logic levels*; a level indicates a gate that is encountered as we trace from the output along a path of interconnections to an input.

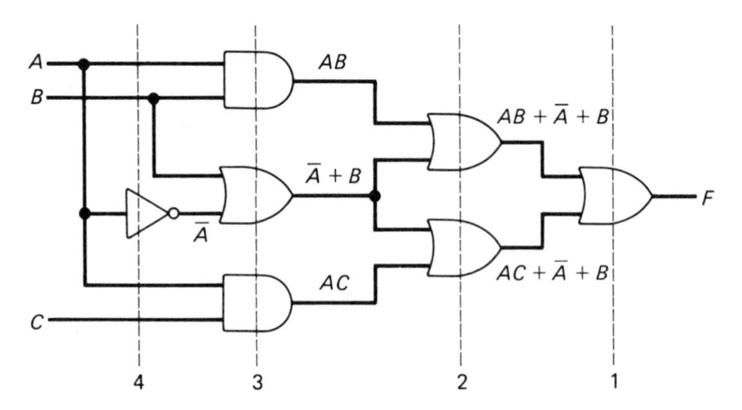

Figure 5.3 *Logic circuit showing gate levels*

A systematic way of analyzing a circuit is to proceed from the inputs and trace through the circuit to the output, labeling (or mentally noting) the outputs of all the gates of each level. The last gate level noted (level 1) is the output F, and the circuit is analyzed.

In the case of Fig. 5.3, the first to be labeled is level 4, which is the output $\bar{A}$ of the inverter. Then level 3 and level 2 are noted, in that order, and finally, at level 1 we have

$$F = (AB + \bar{A} + B) + (AC + \bar{A} + B)$$

which may be simplified to

$$F = \bar{A} + B + C \tag{5.2}$$

It is generally easier to analyze circuits with only AND, OR, and NOT gates because of the familiar operations they represent. The analysis procedure is exactly the same, however, regardless of the type of gates in the network. To illustrate the analysis of a network of NAND gates, let us find F in Fig. 5.4.

This is a two-level circuit, so that upon labeling the leftmost level, as shown, we have only to write the expression for F. The result is

$$F = \overline{(\overline{AB})(\overline{BC})}$$

or

$$F = AB + BC \tag{5.3}$$

This example is simple enough to analyze, but circuits of NAND or NOR gates often present difficulties because of the number of times complementation must be

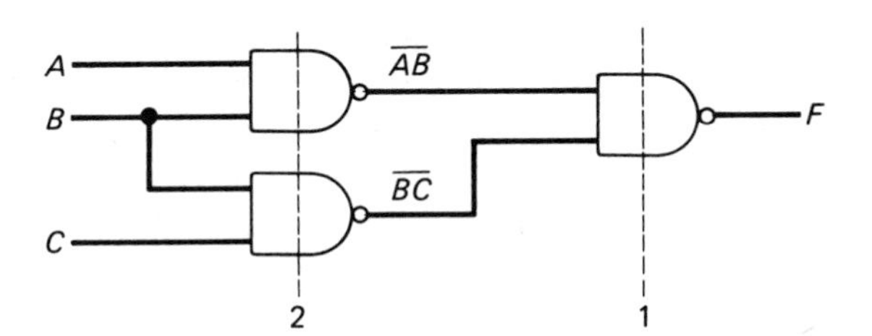

Figure 5.4 *Network of NAND gates*

applied. Either De Morgan's laws must be continually used, or else the output will be a function involving many complementation operations. An alternative approach is to replace NAND and NOR gate circuits by their equivalent AND, OR, and NOT gate circuits and then analyze the result. Of course, this may be done by replacing NAND and NOR gates directly by the equivalent cascaded AND–NOT or OR–NOT gates, which were shown in Figs. 4.11 and 4.17, respectively. But there is another, very elegant, way of converting NAND and NOR circuits to AND, OR, and NOT circuits, which we now consider.

The procedure, in general, is to replace NAND and NOR gates in odd-numbered levels in the circuit by their De Morgan equivalent gates, shown in Figs. 4.16 and 4.22, respectively. Then, since the nodes on the gate symbols represent inversions, two nodes in one line between gates (two inversions) cancel each other and may be removed. To illustrate the process with the circuit in Fig. 5.4, let us replace the NAND gate at level 1 by its De Morgan equivalent, obtaining Fig. 5.5(a). Then removing the two pairs of nodes in the two lines between the two levels

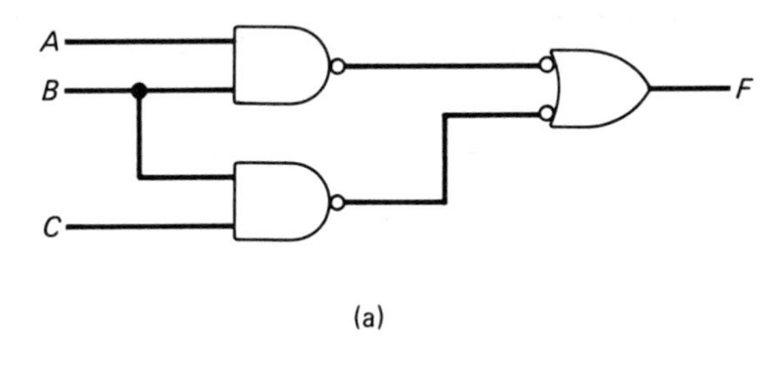

(a)

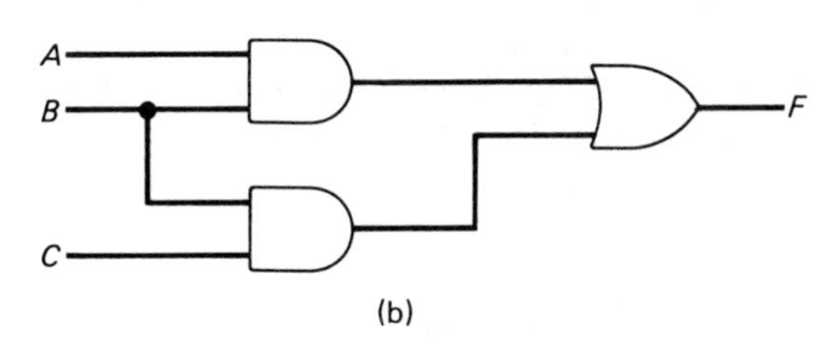

(b)

Figure 5.5 *(a) NAND gate circuit and (b) its equivalent AND–OR gate circuit*

yields the AND–OR circuit shown in Fig. 5.5(b). As the reader may easily verify, the output of this circuit is Eq. (5.3), as before.

As a more complicated example, let us change the circuit in Fig. 5.6 to an equivalent one made up of AND, OR, or NOT gates and then find F. Replacing the NAND and NOR gates of levels 1 and 3 by their De Morgan equivalents, we obtain the circuit in Fig. 5.7(a). We may simply remove the nodes at levels 1 and 2, since their effects cancel. However, at level 3 we must note the inputs before removing the nodes. In every case, because of the nodes, the input is $\bar{A}$, $\bar{B}$, or $\bar{C}$, so that removing the nodes must be accompanied by changing the inputs to their complements. The final result using AND or OR gates is shown in Fig. 5.7(b). Analysis of this circuit yields, in simplified form,

$$F = \bar{A} + \bar{B} + \bar{C} \tag{5.4}$$

As a last example, let us consider the circuit in Fig. 5.8, which contains AND, OR, NOT, NAND, and NOR gates. Replacing the NAND and NOR gates at levels 1 and 3 by their De Morgan equivalents, we obtain the circuit in Fig. 5.9(a). Removing the canceling nodes at levels 1 and 2, and the node at level 3 and the

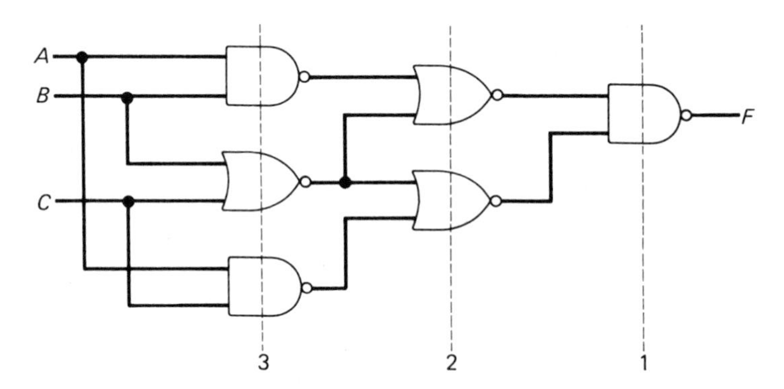

Figure 5.6 *NAND–NOR circuit*

inverter in line with it, yields Fig. 5.9(b). There is still a node at an input of an OR gate at level 3, which indicates that the input C is inverted. Since C is also inverted by the NOT gate, we may remove the node and the NOT gate and replace C by $\bar{C}$, as shown in Fig. 5.9(c). We now have only AND and OR gates and, of course, the indicated inversion. The output F may now be shown to be

$$F = AB \tag{5.5}$$

In most cases it is probably as easy to perform the analysis on the original circuit as it is to transform the circuit to an equivalent AND–OR–NOT circuit and

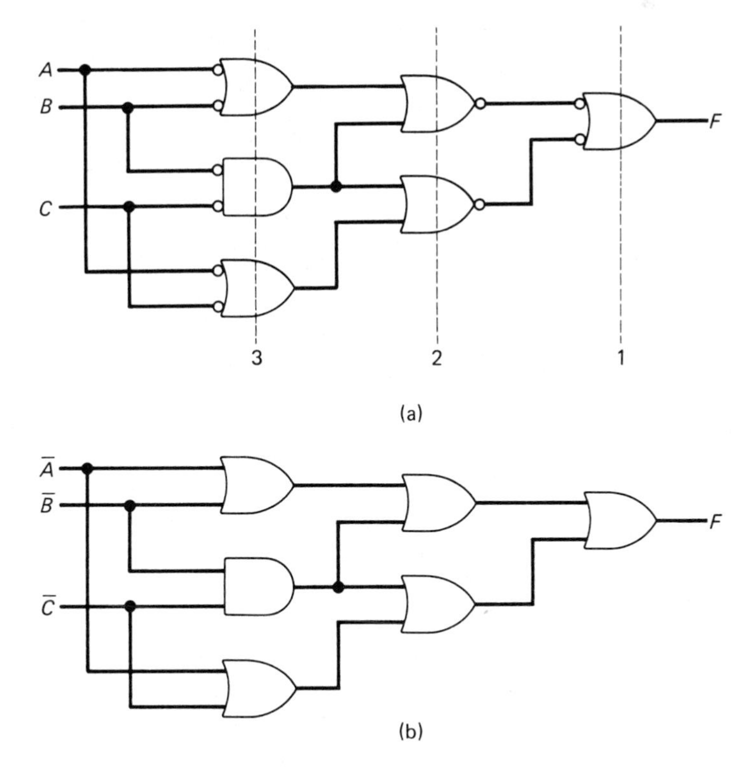

Figure 5.7 *Circuits equivalent to those in Fig. 5.6*

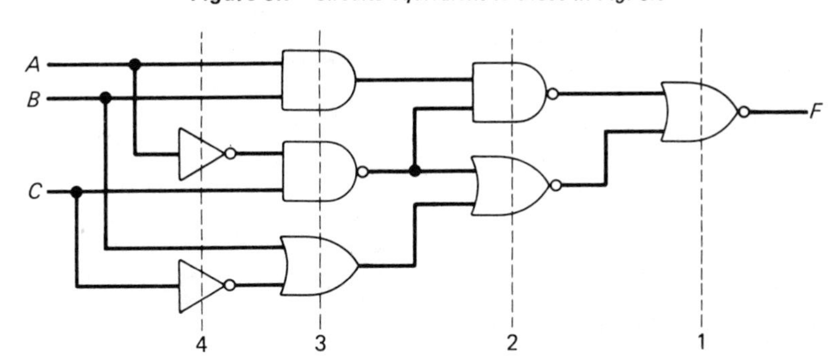

Figure 5.8 *AND–OR–NOT–NAND–NOR circuit*

analyze the result. However, in many cases, particularly if all the gates are NAND gates and/or NOR gates, we may visualize the transformations without actually making them. In this case, an example of which is Fig. 5.4, we may perform the analysis on the original circuit using only AND, OR, and NOT operations.

As we will see in the remainder of this chapter, gate transformations are also convenient in many cases in network synthesis.

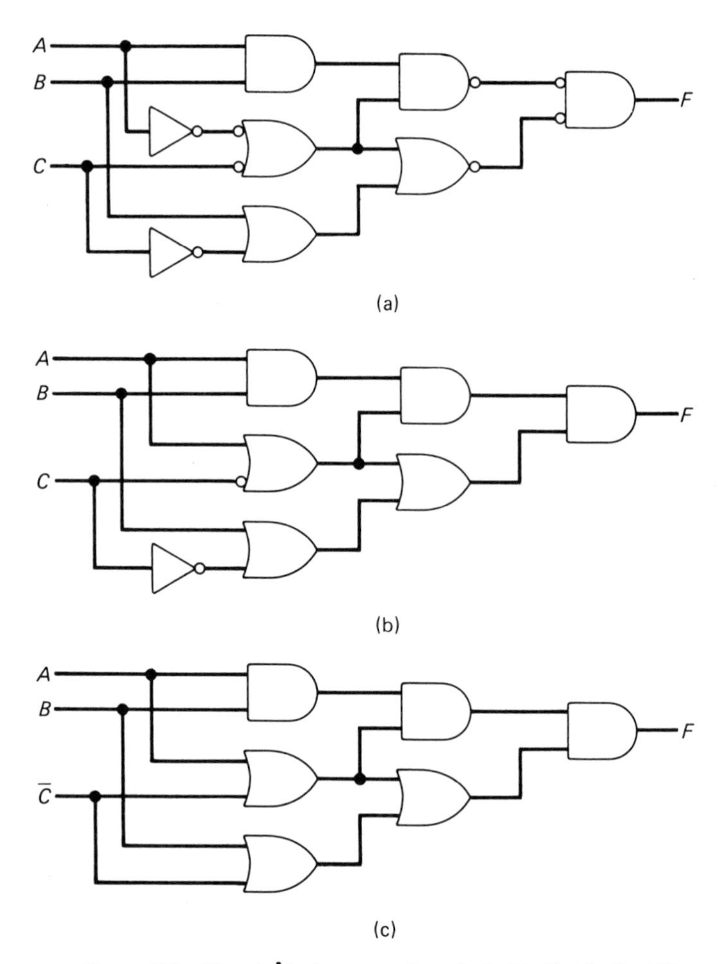

(a)

(b)

(c)

Figure 5.9 *Steps in finding a circuit equivalent to that in Fig. 5.8*

EXERCISES

5.1.1 Analyze the given circuit in its original form.

Ans: $F = B + \bar{C}$

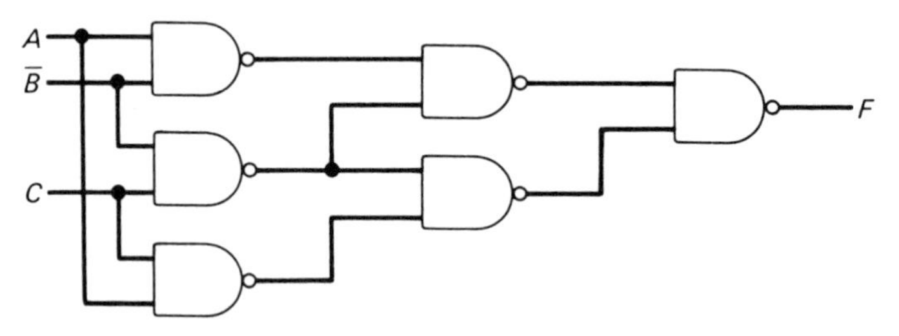

Exercise 5.1.1

5.1.2 Check the result of Ex. 5.1.1 by first transforming the circuit to an equivalent AND–OR–NOT circuit.

5.1.3 Change all the NAND gates in Ex. 5.1.1 to NOR gates and repeat Exs. 5.1.1 and 5.1.2.

Ans: $F = B\bar{C}$

5.2 Network Synthesis Using AND, OR, and NOT Gates

The synthesis of a combinational circuit begins, as we have said, with the output function F, which may be defined by a Boolean expression or by a truth table. In either case the input variables will be specified. Therefore the inputs and output (or outputs) are given, and our task is to find, or *synthesize*, or *realize*, the network.

Every occurrence of a variable in the function F, whether complemented or uncomplemented, is called a *literal*, and the number of literals is a measure of how complicated the circuit will be. For example, the circuit in Fig. 5.3 was shown to have an output

$$F = AB + \bar{A} + B + AC + \bar{A} + B \tag{5.6}$$

which was given in simplified form in Eq. (5.2) by

$$F = \bar{A} + B + C \tag{5.7}$$

The latter version contains three literals, whereas the former contains eight. The implementation of Eq. (5.6) required seven gates, but an equivalent circuit implementing Eq. (5.7) requires only two gates, as we see in Fig. 5.10.

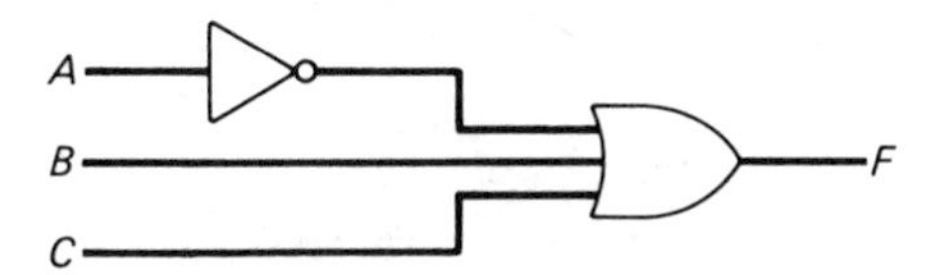

Figure 5.10 *Implementation of Eq. (5.7)*

Another measure of the complexity of a circuit, which is also related to the number of literals in its output, is the total number of gate inputs. In Fig. 5.3, for example, there are thirteen inputs, whereas the equivalent circuit in Fig. 5.10 involves only four inputs. Finally, as a practical matter, a circuit with many gate levels has a higher overall propagation delay time (see Sec. 4.7) than one with fewer gate levels, and this could be a critical consideration in a complex circuit. The circuit in Fig. 5.3 has four gate levels, whereas its equivalent in Fig. 5.10 has only two.

From the standpoint of mathematical elegance it is obviously better to synthesize a function with as few elements and gate levels as possible. This is true from the point of view of cost as well, although with the integrated-circuit techniques of today, which make hundreds and thousands of gates available on a single chip, ideas about saving cost by circuit simplification are not as applicable as they once were. Nevertheless, in many cases it is desirable to simplify the output function before synthesizing it, and we have several methods at our disposal for doing so. One method is to apply the Boolean theorems of Chap. 3 to the output function whenever possible. Another method, which we will consider in Sec. 5.4, is the *Karnaugh map* procedure. There are many other techniques which the interested reader may find in the literature. In this section we will assume that the output functions are given in the desired form and will not be concerned with simplication procedures.

It is always possible to synthesize a given output F with a circuit having only two levels, not counting inverters. This is because the function may be expressed as a sum of products or as a product of sums. For example, suppose we have the sum of products

$$F = AB\bar{C} + \bar{B}D + \bar{C}D \tag{5.8}$$

This function can be implemented, if both the complemented and the uncomplemented inputs (commonly referred to as *double rail* inputs) are available, with a single OR gate at level 1 and three AND gates at level 2. The OR gate will accomplish the addition of the three terms in the sum, and one AND gate will perform each of the three products. The result is shown in Fig. 5.11, where inverters have been added to make $\bar{B}$ and $\bar{C}$ available.

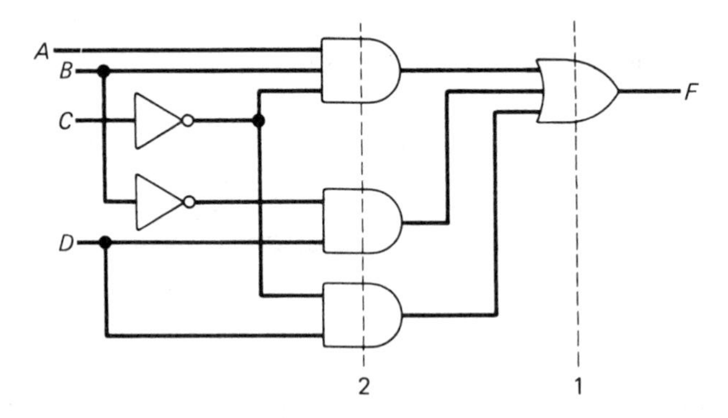

Figure 5.11 *Implementation of Eq. (5.8)*

In the general case of an output expressed in SOP form, aside from the inverters, only one OR gate is needed at level 1 to take care of the sum operation, and an AND gate is required at level 2 for each product operation. The result is sometimes called an AND–OR network, since the AND gates are encountered first and the OR gate second in proceeding from input to output. In the case of multiple outputs level 1 will contain an OR gate for each output.

If we express the output in POS form, such as

$$F = (A + B + C)(B + D) \tag{5.9}$$

then the so-called OR–AND network is always possible. This is a network containing an AND gate for each output at level 1 to accomodate the product operations, and OR gates at level 2 to account for the sum operations. Of course, any complementations necessary will require inverters. As an example, F in Eq. (5.9) is synthesized in Fig. 5.12.

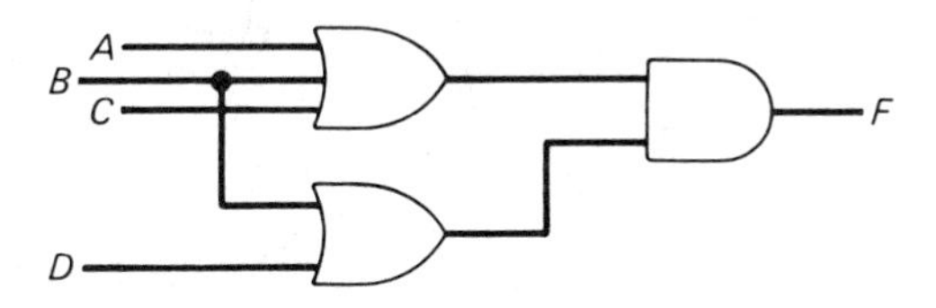

Figure 5.12 *Implementation of Eq. (5.9)*

The OR–AND or AND–OR circuits may always be obtained and from the standpoint of delay time are generally the best solutions, since they have only two levels, aside from any necessary inverters. However, it may be possible to express the output in another way in which there are fewer literals. In this case the circuit may be constructed with simpler hardware. For example, let us write Eq. (5.9) in the equivalent form

$$F = B + D(A + C) \tag{5.10}$$

There are now only four literals as opposed to five in Eq. (5.9). The circuit realization is shown in Fig. 5.13, where it may be seen that the hardware is slightly simpler, the three-input OR gate of Fig. 5.12 being somewhat more complex than a two-input OR gate.

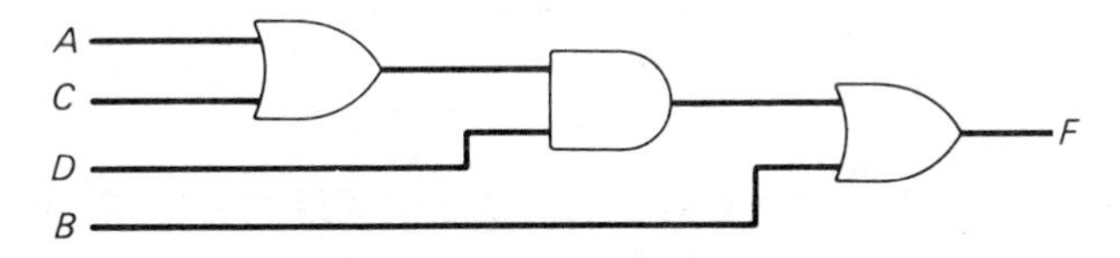

Figure 5.13 *Alternative implementation of Eq. (5.9)*

The decision as to which of several implementations is better may depend on several factors. If propagation delay time is an important consideration, the circuit with fewer levels might be chosen. In this respect, Fig. 5.12 (with two levels) is preferable to Fig. 5.13 (with three levels). On the other hand, if it is important to minimize the complexity of the hardware, Fig. 5.13 is preferable.

The synthesis of the circuit in Fig. 5.13 illustrates a general method. Noting that the function is a sum of products or a product of sums, we begin at level 1 with an

OR gate or an AND gate. In the case of Eq. (5.10) we see that F is a sum of two inputs, B and $D(A + C)$. Therefore an OR gate with these two inputs is required at level 1. Next we analyze the inputs as sums of products or products of sums and proceed to level 2, and so on. In the case of Eq. (5.10), B has been observed at level 1 as an input. The other input $D(A + C)$ is a product of two inputs, D and $A + C$, requiring an AND gate at level 2. Finally, the input $A + C$ is realized with an OR gate at level 3. In the general case the odd levels will contain OR (AND) gates and the even levels will contain AND (OR) gates if the function is a sum of products (product of sums). Inverters may also be required, of course.

In this section we have limited ourselves to AND–OR–NOT circuits. We will consider NAND and NOR synthesis in the next section.

EXERCISES

5.2.1 Synthesize the function

$$F = ABC + AB\bar{D} + AE$$

with a two-level network (not counting the inverters) of AND, OR, or NOT gates.

Ans:

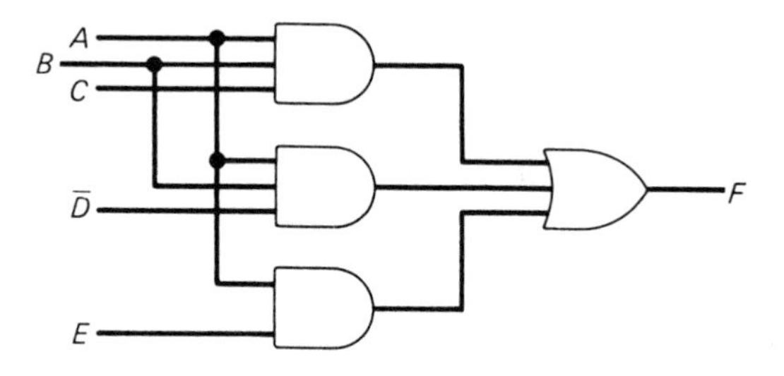

Exercise 5.2.1

5.2.2 Rewrite F in Ex. 5.2.1 so that only five literals appear and synthesize it with a network of eight inputs.

Ans: $F = A[B(C + \bar{D}) + E]$

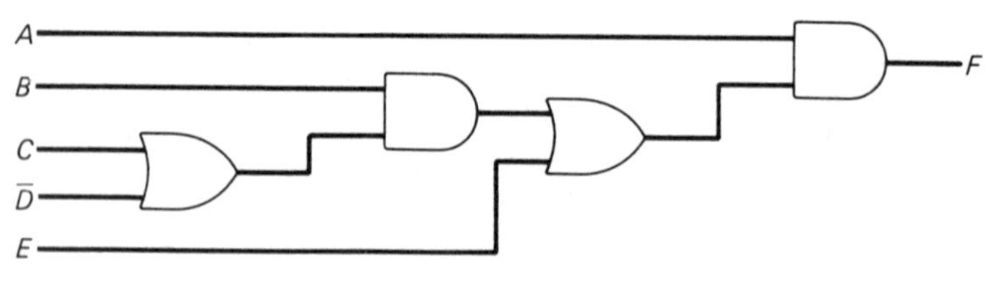

Exercise 5.2.2

5.2.3 Find an OR–AND implementation of

$$F = AB\bar{C} + A\bar{B}C + \bar{A}BC + \bar{A}\bar{B}C + \bar{A}\bar{B}\bar{C}$$

(*Suggestion:* Find $\bar{F}$ in SOP form and complement it.)

Ans:

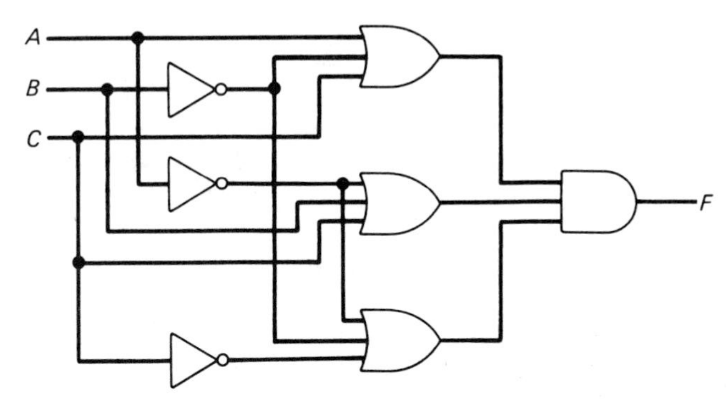

Exercise 5.2.3

5.3 NAND and NOR Gate Synthesis

In most logic families the basic gate is either a NAND gate or a NOR gate. Also, since NAND or NOR gates may be used to implement the complete set of Boolean operations, a designer does not need other gates. For these reasons, as well as others, we must consider systematic ways to synthesize networks with only NAND gates, or with only NOR gates.

In the case of NAND gate synthesis, we have already considered (in Sec. 4.4) a systematic method, which we may call the double complement method. Since this procedure was illustrated previously, we simply reiterate it here for the general case.

If the function F to be synthesized is expressed as an SOP, then the double complement $\bar{\bar{F}}$ is the complement of the complement of an SOP, which by De Morgan's law is the complement of a POS. This is the form we need for direct NAND gate synthesis.

For example, let F be given by

$$F = AB + BC \tag{5.11}$$

Then we have

$$\begin{aligned}\bar{\bar{F}} &= \overline{\overline{AB + BC}} \\ &= \overline{(\overline{AB})(\overline{BC})} \\ &= (\overline{AB}) \text{ NAND } (\overline{BC})\end{aligned}$$

Also, since

$$\overline{AB} = A \text{ NAND } B$$

and

$$\overline{BC} = B \text{ NAND } C$$

we may synthesize Eq. (5.11) as shown in Fig. 5.14.

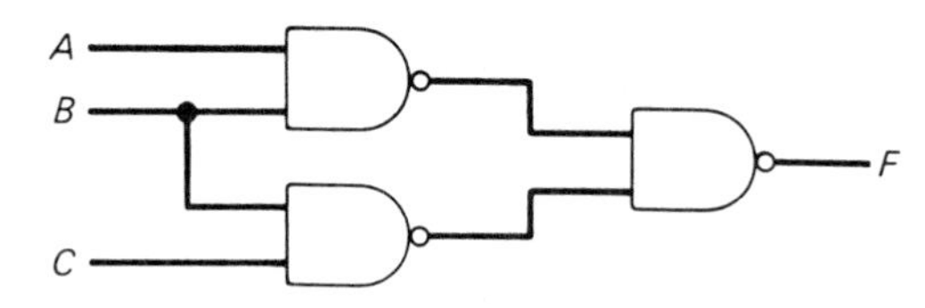

Figure 5.14 *NAND gate implementation of Eq. (5.11)*

A direct NOR gate synthesis may be carried out by the double complement procedure by starting with the output function in POS form. This was discussed earlier in Sec. 4.5 and will be illustrated here with the function

$$F = AB + \bar{A}\bar{B} \tag{5.12}$$

The complement of this function is the SOP of the minterms that are missing and therefore is given by

$$\bar{F} = A\bar{B} + \bar{A}B$$

We thus have

$$F = \overline{A\bar{B} + \bar{A}B}$$

By De Morgan's law this may be written as a POS, and applying double complementation will lead to the complement of an SOP, the form suitable for implementation with a NOR gate. However, in this case we already have the complement of an SOP, so we may write

$$F = (A\bar{B}) \text{ NOR } (\bar{A}B)$$

Finally, we note that

$$A\bar{B} = \bar{A} \text{ NOR } B$$

and

$$\bar{A}B = A \text{ NOR } \bar{B}$$

Therefore the NOR gate realization is as shown in Fig. 5.15. The complemented functions may be obtained also with NOR gates, as was shown in Sec. 4.5.

We may also perform NAND and NOR synthesis by using, in reverse, the idea discussed in Sec. 5.1 of replacing NAND and NOR gates by AND and OR gates. That is, we synthesize the given function using AND, OR, and NOT gates and transform the result to a NAND or a NOR circuit using the De Morgan equivalent gates. This procedure is sometimes called the *transform method.*

Let us first consider the general case of NAND gate synthesis. If the function to

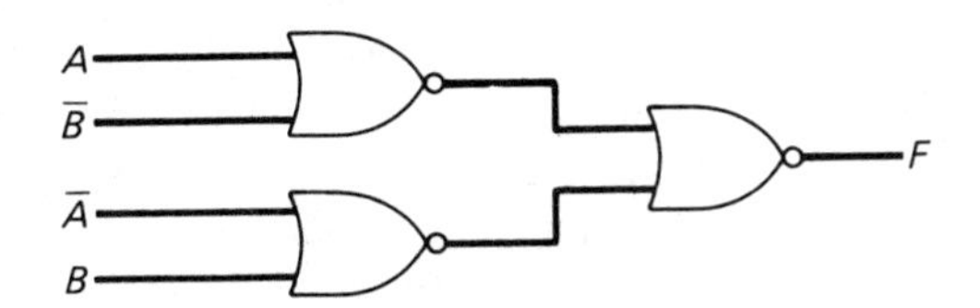

Figure 5.15 *NOR gate implementation of Eq. (5.12)*

be synthesized is in the SOP form, an AND–OR–NOT synthesis results in OR gates at the odd-numbered levels and AND gates at the even-numbered levels. A possible exception is the last level (or the input level), where NOT gates may appear. Placing nodes, indicating complementation, on the inputs of the OR gates (odd levels) and on the outputs of the AND gates (even levels) does not change the output, since these nodes form canceling pairs. An exception is the last level, where nodes placed on the input of an OR gate (in the odd level case) must be compensated for by complementing the input variable. The circuit, by virtue of De Morgan's equivalent NAND gate, is then a NAND circuit. Of course, any complemented inputs must be obtained with a NAND gate or, if desired, a NOT gate.

To illustrate the procedure, let us realize with NAND gates the function F of Eq. (5.11). Its AND–OR realization is shown in Fig. 5.16(a). Placing nodes on the

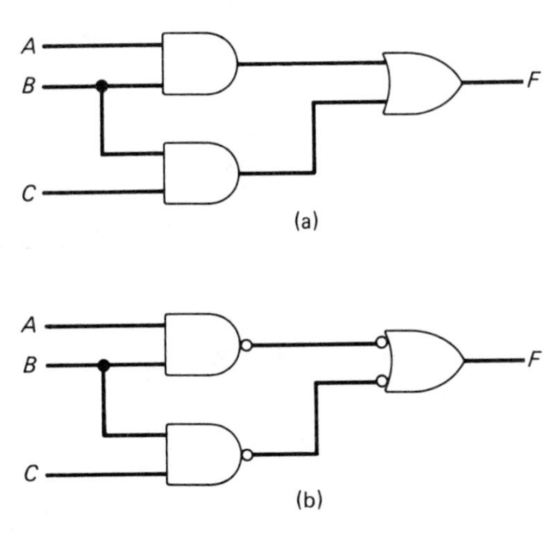

Figure 5.16 *Steps in the NAND gate synthesis of Eq. (5.11)*

inputs of the NOR gate and the outputs of the NAND gates yields Fig. 5.16(b), which is a NAND circuit. Using the more familiar form of the output NAND gate, we have the result previously given in Fig. 5.14.

As another, more involved, example let us synthesize with NAND gates the function

$$F = AB(C + \bar{D}) + AE \tag{5.13}$$

The AND–OR realization is shown in Fig. 5.17(a), to which we shall apply the transform method. The result is shown in Fig. 5.17(b), with the equivalent result given in Fig. 5.17(c). The function in Eq. (5.13) was synthesized in two other ways in Exs. 5.2.1 and 5.2.2, and the reader may wish to compare the three results.

In the case of NOR gate synthesis, if the output is expressed in POS form, its AND–OR gate synthesis will contain AND gates at odd levels and OR gates at even levels. Application of the transform method exactly as described in NAND gate synthesis will lead to a network of NOR gates. To illustrate the procedure, let us reconsider the function of Eq. (5.13), written in the POS form

$$F = A[B(C + \bar{D}) + E] \tag{5.14}$$

This form was synthesized earlier in Ex. 5.2.2, and the resulting circuit is in the proper form for conversion to a NOR gate circuit. Applying the transform method

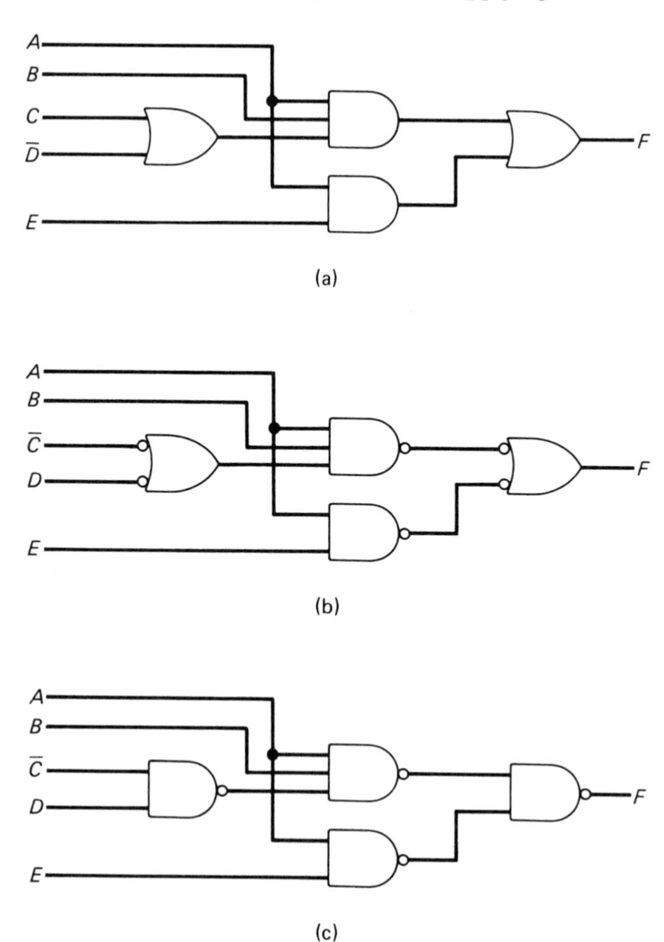

Figure 5.17 *Steps in the NAND gate synthesis of Eq. (5.13)*

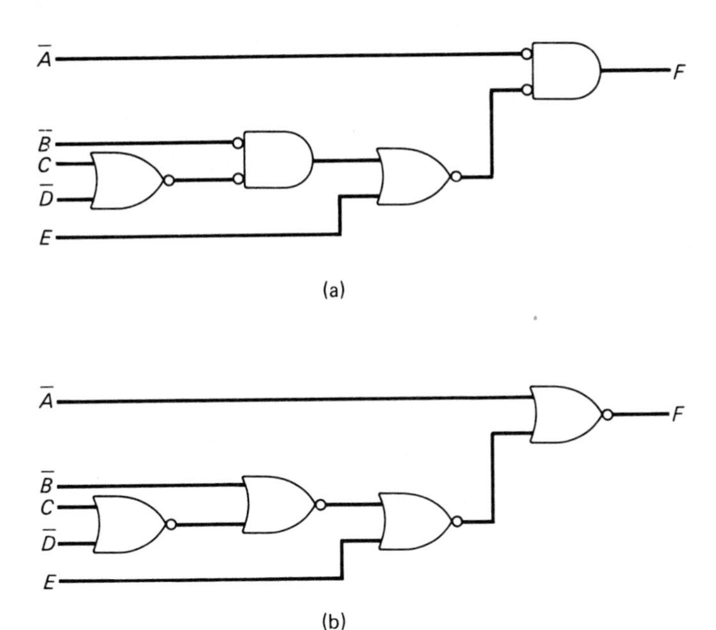

Figure 5.18 *Steps in the NOR gate synthesis of Eq. (5.14)*

to the circuit yields Fig. 5.18(a), which is equivalent to the NOR gate circuit in Fig. 5.18(b).

In the next section we will digress briefly to consider the *Karnaugh map*, which is a procedure that may be used to minimize the number of literals in a Boolean function or to express the function in a standard form for NOR or NAND synthesis. In Chap. 6 we will apply synthesis techniques to obtain a number of important combinational circuits, and we will find the Karnaugh map method very useful.

EXERCISES

5.3.1 Obtain a NAND gate synthesis of

$$F = A\bar{B} + \bar{A}B + C$$

Ans:

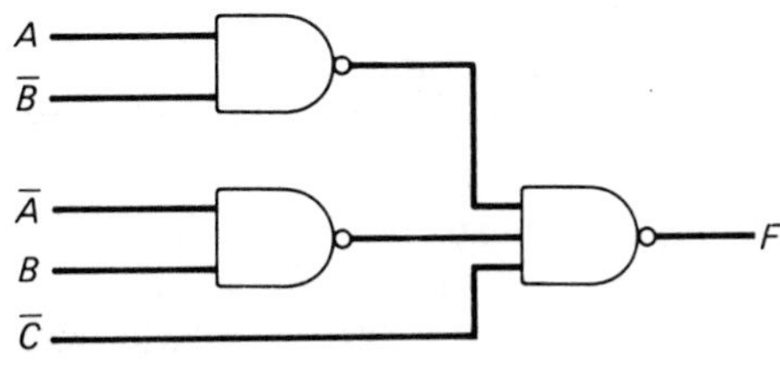

Exercise 5.3.1

5.3.2 Obtain a NOR gate synthesis of

$$F = (A + \bar{B})(\bar{A} + B)C$$

Ans:

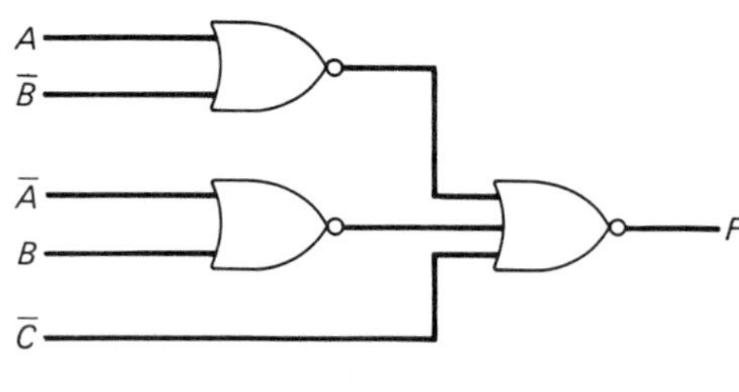

Exercise 5.3.2

5.4 Karnaugh Maps

The Boolean function F to be synthesized may be simplified beforehand so that the resulting circuit is as concise as possible. The simplification may be performed by means of the Boolean theorems considered in Chap. 3, but for a complex function it is sometimes difficult to know when the process has been carried as far as possible. A very powerful method of going directly to the simplest, or *minimal*, expression of F is the *Karnaugh map* procedure, which we consider in this section. Although Karnaugh maps may be applied to more complex functions, we will restrict ourselves to functions of three and four variables. Functions of two variables are easy enough to deal with using the Boolean theorems, and the Karnaugh maps become rather involved and impractical for functions of a large number of variables.

A Karnaugh map is simply an array of rectangles corresponding to every possible minterm of the function in question. If a minterm is present in the function, its corresponding rectangle contains a 1; otherwise, the rectangle is left blank, indicating that a 0 belongs in it. The Karnaugh map is thus simply another form of a truth table. However, the rectangles in the map are arranged, as we shall see, so that a simplified form of the function may be obtained by inspection of the map.

The Karnaugh map for a three-variable function $F(A, B, C)$ has eight rectangles, arranged as shown in Fig. 5.19(a). An alternative form, which is useful if F is mapped from a truth table, is shown in Fig. 5.19(b). For example, if minterm $\bar{A}\bar{B}\bar{C}$ is present in F, a 1 is placed in the rectangle constituting the first column ($\bar{A}\bar{B}$) and first row ($\bar{C}$) of Fig. 5.19(a). Similarly, since $\bar{A}\bar{B}\bar{C}$ corresponds in the truth

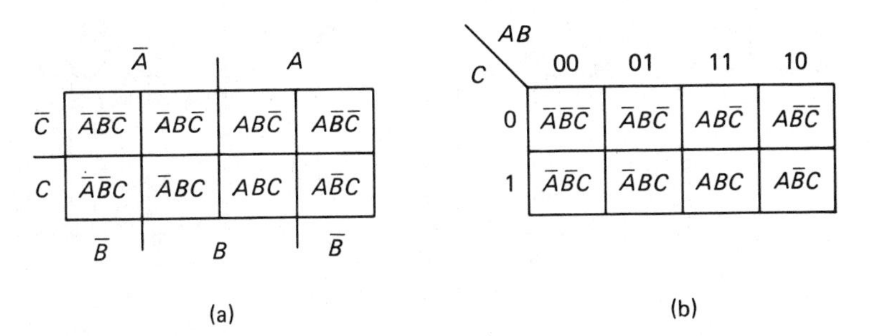

Figure 5.19 *Two versions of the general three-variable Karnaugh map*

table to 0 0 0, the 1 is placed in position 0 0 (for AB) and 0 (for C) in Fig. 5.19(b). This is, of course, the same rectangle as in Fig. 5.19(a).

As an example, the function

$$F = ABC + \bar{A}\bar{B}C + \bar{A}BC + \bar{A}B\bar{C} \tag{5.15}$$

is *mapped* as shown in Fig. 5.20. The rectangles corresponding to minterms ABC,

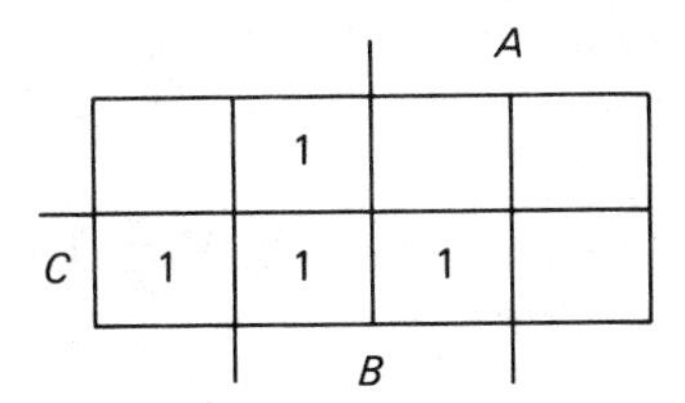

Figure 5.20 *Karnaugh map of Eq. (5.15)*

$\bar{A}\bar{B}C$, $\bar{A}BC$, and $\bar{A}B\bar{C}$ contain 1's, and all the other rectangles are blank, indicating 0's.

The lines extending beyond the map in Figs. 5.19(a) and 5.20 serve to separate A from $\bar{A}$, B from $\bar{B}$, and C from $\bar{C}$. The $\bar{B}$ rectangles are those in the first and last column, so that the map is "continuous," in that the left column is "adjacent" to the right column. It should also be noted that in Fig. 5.19(b) the numbering of the rectangles is such that combinations in adjacent rectangles differ only in the value of one variable. Therefore two-rectangle groups (blocks of two *adjacent* rectangles) are independent of one variable, and four-rectangle groups are independent of two variables.

As examples, in Fig. 5.21(a) the two encircled 1's constitute

$$\begin{aligned} F_1 &= \bar{A}BC + \bar{A}\bar{B}C \\ &= \bar{A}C \end{aligned} \tag{5.16}$$

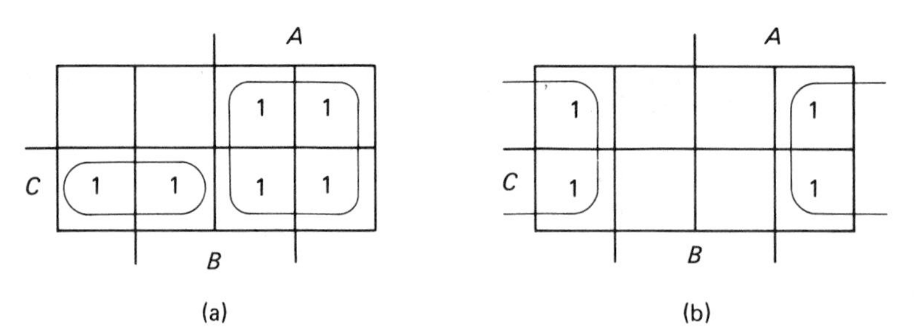

Figure 5.21 *Maps of two- and four-rectangle terms*

and the four encircled 1's constitute

$$F_2 = AB\bar{C} + A\bar{B}\bar{C} + ABC + A\bar{B}C$$
$$= A \qquad (5.17)$$

These are easy to see by noting that in the case of Eq. (5.16), the two 1's fill up the rectangles common to $\bar{A}$ and C, and in the case of Eq. (5.17), the four 1's completely fill the rectangles labeled A. As another example, the map shown in Fig. 5.21(b) is that of

$$F_3 = \bar{B}$$

since the 1's completely fill up region $\bar{B}$. This also illustrates the continuity of the left and right columns of the map.

The general four-variable Karnaugh map is shown in two versions in Fig. 5.22(a) and (b). These correspond, respectively, to Fig. 5.19(a) and (b) in the three-variable case. Again, the overlapping lines in Fig. 5.22(a) separate the complemented and uncomplemented states, and the numbers in Fig. 5.22(b) identify the minterms. Moreover, the map is "continuous" in that the first row and last row are "adjacent" (parts of $\bar{D}$), and the first and last columns are "adjacent" (parts of $\bar{B}$).

In the four-variable case, two-rectangle groups of 1's represent functions that are independent of one variable, four-rectangle groups are independent of two variables, and eight-rectangle groups are independent of three variables. For example, in Fig. 5.23 the two encircled 1's constitute

$$F_1 = AB\bar{C}\bar{D} + A\bar{B}\bar{C}\bar{D}$$
$$= A\bar{C}\bar{D} \qquad (5.18)$$

the four encircled 1's constitute

$$F_2 = \bar{A}\bar{B}\bar{C}\bar{D} + \bar{A}B\bar{C}\bar{D} + \bar{A}\bar{B}\bar{C}D + \bar{A}B\bar{C}D$$
$$= \bar{A}\bar{C} \qquad (5.19)$$

	$\bar{A}$		A		
$\bar{C}$	$\bar{A}\bar{B}\bar{C}\bar{D}$	$\bar{A}B\bar{C}\bar{D}$	$AB\bar{C}\bar{D}$	$A\bar{B}\bar{C}\bar{D}$	$\bar{D}$
$\bar{C}$	$\bar{A}\bar{B}\bar{C}D$	$\bar{A}B\bar{C}D$	$AB\bar{C}D$	$A\bar{B}\bar{C}D$	D
C	$\bar{A}\bar{B}CD$	$\bar{A}BCD$	$ABCD$	$A\bar{B}CD$	D
C	$\bar{A}\bar{B}C\bar{D}$	$\bar{A}BC\bar{D}$	$ABC\bar{D}$	$A\bar{B}C\bar{D}$	$\bar{D}$
	$\bar{B}$	B	B	$\bar{B}$	

(a)

CD \ AB	00	01	11	10
00	$\bar{A}\bar{B}\bar{C}\bar{D}$	$\bar{A}B\bar{C}\bar{D}$	$AB\bar{C}\bar{D}$	$A\bar{B}\bar{C}\bar{D}$
01	$\bar{A}\bar{B}\bar{C}D$	$\bar{A}B\bar{C}D$	$AB\bar{C}D$	$A\bar{B}\bar{C}D$
11	$\bar{A}\bar{B}CD$	$\bar{A}BCD$	$ABCD$	$A\bar{B}CD$
10	$\bar{A}\bar{B}C\bar{D}$	$\bar{A}BC\bar{D}$	$ABC\bar{D}$	$A\bar{B}C\bar{D}$

(b)

Figure 5.22 *Two versions of the general four-variable Karnaugh map*

and the eight encircled 1's constitute

$$
\begin{aligned}
F_3 &= \bar{A}\bar{B}CD + \bar{A}BCD + ABCD + A\bar{B}CD \\
&\quad + \bar{A}\bar{B}C\bar{D} + \bar{A}BC\bar{D} + ABC\bar{D} + A\bar{B}C\bar{D} \\
&= C
\end{aligned}
\qquad (5.20)
$$

These results also follow from the fact that the 1's of F_1 completely fill the rectangles common to A, $\bar{C}$, and $\bar{D}$, those of F_2 fill the rectangles common to $\bar{A}$ and $\bar{C}$, and those of F_3 fill all the rectangles labeled C.

Once the Karnaugh map of a function is obtained, the function may be expressed as a sum of product terms by representing *all* the 1's in the map by terms in the function. The minimal expression is the one for which all the 1's are *covered* in such a way that every 1 is in the largest possible rectangle of two or four smaller rectangles in the three-variable case, and of two, four, or eight smaller rectangles in the four-variable case, and the number of "coverings" is a minimum. The rectangles may overlap, but all the 1's must be covered.

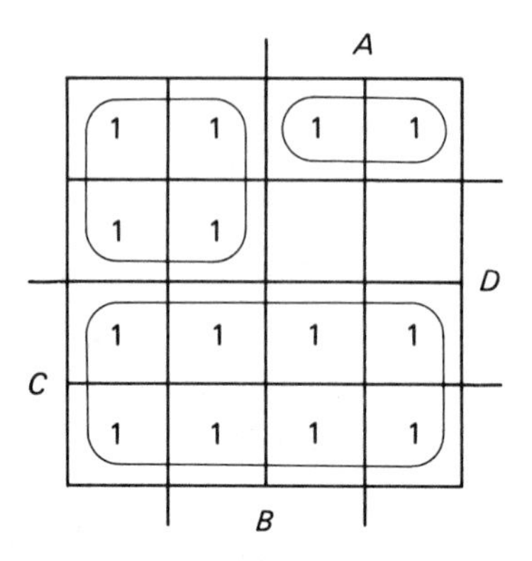

Figure 5.23 *Example of a Karnaugh map*

As an example, let us consider the Karnaugh map in Fig. 5.20, redrawn in Fig. 5.24(a). One form of the function F represented by the map was given earlier in Eq.

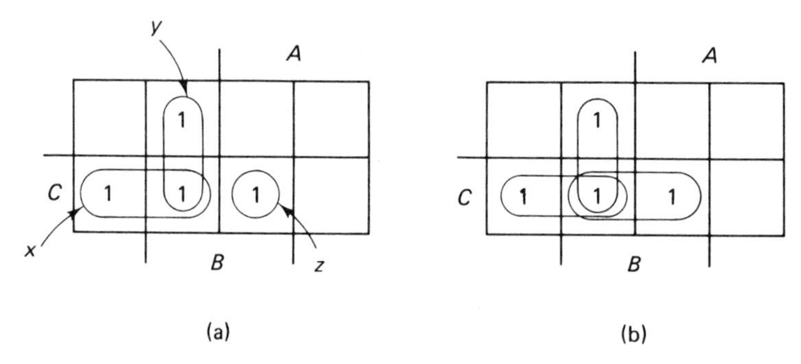

Figure 5.24 *Three-variable maps*

(5.15). One covering of the map is shown in Fig. 5.24(a) by groupings x, y, and z, corresponding, respectively, to terms $\bar{A}C$, $\bar{A}B$, and ABC. Therefore a form of the function is

$$F = \bar{A}C + \bar{A}B + ABC$$

This result is simpler than Eq. (5.15), but is it the simplest possible form? The answer is no, because the 1 in region z is not in the largest possible rectangle of two or four smaller rectangles. The minimal F will come from the covering shown in Fig. 5.24(b), in which x and y have remained the same but z is extended to cover a two-rectangle region. From this map we have the minimal function

$$F = \bar{A}C + \bar{A}B + BC$$

It may help in drawing or reading the Karnaugh map of a given function to visualize the rectangles as sets of points. Then the terms in the function may be interpreted as sets. For example, the sets A and B are shown shaded in Figs. 5.25(a) and (b). The term AB, or A AND B, represents the set of points (or group of rectangles identified with 1's) in A and in B, that is, common to A and B, as shown in Fig. 5.25(c). The term $A + B$, or A OR B, represents the points in *A or B or* both, as shown in Fig. 5.25(d). The complement $\overline{A + B}$, shown in Fig. 5.25(e), is the set *not* in $A + B$, or the set *outside* $A + B$. The set $(A + B)C$, or (A OR B) AND C, is the set common to $A + B$ and C, shown in Fig. 5.25(f). Any Boolean function may be represented by sets in this manner, and, of course, the idea may be extended to Karnaugh maps of four-variable functions, as well.

The Karnaugh maps, when viewed as sets of shaded areas as in Fig. 5.25, are sometimes called *Venn diagrams*, although in the latter case the sets are usually represented by shapes other than rectangles. The only difference between Karnaugh maps and Venn diagrams, when drawn in rectangular shapes, is that 1's are placed in the Karnaugh map rectangles where the Venn diagram rectangles are shaded. In set theory the operations AND and OR are often called *intersection* and *union* and are symbolized $\cap$ and $\cup$, respectively. However, we will continue to use AND and OR and $+$ and $\cdot$.

As a last example illustrating a Karnaugh map of a four-variable function and its relationship to point sets, let us obtain the minimal expression for the function

$$F = \overline{AB}(\overline{C + D}) + \bar{A}B\bar{C} + AB\bar{C}\bar{D} \tag{5.21}$$

We could carry out the complementation and multiplication and find the rectangles

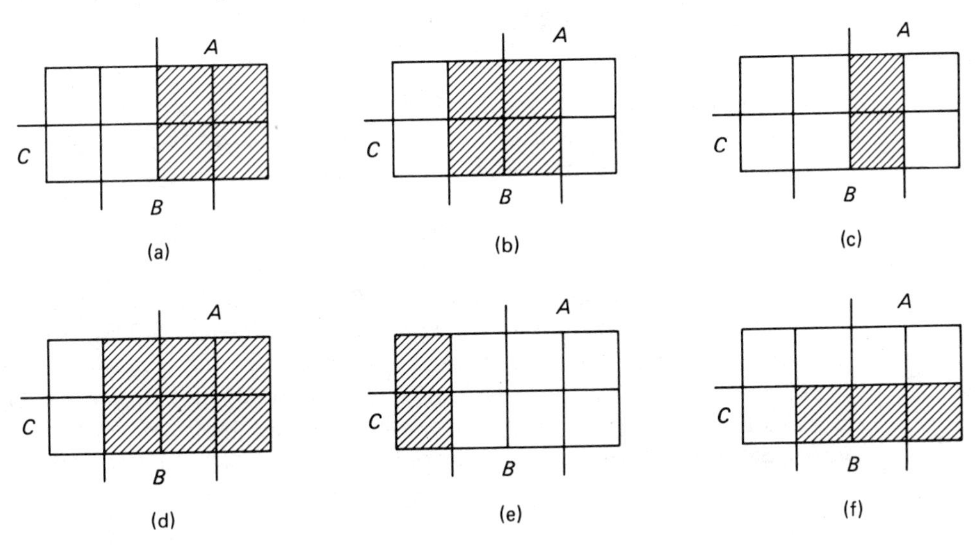

Figure 5.25 *Sets (a) A, (b) B, (c) AB, (d) A + B, (e)* $\overline{A + B}$*, and (f) (A + B)C*

in the Karnaugh map corresponding to each term in the resulting sum. However, let us apply the idea of sets to find the location of the 1's. Referring to the general four-variable map of Fig. 5.22(a), we see that set $\overline{AB}$, the set of rectangles outside those common to A and B, is made up of the rectangles of the first, second, and fourth columns. The set $\overline{C + D}$, the set outside C or D or both, is the first row of rectangles. Therefore the set $\overline{AB}(\overline{C + D})$, the set common to these two sets, is the first, second, and fourth rectangles in the first row. Therefore 1's will go in these three places. In a similar manner we may find $\bar{A}B\bar{C}$ as the first two rectangles in the second column (common to NOT A, B, and NOT C), and $AB\bar{C}\bar{D}$ as the third rectangle in the first row (common to A, B, NOT C, and NOT D). The Karnaugh map is now complete and is shown in Fig. 5.26.

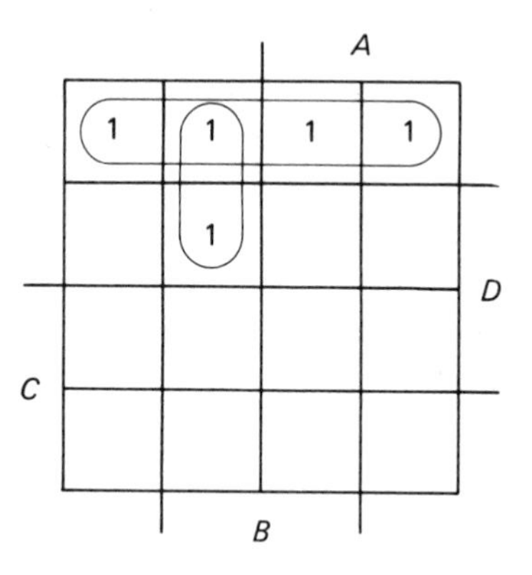

Figure 5.26 *Karnaugh map of Eq. (15.21)*

The minimal function for Fig. 5.26 is now easily seen to arise when the four 1's in the first row are covered in one term ($\bar{C}\bar{D}$) and the two 1's in the second column are covered in one term ($\bar{A}B\bar{C}$). The resulting function is

$$F = \bar{C}\bar{D} + \bar{A}B\bar{C}$$

Since the complement $\bar{F}$ of a function F has 1's in its Karnaugh map where that of F is blank, we may also write $\bar{F}$ by inspection of the Karnaugh map for F. Since this yields $\bar{F}$ in SOP form, applying De Morgan's law will then result in F in POS form. Karnaugh maps are often used this way to obtain the POS form, but in many cases $\bar{F}$ may have fewer minterms than F, so that the procedure yields F with much less effort. For example, in the case of the Karnaugh map of Fig. 5.23 we may write

$$\bar{F} = A\bar{C}D$$

and therefore

$$F = \bar{A} + C + \bar{D}$$

EXERCISES

5.4.1 Find the Karnaugh maps of

(a) $A + B\bar{C} + \bar{B}C$

(b) $A(B + C) + \bar{A}C$

(c) $\bar{A}\bar{C} + \overline{A + \bar{C}} + \bar{A}BC + \bar{A}\bar{B}C$

Ans:

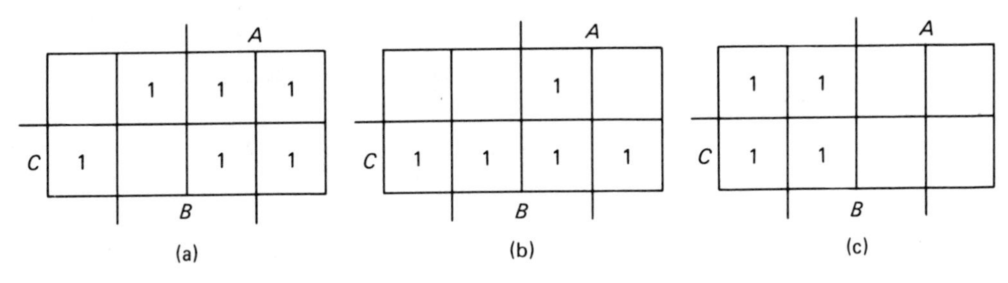

Exercise 5.4.1

5.4.2 Show that the function in Ex. 5.4.1(a) is minimal, and find minimal functions that are equivalent to those of Ex. 5.4.1(b) and (c).

Ans: (b) $AB + C$; (c) $\bar{A}$

5.4.3 Find the Karnaugh maps of

(a) $\bar{A}C + \bar{A}\bar{B}\bar{C} + A\bar{B}C + A\bar{B}\bar{C}\bar{D} + A\bar{B}\bar{C}D$

(b) $\overline{AB + C} + \bar{A}(\overline{B + \bar{C}}) + AB\bar{C}$

Ans:

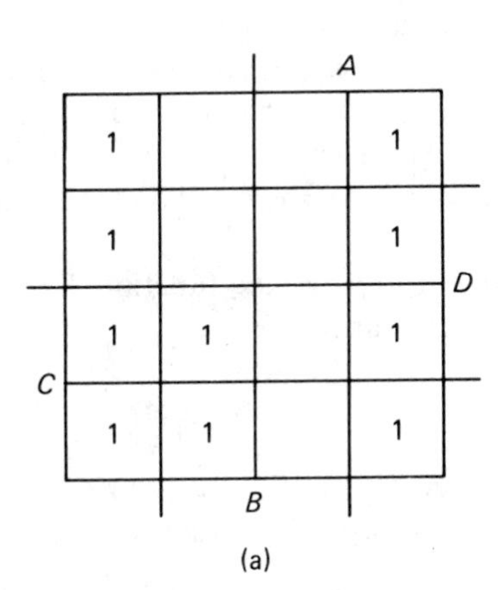

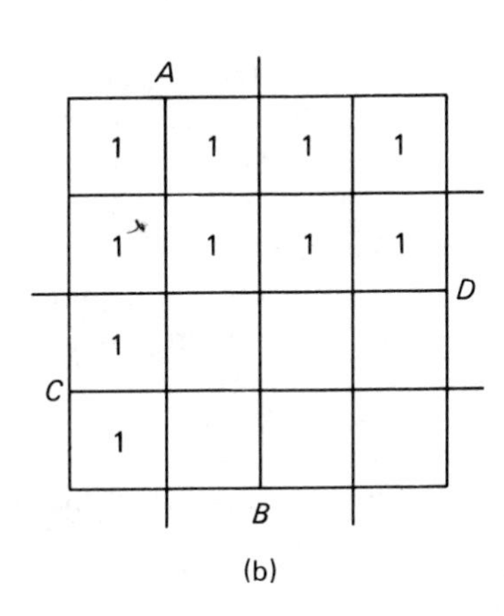

Exercise 5.4.3

5.4.4 Find minimal functions in Ex. 5.4.3(a) and (b).

Ans: (a) $\bar{B} + \bar{A}C$; (b) $\bar{C} + \bar{A}\bar{B}$

5.4.5 Find the minimal function F represented by the given Karnaugh map. (*Suggestion:* Find $\bar{F}$ and complement it.)

Ans: $\bar{A} + \bar{C} + \bar{D}$

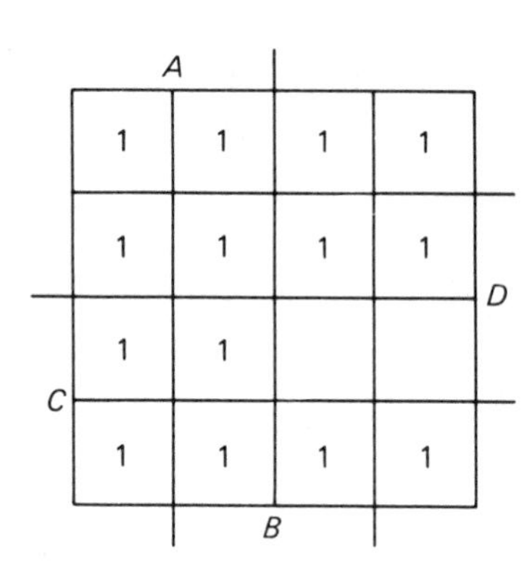

Exercise 5.4.5

5.5 Don't Care Conditions

The combinational circuits considered thus far have been fully specified, by which we mean that a specific output is defined for each combination of input variables. There are cases, however, where we *don't care* how the circuit responds to some input combinations, because either these combinations are never applied or, if they are, the outputs are of no concern to us. Inputs for which the outputs are irrelevant are called *don't care* conditions, or simply *don't cares.*

The presence of don't cares in the design specifications gives us flexibility in selecting the output function that is to be synthesized, since for a don't care condition the output may be either 0 or 1. The Karnaugh map is especially useful for taking advantage of this flexibility, since we can often tell at a glance if selecting a don't care to be 1 will result in a larger grouping of rectangles, and thus a simpler function.

Suppose, for example, we wish to realize the function F defined by Tab. 5.1, where the x's represent don't cares. That is,

$$F = \bar{A}\bar{B}C + \bar{A}BC + A\bar{B}C + AB\bar{C}$$

with the stipulation that we don't care what F is for inputs $\bar{A}\bar{B}\bar{C}$ and ABC. Using Boolean theorems or a Karnaugh map, we may reduce F to the simplified form

$$F = \bar{A}C + \bar{B}C + AB\bar{C} \tag{5.22}$$

Table 5.1 *Truth table for F*

A	B	C	F
0	0	0	x
0	0	1	1
0	1	0	0
0	1	1	1
1	0	0	0
1	0	1	1
1	1	0	1
1	1	1	x

Can we further simplify F by taking into account the don't care conditions?

To answer this question, let us draw the Karnaugh map for F given in Eq. (5.22) and add x's for the don't care conditions. The result is shown in Fig. 5.27(a).

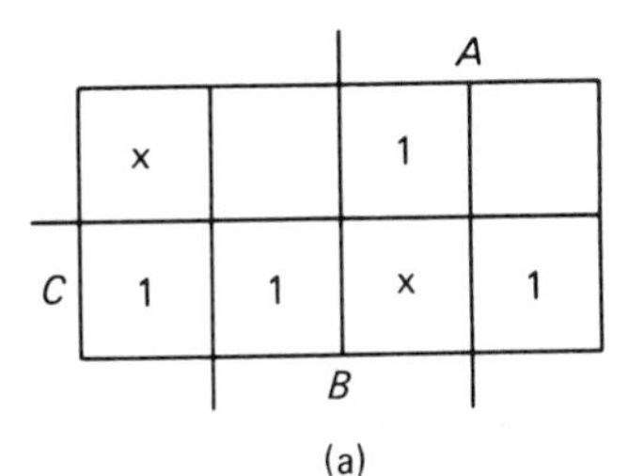

(a)

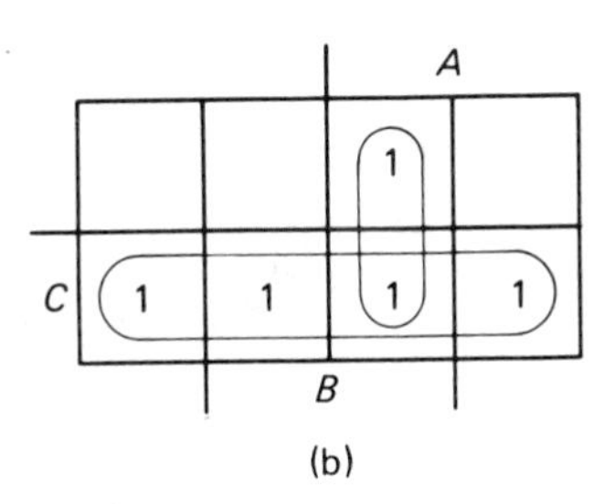

(b)

Figure 5.27 *Karnaugh maps for Tab. 5.1*

In this simple example it is clear that if the x for the don't care condition ABC is replaced by a 1 and that for condition $\bar{A}\bar{B}\bar{C}$ is removed (replaced by a 0), then the map, shown in Fig. 5.27(b), defines the simplest possible function, which is

$$F = AB + C \tag{5.23}$$

Therefore implementing this function will satisfy all the requirements of Tab. 5.1 with $F(0, 0, 0) = 0$ and $F(1, 1, 1) = 1$.

As another example, suppose a circuit is required to receive the binary equivalents of the decimal digits 0, 1, . . . , 9 and generate an output signal of 1 when a received digit is equivalent to an even decimal digit. It is assumed that only binary equivalents of decimal digits are possible inputs. Thus numbers such as 1010 and 1011 (equivalents of decimal 10 and 11) never appear as inputs. Letting the inputs be represented by $(ABCD)_2$ and the output by F, we may construct the truth table shown in Tab. 5.2. The last six entries in the output column are irrelevant and are indicated as don't cares, since the inputs in these cases never appear.

Table 5.2 *Truth table for generating even decimal digits*

Decimal	*A*	*B*	*C*	*D*	*F*
0	0	0	0	0	1
1	0	0	0	1	0
2	0	0	1	0	1
3	0	0	1	1	0
4	0	1	0	0	1
5	0	1	0	1	0
6	0	1	1	0	1
7	0	1	1	1	0
8	1	0	0	0	1
9	1	0	0	1	0
10	1	0	1	0	x
11	1	0	1	1	x
12	1	1	0	0	x
13	1	1	0	1	x
14	1	1	1	0	x
15	1	1	1	1	x

The Karnaugh map for Tab. 5.2 is shown in Fig. 5.28(a), from which we see that replacing the x's in rows one and four by 1's results in a drastically simplified function. The Karnaugh map in this case is shown in Fig. 5.28(b); it represents the function

$$F = \bar{D} \tag{5.24}$$

Of course, this is not an astounding result, since every binary digit with $\bar{D} = 1$ ($D = 0$) contains a 0 in its units position and is therefore even. However, without the don't cares, the function, from Fig. 5.28(a), would be

$$F = \bar{A}\bar{D} + \bar{B}\bar{C}\bar{D}$$

which is more complicated.

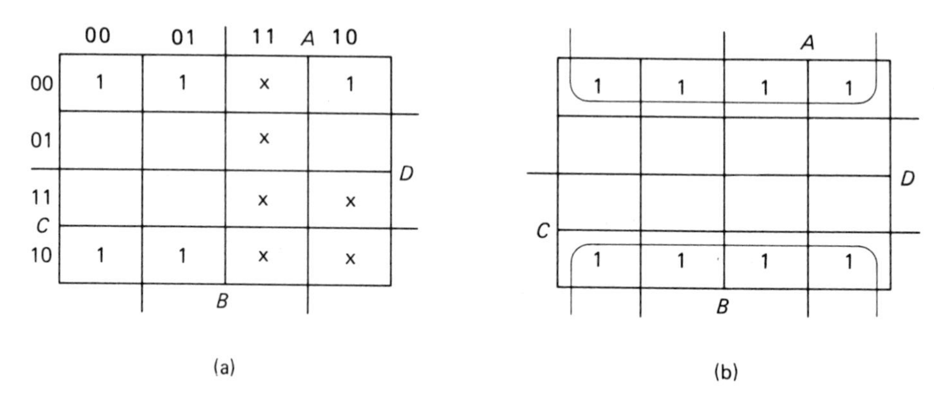

Figure 5.28 *Karnaugh maps for Tab. 5.2*

EXERCISES

5.5.1 Find the simplest expression for a function satisfying

$$F = \bar{A}\bar{B}\bar{C} + AB\bar{C} + ABC$$

with don't cares $\bar{A}B\bar{C}$, $A\bar{B}\bar{C}$, and $\bar{A}\bar{B}C$.

Ans: $AB + \bar{C}$

5.5.2 Find the simplest expression for a function satisfying

$$F = A\bar{B}\bar{C} + A\bar{C}D + A\bar{B}D + \bar{A}CD + \bar{A}\bar{B}\bar{C}D$$

with don't cares $\bar{A}B\bar{C}\bar{D}$, $\bar{A}B\bar{C}D$, $ABCD$, $\bar{A}BC\bar{D}$, and $A\bar{B}C\bar{D}$

Ans: $A\bar{B} + D$

5.5.3 Alter Tab. 5.2 so that the output signal of 1 is generated when a received digit is equivalent to an odd decimal integer. (The don't cares are still the same.) Find the minimal expression for F.

Ans: D

PROBLEMS

5.1 **(a)** Find F.

(b) Simplify F to contain only three variable appearances.

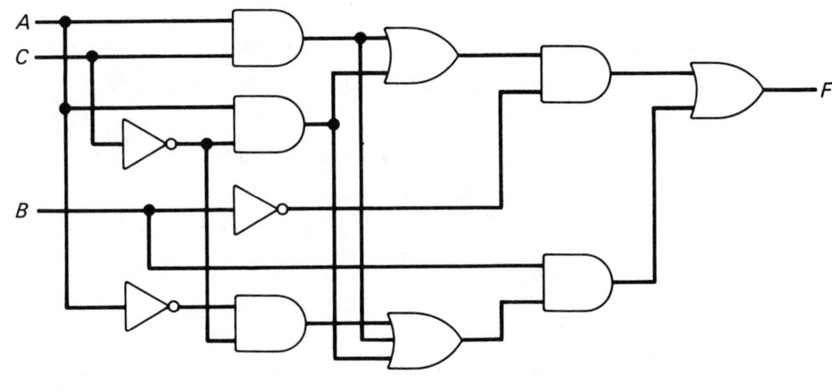

Problem 5.1

5.2 Find F.

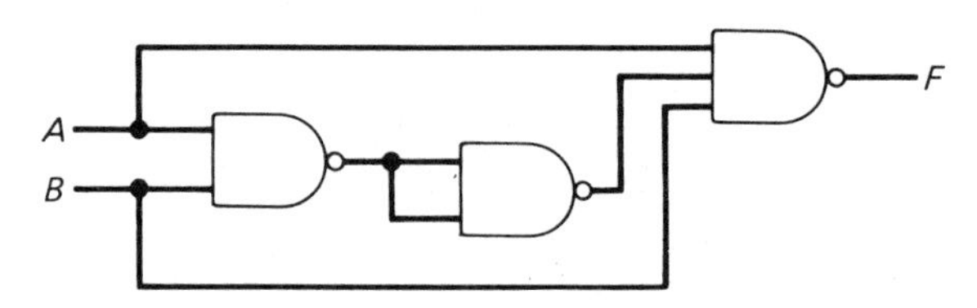

Problem 5.2

5.3 Change the circuit of Prob. 5.2 to an equivalent AND–OR–NOT circuit.

5.4 Find F.

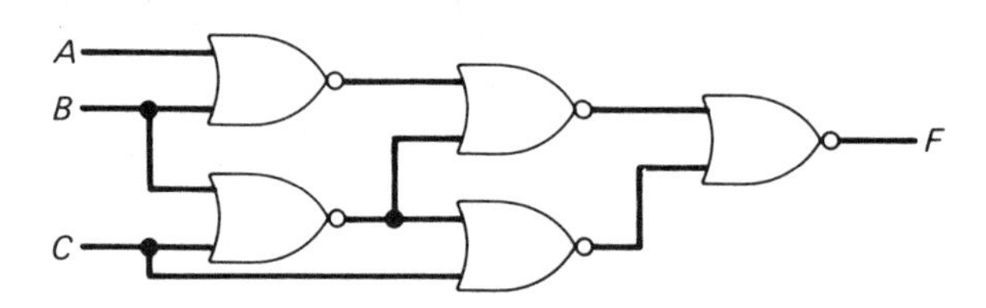

Problem 5.4

5.5 Check the answer to Prob. 5.4 by changing the circuit to an equivalent AND–OR–NOT circuit and analyzing the result.

5.6 Replace the NAND gates of Prob. 5.2 by NOR gates and find F.

5.7 Replace the NOR gates of Prob. 5.4 by NAND gates and find F.

5.8 Synthesize in its present from with an AND–OR–NOT circuit the function

$$F = \overline{AB} \cdot \overline{CD} + D \cdot \overline{CD}$$

5.9 Show that the function of Prob. 5.8 can be written in the form

$$F = \bar{A}\bar{D} + \bar{B}\bar{D} + \bar{C}D$$

and obtain an AND–OR–NOT realization of two levels (not counting inverters).

5.10 Using the transform method, convert the answer obtained in Prob. 5.8 to a circuit of NAND gates.

5.11 Repeat Prob. 5.10 for the answer obtained in Prob. 5.9.

5.12 Obtain a NOR circuit realization of

$$F = (A + \bar{B}\bar{C})(C + D)$$

5.13 Rewrite the function

$$F = \bar{A}B\bar{C} + \bar{A}BD + \bar{A}E + A$$

so that only five literals appear, and synthesize it with a network of AND, OR, and NOT gates having only eight inputs. (Include any required inverters.)

5.14 Change the circuit obtained in Prob. 5.13 to one containing only NAND gates.

5.15 Prove Theorems 12a, 13a, and 13b of Sec. 3.6 by using Karnaugh maps. (*Suggestion:* Two functions are equal if they have the same Karnaugh maps.)

5.16 Solve Prob. 3.1 using Karnaugh maps instead of perfect induction.

5.17 Solve Prob. 3.4 using Karnaugh maps instead of perfect induction.

5.18 Find the minimal equivalent of

$$F = (A + B)(A + \bar{B} + C)(\bar{A} + C)$$

using a Karnaugh map.

5.19 Find the minimal expression for the function defined by the given Karnaugh map.

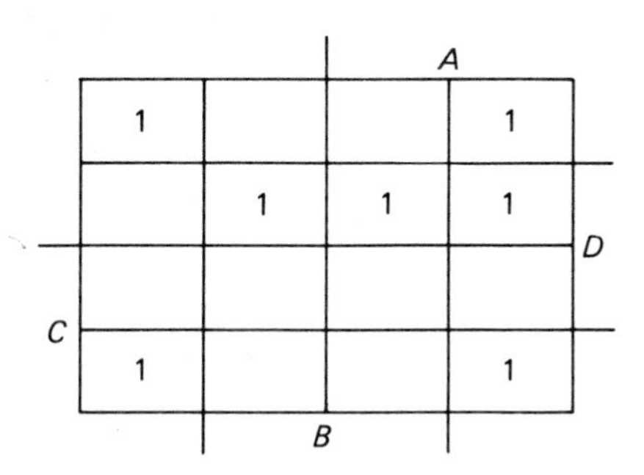

Problem 5.19

5.20 A digital circuit is to have an output F and inputs A, B, C, and D, so that $F = 1$ if and only if the decimal equivalent of $(ABCD)_2$ is divisible by either 3 or 4. Find the minimal expression for F.

5.21 Solve Prob. 5.20 if the decimal equivalents 5, 7, and 13 are don't cares.

5.22 Solve Prob. 5.21 if, in addition, the decimal equivalen's 1, 11, and 14 are don't cares.

6

APPLICATIONS OF COMBINATIONAL NETWORKS

In Chap. 5 we defined combinational networks and considered general procedures in their analysis and synthesis. In this chapter we describe a number of important special examples of combinational circuits, such as decoders, encoders, multiplexers, and adders. In most cases we will give specific circuits of logic gates and show that they perform the stated functions. An exception is the *read only memory* (ROM), which we describe in Sec. 6.4 in terms of a decoder and certain *coupling* elements.

6.1 Encoders and Decoders

As was pointed out in Sec. 4.6, an encoder is a device that translates a signal, such as a decimal number or a set of alphabetical characters, into a code. Normally the symbols translated are relatively complex ones, such as "A," "B," and "=," and those used in the code are relatively simple, such as 0 and 1. Thus the encoder converts the data into a form that can be interpreted by a digital circuit.

On the other hand, a decoder (also mentioned in Sec. 4.6) is a device that converts a code into the uncoded signal it represents. For example, a decoder may translate a message made up of 0's and 1's into its equivalent in terms of A, B, =, etc. Thus decoding is the opposite of encoding. Encoders and decoders are important examples of combinational circuits. We considered some examples in Chap. 4, and we will look at others in this section.

As a first example, suppose we wish to represent the decimal digits by their binary equivalents. That is, we wish to design an encoder that will accept the decimal digit N as an input (from a keyboard, perhaps) and yield its binary equivalent $(ABCD)_2$ as the output. The truth table for N is shown in Tab. 6.1.

Table 6.1 *Truth table for decimal digits*

N	A	B	C	D
0	0	0	0	0
1	0	0	0	1
2	0	0	1	0
3	0	0	1	1
4	0	1	0	0
5	0	1	0	1
6	0	1	1	0
7	0	1	1	1
8	1	0	0	0
9	1	0	0	1

The encoder is to have outputs A, B, C, and D, which are to be actuated (that is, made 1) or not, depending on the input N. From the columns of Tab. 6.1 we see that $A = 1$ when 8 or 9 is the input. Thus A is the output of an OR gate with inputs 8 and 9. Similarly, we see that B is the output of an OR gate with inputs 4, 5, 6, and 7, C is the output of an OR gate with inputs 2, 3, 6, and 7, and D is the output of an OR gate with inputs 1, 3, 5, 7, and 9. The circuit is shown in Fig. 6.1, where we see that each input will activate one or more outputs, resulting in its binary equivalent. (An exception is 0, which activates no output lines, since its binary equivalent is 0000.)

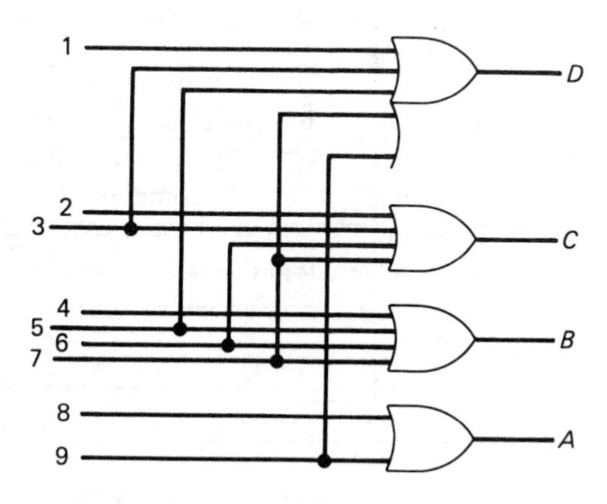

Figure 6.1 *Encoder for decimal to binary*

Two examples of decoders were given in Sec. 4.6. A 1-of-2^n decoder, which selects one output out of 2^n possibilities, was given in Fig. 4.24 for the case $n = 2$. A circuit for displaying decimal equivalents of binary inputs was shown in Fig. 4.26.

To illustrate decoders further, let us design a circuit to decode the binary numbers $(ABCD)_2$ of Tab. 6.1. Such a circuit must have ten outputs, one for each of the decimal digits, and an output N must be 1 when the input is the set of bits corresponding to N and must be 0 otherwise. To make the problem somewhat different from that of Fig. 4.26, let us further require that a *select signal* S be 1 when we want an output to appear, and when $S = 0$, all outputs are 0. The signal S is also called an *enable* input, because it *enables* the gate (or allows it to be activated) whose output is to be displayed.

From Tab. 6.1 we see that the decimal numbers are given in terms of A, B, C, and D by

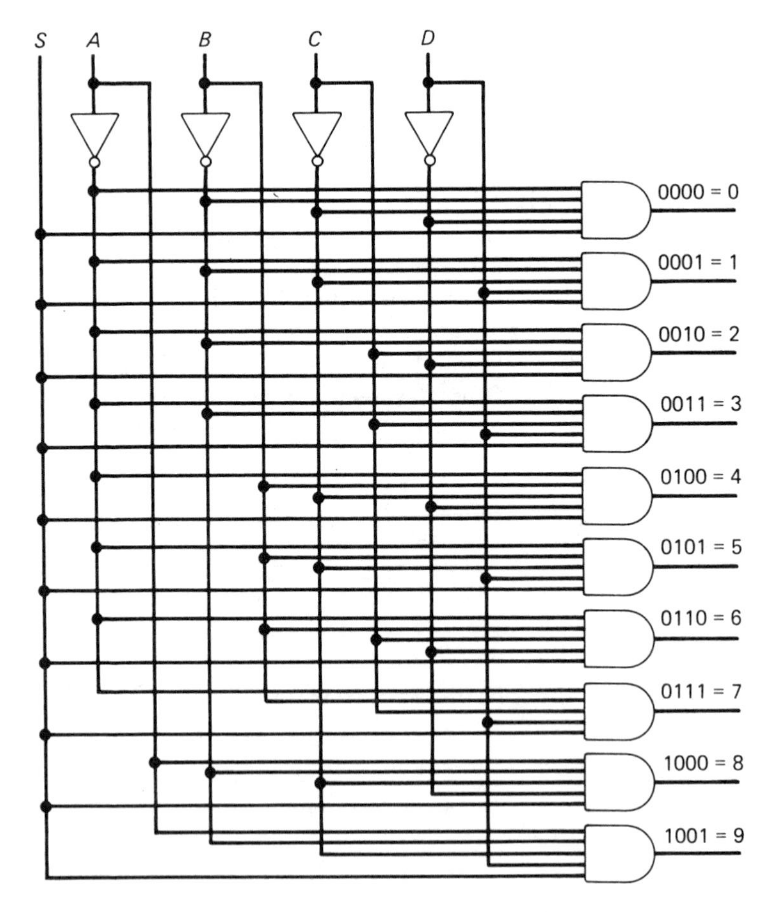

Figure 6.2 *Binary-to-decimal decoder*

$$0 = \bar{A}\bar{B}\bar{C}\bar{D}$$
$$1 = \bar{A}\bar{B}\bar{C}D$$
$$2 = \bar{A}\bar{B}C\bar{D}$$
$$\cdots$$
$$8 = A\bar{B}\bar{C}\bar{D}$$
$$9 = A\bar{B}\bar{C}D$$

Thus the digits 0, 1, 2, . . . , 9 are outputs of AND gates whose inputs are the bits A, B, C, and D in complemented or uncomplemented form. The select signal S must also be an input of every gate. The resulting decoder is shown in Fig. 6.2, which the reader should check with Tab. 6.1. For example, when $SABCD = 10011$, the gate labeled 3 will be the only gate with an output of 1. Also, if $SABCD =$ 00111, no gate will have an output of 1, because the enable signal is $S = 0$. Finally, if $SABCD = 11011$, none of the gates will be activated, because 1011_2 is the decimal number 11 and is not a valid input.

EXERCISES

6.1.1 Design an encoder to convert the decimal integer N to the binary integer $(ABC)_2$, as defined by the given table. This code is known as a *Gray* code and has the property that any two consecutive digits, including 5 and 0, differ in only one bit position. (We will consider the Gray code in more detail in Chap. 8.)

N	A	B	C
0	0	0	0
1	0	0	1
2	0	1	1
3	0	1	0
4	1	1	0
5	1	0	0

Ans:

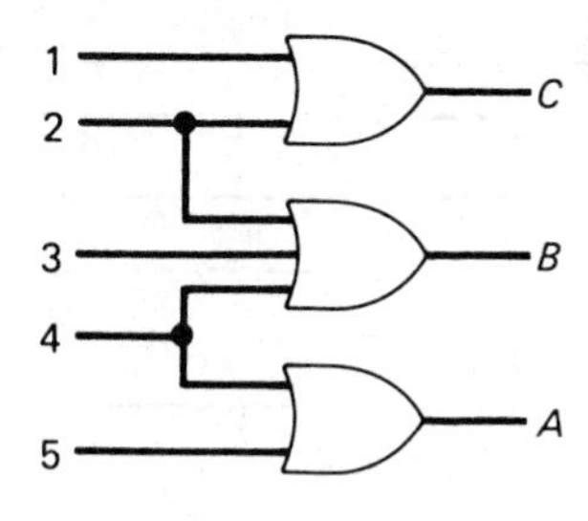

Exercise 6.1.1

6.1.2 Design a decoder to convert the binary integer $(ABC)_2$ to the decimal integer N, as defined by the table of Ex. 6.1.1.

Ans:

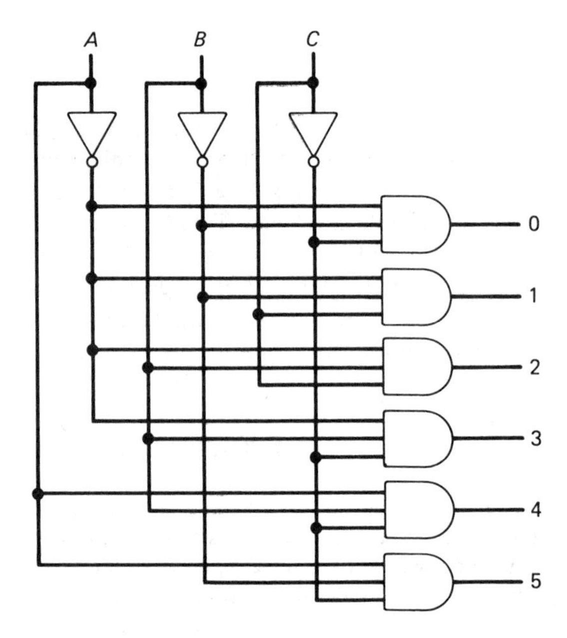

Exercise 6.1.2

6.1.3 Let X, Y, Z, and W be characters that are to be encoded as shown in the accompanying table. Design an encoder with switches labeled X, Y, Z, and W such that when a switch is closed, an input of 1 is connected to a digital circuit that yields the appropriate coded output.

	Code			
Character	*A*	*B*	*C*	*D*
X	1	0	0	1
Y	1	0	1	0
Z	0	0	1	1
W	1	1	0	0

Ans:

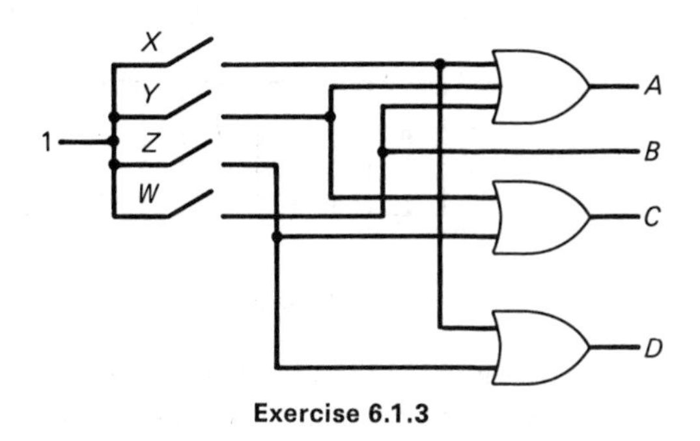

Exercise 6.1.3

6.2 Multiplexers and Demultiplexers

Another useful combinational circuit is a *multiplexer*, which is a device that selects data from one of two or more input lines and transmits it on a single output line. In this manner several messages may be sent over a single output line by sending in succession the first character of each message, then the second character, then the third, and so on, until each character of each message has been sent. The multiplexed signal is then separated at the receiving end by a *demultiplexer*, which channels the various characters of each message into one continuous signal in one output line. Thus a multiplexer has, say, N input lines and one output line, while a demultiplexer has one input and N output lines.

As a simple illustration consider the circuit in Fig. 6.3, which is a two-line to

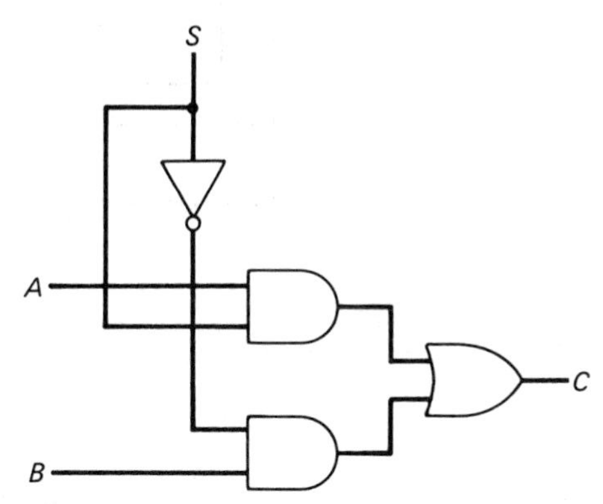

Figure 6.3 *2-to-1 multiplexer*

one-line multiplexer that transmits signals A and B over the output line C. The select signal S determines which of A or B is the output C. When $S = 1$ we have $C = A$, and when $S = 0$ we have $C = B$. The select signal may be alternately 1 and 0 as shown in Fig. 6.4, so that first A is the output, then B, then A, and so on. The duration of the 1 pulse must be long enough for the code representing each character to be transmitted.

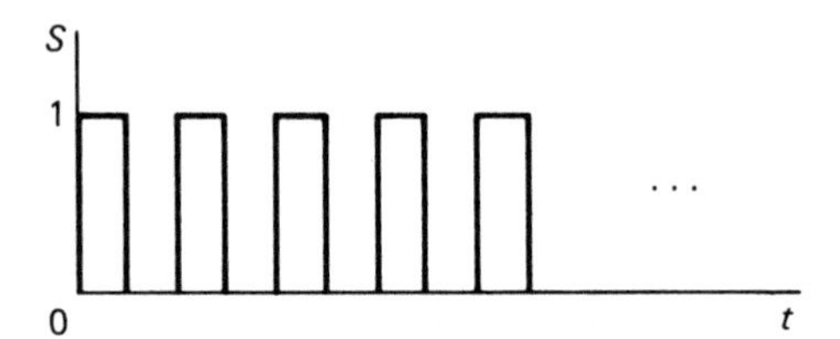

Figure 6.4 *Select signal*

Let us now design a demultiplexer whose input is the signal C of Fig. 6.3 and whose purpose is to separate C into two signals A and B. That is, signals A and B are to be recovered from the multiplexed signal C. We again require a select signal S like that in Fig. 6.4 to provide alternately A, then B, then A, and so on. The result, shown in Fig. 6.5, yields the demultiplexed signals A and B, as the reader may verify.

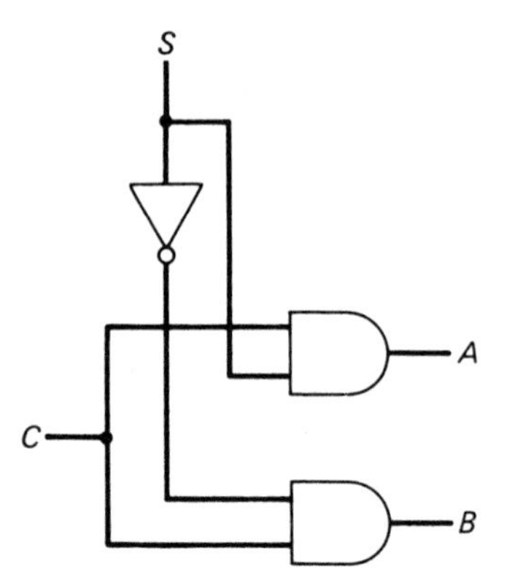

Figure 6.5 *Two-channel demultiplexer*

EXERCISES

6.2.1 Obtain a 4-to-1 multiplexer that has inputs X_0, X_1, X_2, and X_3 and multiplexed output F. Let the select signal be represented by $(AB)_2$ where 00, 01, 10, and 11 enable, respectively, the gate whose output is X_0, X_1, X_2, and X_3.

Ans:

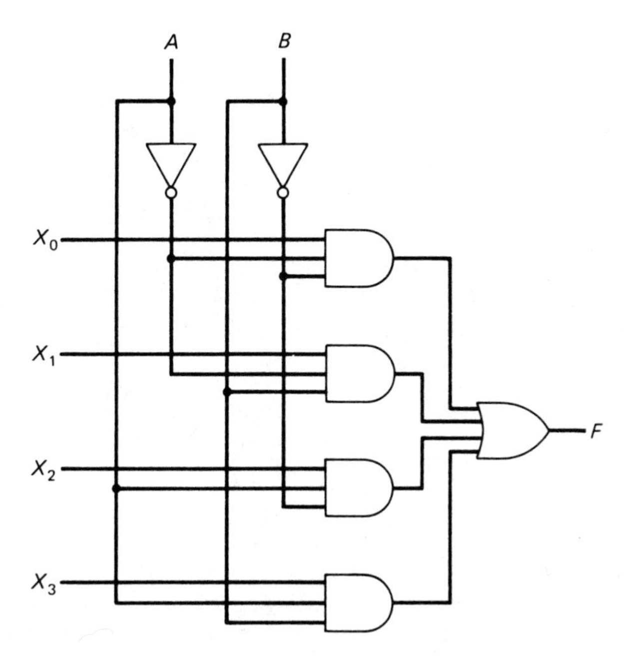

Exercise 6.2.1

6.2.2 Represent the multiplexer of Ex. 6.2.1 by the symbol as shown and find the output F.

Ans: $X_0\bar{A}\bar{B} + X_1\bar{A}B + X_2A\bar{B} + X_3AB$

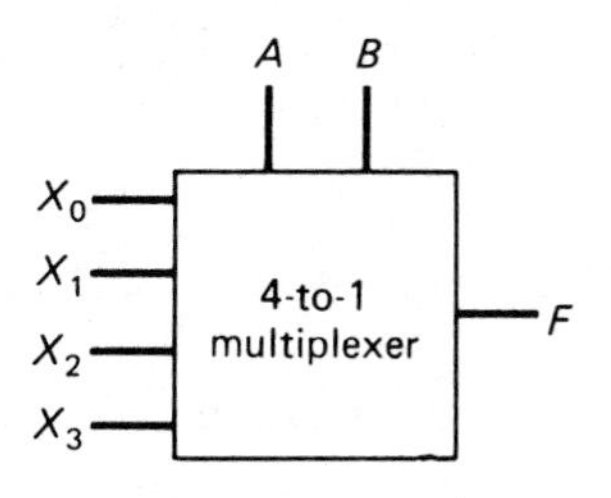

Exercise 6.2.2

6.2.3 Note in Ex. 6.2.2 that the 4-to-1 multiplexer may be used to generate various functions $F(A, B)$ by assigning values to the inputs X_0, X_1, X_2, and X_3. Find the inputs to yield an output of

(a) A

(b) $\bar{B}$

(c) A XOR B

Ans: (a) 0011; (b) 1010; (c) 0110

6.3 Adders

The arithmetic operations performed directly by a computer are simply addition and subtraction. All other operations, such as multiplication, division, or taking a logarithm or a power can be performed by repeated addition or subtraction. As an example of circuit synthesis we will develop in this section a digital circuit that performs addition. As we will see in Chap. 10, subtraction may also be carried out by performing addition.

Let us consider first the *half adder* (HA), which is a device that adds two bits of binary data. In other words, the half adder performs the operations

$$\begin{aligned} 0 + 0 &= 0 \\ 0 + 1 &= 1 \\ 1 + 0 &= 1 \\ 1 + 1 &= 0, \text{ carry } 1 \end{aligned} \tag{6.1}$$

The last operation is, of course, $1 + 1 = 10$, which is 0 with a carry of 1 to the next bit position. We may express Eqs. (6.1) in the form of a truth table, as shown in Tab. 6.2, where S represents the sum $A + B$ and C represents the carry. Of course, $C = 0$ in every case except the last equation of (6.1).

Table 6.2 *Truth table for a half adder*

A	B	S	C
0	0	0	0
0	1	1	0
1	0	1	0
1	1	0	1

From the truth table we see that

$$\begin{aligned} S &= \bar{A}B + A\bar{B} \\ &= A \oplus B \end{aligned} \tag{6.2}$$

and

$$C = AB \tag{6.3}$$

Therefore a half adder may be made with an EXCLUSIVE-OR gate and an AND gate, as shown in Fig. 6.6(a). The symbol for the half adder is given in Fig. 6.6(b). The inputs are A and B, and the outputs are S and C.

The half adder adds only two bits at a time, so that it cannot be used to add two bits and a carry bit from a previous step, as is generally required in adding two binary numbers. If we designate the carry from the previous step as a *carry-in* C_i to the next step (the next column of bits), then to continue the addition process we need a device to add the two bits, say, A and B, of the next column and the

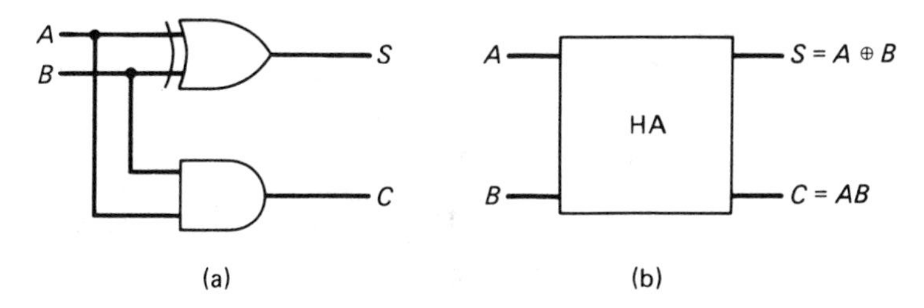

Figure 6.6 *(a) Half-adder circuit and (b) half-adder symbol*

carry-in C_i. This addition produces the sum

$$S = A + B + C_i \tag{6.4}$$

and, of course, a carry of its own, which we designate as a *carry-out* C_0. An adder that performs this operation is called a *full adder* and is symbolized as shown in Fig. 6.7. The inputs are the bits A and B of the column being added and the carry-in bit C_i from the addition of the previous column. The outputs are the sum S of Eq. (6.4) and the carry-out bit C_0.

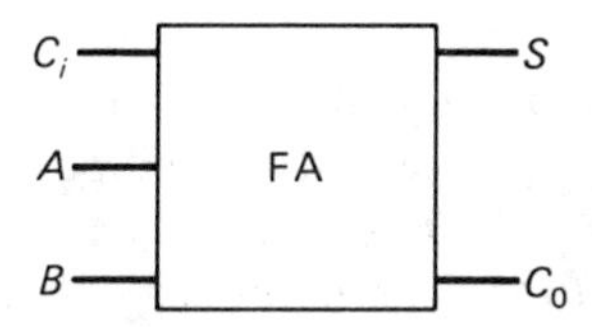

Figure 6.7 *Full adder symbol*

The truth table for the full adder is determined by the eight possible combinations of the inputs A, B, and C_i and the corresponding values of S and C_0 as determined by Eq. (6.4). The result is given in Tab. 6.3, from which we may write

$$S = \bar{A}\bar{B}C_i + \bar{A}B\bar{C}_i + A\bar{B}\bar{C}_i + ABC_i \tag{6.5}$$

and

$$C_0 = \bar{A}BC_i + A\bar{B}C_i + AB\bar{C}_i + ABC_i \tag{6.6}$$

Therefore to design a full adder we need only synthesize these two functions.

Table 6.3 *Truth table of a full adder*

A	B	C_i	S	C_0
0	0	0	0	0
0	0	1	1	0
0	1	0	1	0
0	1	1	0	1
1	0	0	1	0
1	0	1	0	1
1	1	0	0	1
1	1	1	1	1

We may write Eq. (6.5) in the form

$$\begin{aligned} S &= (\bar{A}\bar{B} + AB)C_i + (\bar{A}B + A\bar{B})\bar{C}_i \\ &= \bar{D}C_i + D\bar{C}_i \\ &= D \oplus C_i \end{aligned}$$

where

$$\begin{aligned} D &= \bar{A}B + A\bar{B} \\ &= A \oplus B \end{aligned}$$

Therefore we have

$$S = A \oplus B \oplus C_i \tag{6.7}$$

In the case of the carry-out, Eq. (6.6) may be simplified to

$$C_0 = \bar{A}BC_i + A\bar{B}C_i + AB$$

or

$$C_0 = AB + C_i(A \oplus B) \tag{6.8}$$

These results suggest that we may use half adders to construct a full adder. From Fig. 6.6(b) we see that if the inputs of a half adder are C_i and $A \oplus B$, then the outputs are $A \oplus B \oplus C_i$ and $C_i(A \oplus B)$, which are S and part of C_0, given in Eqs. (6.7) and (6.8). The terms $A \oplus B$ and AB, which are also needed, may be obtained as the outputs of another half adder. Thus two half adders and an OR gate, to perform the OR operation of Eq. (6.8), may be used to realize a full adder. The result, as the reader may verify, is shown in Fig. 6.8.

As an example, suppose we want to add the binary numbers 110 and 111. Since these are three-bit numbers, we need a *three-bit* full adder, or three full adders connected so as to add each pair of bits and obtain the appropriate carry. The process is shown in Fig. 6.9, where the bits in 110 are the A bits and those in 111 are the B bits. The carry-out is shown in each case and used as a carry-in to the next stage. The result, as the reader may verify by tracing through the circuit, is the sum 1101.

The addition of any two binary numbers can be performed by using enough

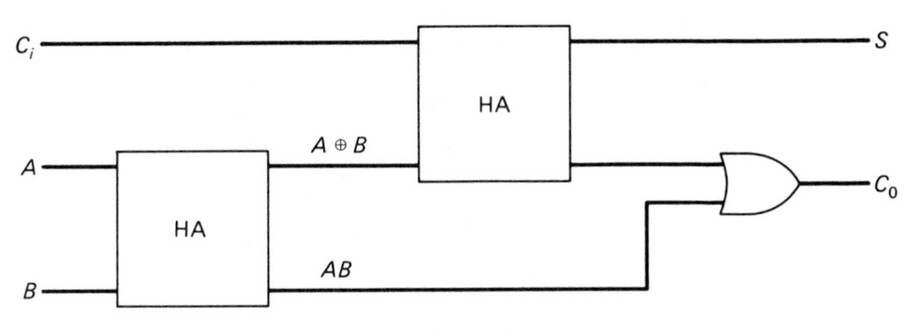

Figure 6.8 *Full adder*

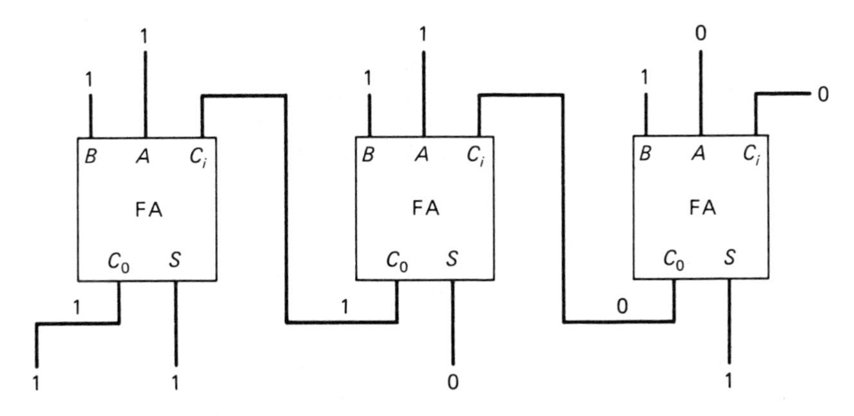

Figure 6.9 *Three-bit full adder*

full adders, as suggested by Fig. 6.9. However, one half adder can be used repeatedly if provision is made for applying the input bits and storing the output bits in the proper sequence, as in *registers* (to be considered in Chapter 9).

EXERCISES

6.3.1 Show that the given circuit is a NAND gate implementation of a half adder.

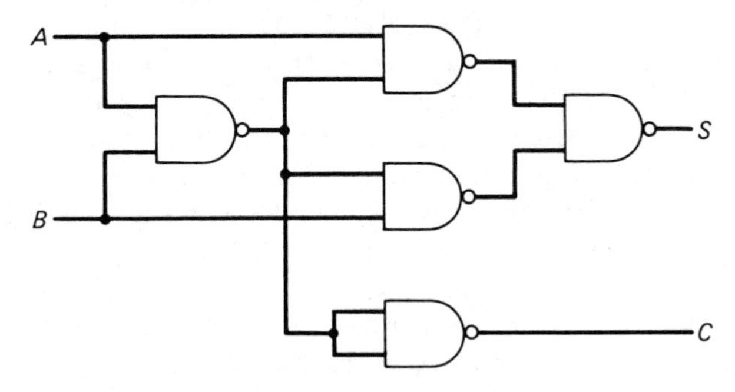

Exercise 6.3.1

6.3.2 Attach another full adder to Fig. 6.9 so that four-bit numbers may be added, and trace through the sum 1101 + 1011 = 11000.

Ans:

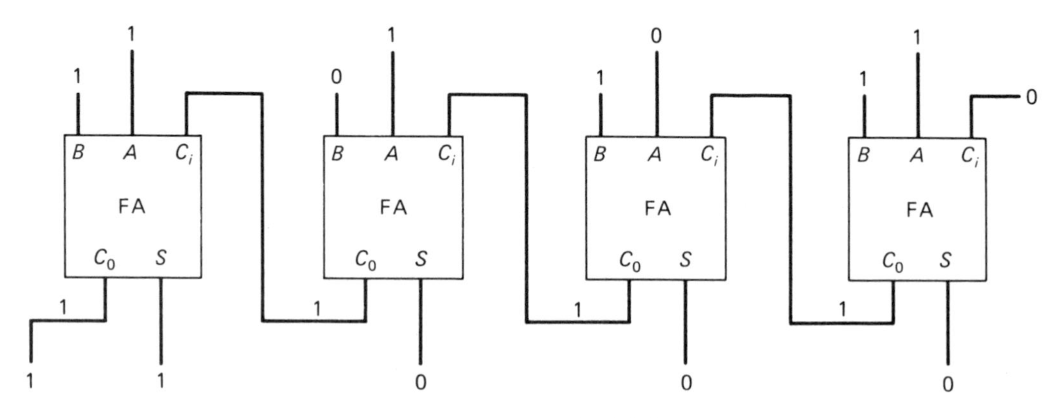

Exercise 6.3.2

6.4 Read Only Memory

As a last example of an important combinational logic circuit, we consider the *read only memory*, or ROM. A ROM is simply a device that generates a specific set of outputs for a given set of inputs in accordance with some rule, such as a Boolean function or a truth table. Indeed, a ROM may be thought of as a device in which a truth table is *stored*, or *programmed*. The term "read only" stems from the fact that the device can be *read*, that is, its output ascertained for a given input, but it cannot be changed to give a different output.

In most cases the ROM is programmed by the manufacturer to perform some function, such as remembering a trigonometric table, and cannot be changed by the user. In other cases the ROM can be programmed by the user to suit his own purposes. A *programmable* ROM (or PROM) comes in two varieties: one that can be programmed only once, so that the data is permanently stored, and one that is reprogrammable (erasable PROM, or EPROM) and can be used over and over for different purposes. However, all ROMS have in common the feature that once programmed and put in use, they can be read only.

A simplified version of the basic ROM is shown in Fig. 6.10, with n input lines and m output lines. The memory is *addressed* by applying an input $A_1A_2 \ldots A_n$ to the decoder, which applies a 1 to the appropriate one of the decoder output lines. If the coupling element between this line and an output line, say, F_1, is intact (as they are all shown to be), then $F_1 = 1$. If the coupling element is opened, or missing, then $F_1 = 0$. Indeed, one method of storing a program in a PROM is to

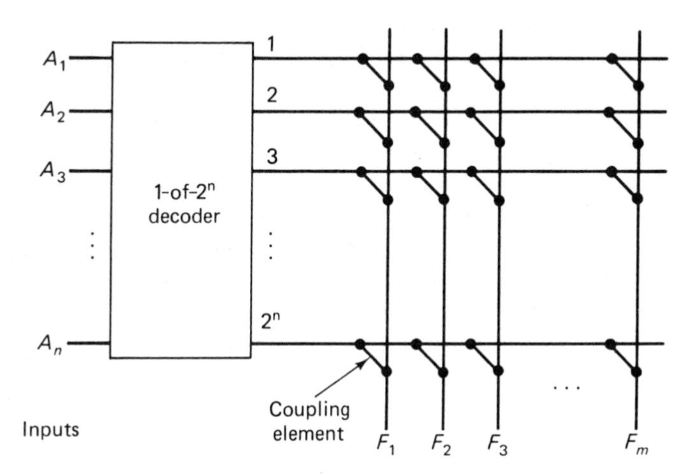

Figure 6.10 *Simplified version of a ROM*

break the contact provided by a coupling element to create a 0 in an appropriate output position.

As an example, let us consider the ROM in Fig. 6.11. There are four intact coupling elements and four whose contacts have been broken. Also, we have numbered the four outputs of the decoder 0, 1, 2, and 3 to identify them with their binary equivalents in the truth table, given in Tab. 6.4.

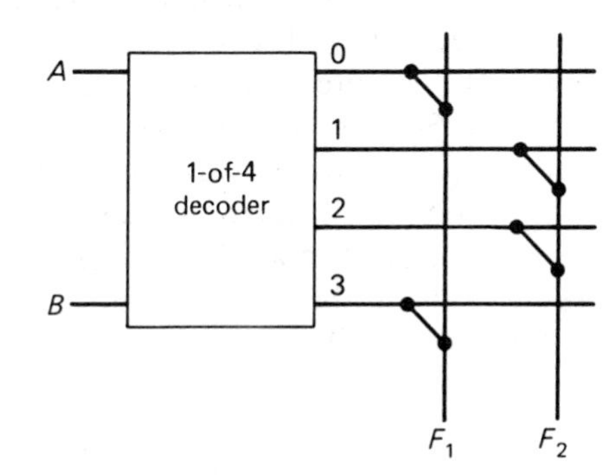

Figure 6.11 *Example of a ROM*

Table 6.4 *Truth table for Fig. 6.11*

Decimal	*A*	*B*	F_1	F_2
0	0	0	1	0
1	0	1	0	1
2	1	0	0	1
3	1	1	1	0

To see that the truth table is correct, we may check each of the decoder output lines and their connection with the F_1 and F_2 lines. For example, line 0, corresponding to binary 00 ($A = 0$, $B = 0$), is connected to F_1 but not to F_2. Thus a 1 at line 0 will result in $F_1 = 1$ and $F_2 = 0$, as indicated.

In Fig. 6.10 we may think of $F_1F_2 \ldots F_m$ as a binary number, or *word*, of m bits, in which case the ROM provides 2^n words of m bits each. This may be seen from the truth table, which has 2^n rows, each row corresponding to a word. The *size* of the ROM is therefore said to be $2^n \times m$ bits. In the case of Fig. 6.11, for example, the size is 4×2 or 8 bits. This is also the number of intersections of the ROM output lines with the decoder output lines.

We symbolize a ROM by a rectangle, as shown in Fig. 6.12. The size, 16×3 in this case, may occasionally be omitted.

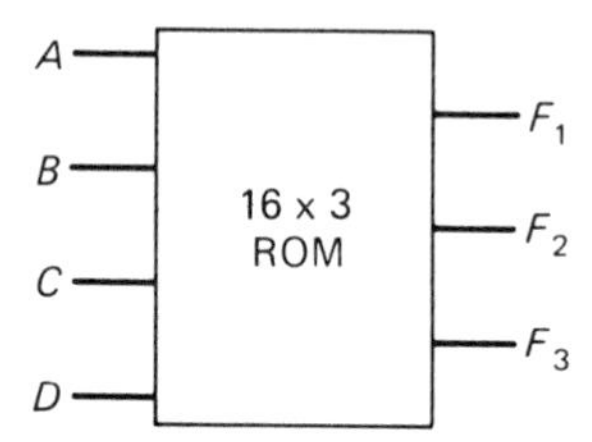

Figure 6.12 *Symbol for a ROM*

As a last example, suppose we wish to program a ROM so that its outputs are the squares of the decimal numbers 0, 1, 2, and 3. In other words, when the output of the decoder is 0, the output of the ROM is $0^2 = 0$, 1 yields $1^2 = 1$, 2 yields $2^2 = 4$, and 3 yields $3^2 = 9$. We first construct the truth table, given as Tab. 6.5, noting that the inputs are two-bit words but the outputs must be four-bit words, since $9 = 1001_2$. From the truth table we construct the ROM, shown in Fig. 6.13.

Another method of storing a truth table such as Tab. 6.5 is illustrated in the problems (see Probs. 6.8 and 6.9). This method, which uses multiplexers, performs the same function as a ROM.

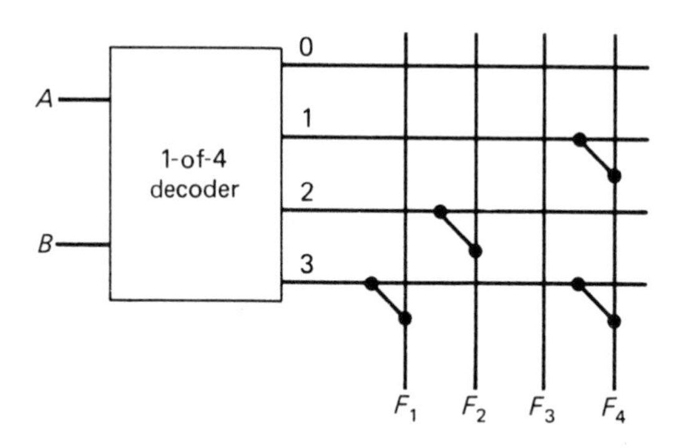

Figure 6.13 *ROM for Tab. 6.5*

Table 6.5 *Truth table for $Y = X^2$*

X			Y				
Decimal	*A*	*B*	F_1	F_2	F_3	F_4	*Decimal*
0	0	0	0	0	0	0	0
1	0	1	0	0	0	1	1
2	1	0	0	1	0	0	4
3	1	1	1	0	0	1	9

EXERCISES

6.4.1 Implement with a ROM the function

$$F = ABC + A\bar{B}C + \bar{A}B\bar{C} + AB\bar{C}$$

(*Suggestion:* The 0 output line corresponds to minterm $\bar{A}\bar{B}\bar{C}$, and so on)

Ans:

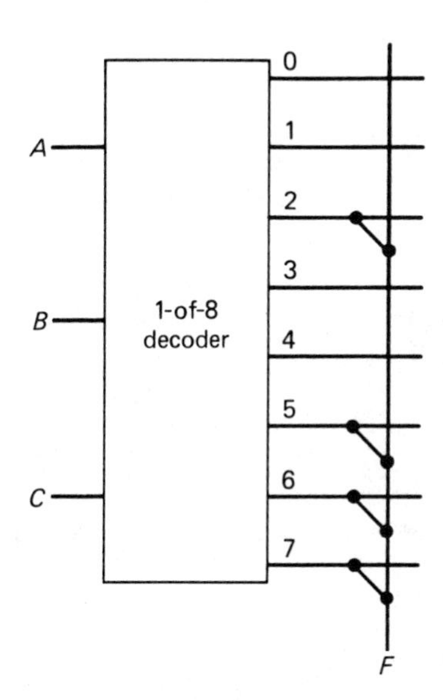

Exercise 6.4.1

6.4.2 Note that the function

$$F = A + B\bar{C}$$

is not in a form that lends itself to programming a ROM. Use the Karnaugh map method to express *F* in the suitable sum-of-minterms form.

Ans: $\bar{A}B\bar{C} + AB\bar{C} + A\bar{B}\bar{C} + ABC + A\bar{B}C$

PROBLEMS

6.1 Obtain a 1-of-4 decoder that accomplishes the same purpose as that of Fig. 4.24 but contains only AND gates and inverters. In addition, have each output gate enabled by a select signal.

6.2 Obtain an encoder to convert the characters shown in the left column of the table to their code shown in the right column. (This code is known as the ASCII code, which we will consider in Chap. 8.)

Character	*Code*						
	A_6	A_5	A_4	A_3	A_2	A_1	A_0
B	1	0	0	0	0	1	0
I	1	0	0	1	0	0	1
T	1	0	1	0	1	0	0
3	0	1	1	0	0	1	1

6.3 Repeat Problem 6.2 for the given table. (This code is the *excess*-3 code, which we will consider in Chap. 8. Note that each coded number is the binary equivalent of $N + 3$.)

N	*Code*			
	A	*B*	*C*	*D*
0	0	0	1	1
1	0	1	0	0
2	0	1	0	1
3	0	1	1	0
4	0	1	1	1
5	1	0	0	0
6	1	0	0	1

6.4 Obtain a decoder for the table of Prob. 6.2.

6.5 Obtain a decoder for the table of Prob. 6.3.

6.6 Obtain a demultiplexer that recovers signals X_0, X_1, X_2, and X_3 from the multiplexed signal F of Ex. 6.2.1. The select signal is represented by $(AB)_2$, as before.

6.7 Show that the truth table for $F(A, B)$ is as shown in (a) if F is the output of the 4-to-1 multiplexer of (b). (*Suggestion:* See Exs. 6.2.1 and 6.2.2.)

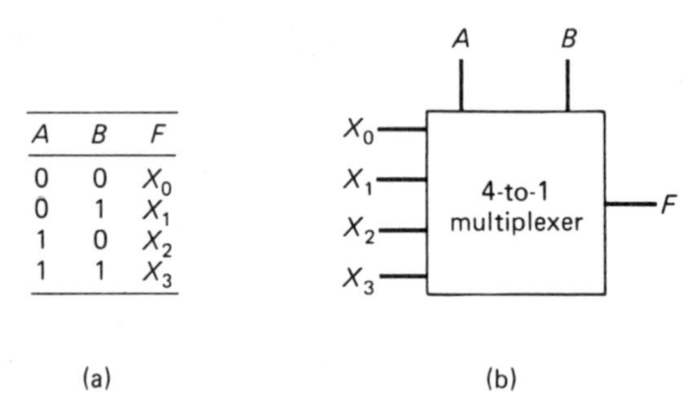

A	B	F
0	0	X_0
0	1	X_1
1	0	X_2
1	1	X_3

Problem 6.7

6.8 Show that if $(AB)_2$ is the binary equivalent of the decimal N, then $(F_1F_2F_3F_4)_2$ is the binary equivalent of $N^2 + 1$, for $N = 0, 1, 2,$ and 3. Thus we may use a number of multiplexers to *store* a function such as $f(N) = N^2 + 1$. (*Suggestion:* See Prob. 6.7.)

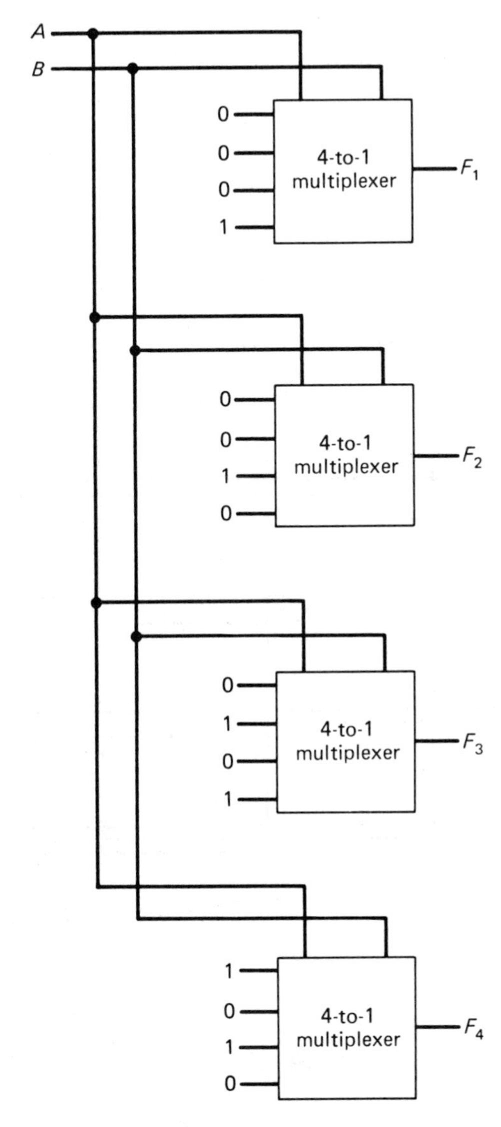

Problem 6.8

6.9 Obtain a network like that of Prob. 6.8 that stores the function

$$f(N) = N^2 + 5$$

for $N = 0, 1, 2,$ and 3.

6.10 Show that the given circuit is a 4-to-1 multiplexer which performs the identical function to that of Ex. 6.2.1 when $C = 1$ and yields $F = 0$ when $C = 0$.

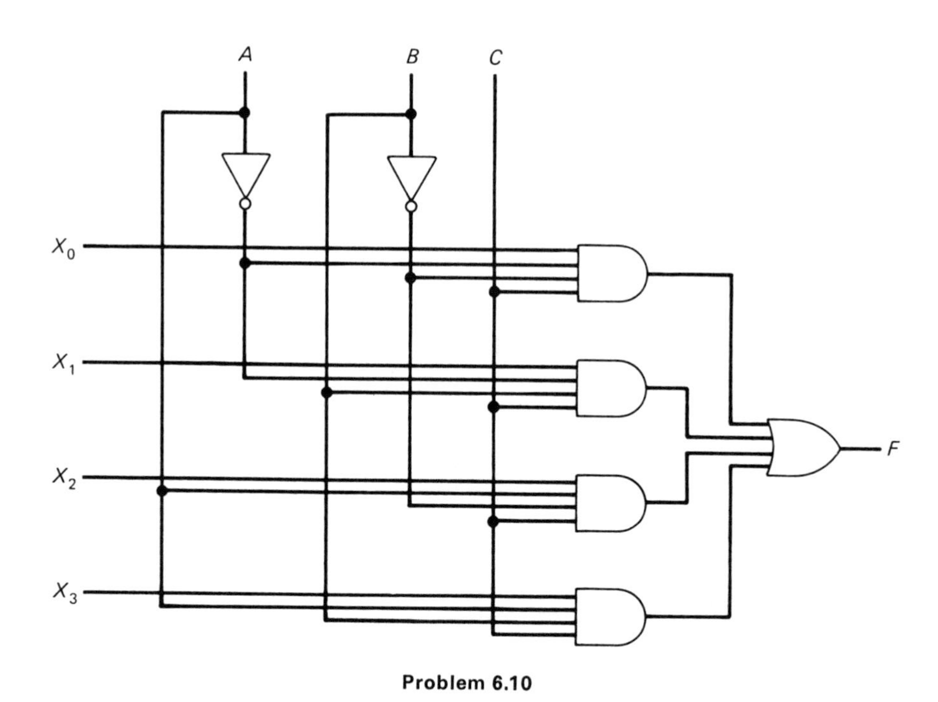

Problem 6.10

6.11 In the figure shown, the block diagrams represent multiplexers of the type shown in Prob. 6.10. Show by constructing a truth table for $F(A, B, C)$ that the circuit is an 8-to-1 multiplexer.

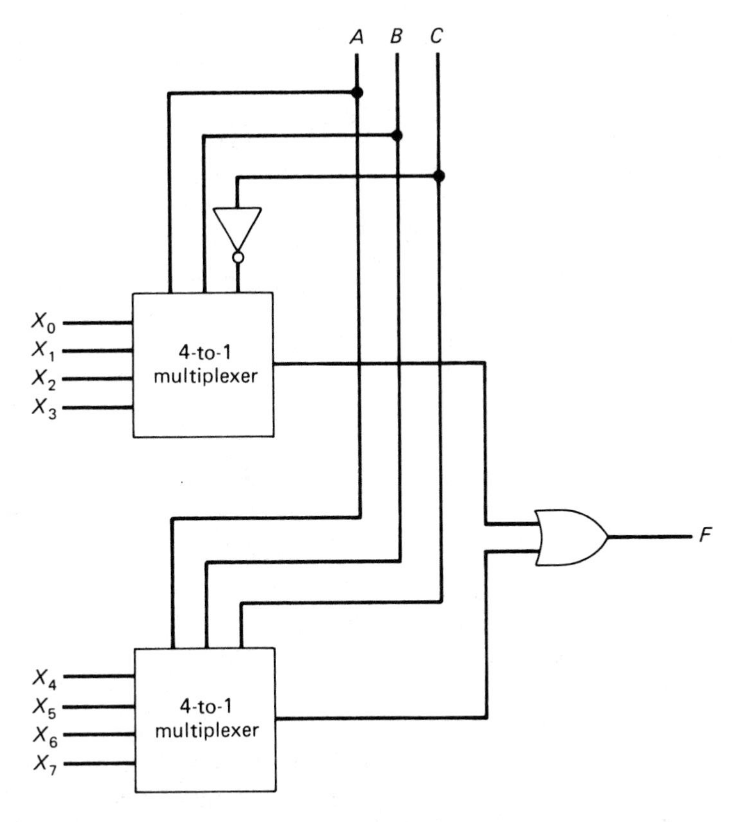

Problem 6.11

6.12 Show that the given circuit is a half adder.

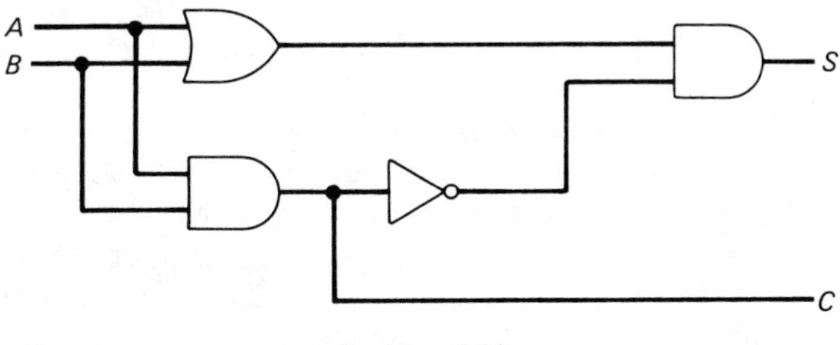

Problem 6.12

6.13 Design a full adder with AND, OR, and XOR gates by synthesizing Eqs. (6.7) and (6.8).

6.14 Draw a simplified version of a ROM that implements Tab. 4.4 (Sec. 4.3).

6.15 Draw a simplified version of a ROM that implements a full adder.

7

FLIP-FLOPS

As we have seen in the previous chapters, combinational circuits can be put to a variety of uses, ranging from extremely simple to highly complex. However, the output of a combinational circuit at any instant is determined solely by its inputs at that instant. That is, it has no capacity for *remembering*, or storing, effects of inputs at previous instants. Therefore, a combinational circuit, cannot be used to perform even the simplest function that depends on past inputs. In this case, in addition to the combinational circuit elements, *memory* devices, or *storage* elements, are needed.

Digital circuits with memory have outputs that depend on present as well as past inputs, and as we have noted, they are called *sequential* circuits. The digital computer is, of course, a prime example, which not only can perform various arithmetical operations but also can store the results for later use.

The basic storage element of a digital circuit is a device called a *flip-flop*, which we consider in some detail in this chapter. A flip-flop is capable of storing a single bit because it has two stable states: a so-called *one* state when it is storing a 1 and a *zero* state when it is storing a 0. The flip-flop has the capacity to remain in either the one or the zero state until some change in its input causes it to change its state. That is, it can be instructed by its input to *flip* to 1 or to *flop* back to 0.

The flip-flop is also a basic building block for larger sequential networks such as counters, shift registers, and memory registers, all of which are fundamental to computers. There are many types of flip-flops, but all are derived from a few simple types, which we will consider. Every flip-flop, as we will see, has one or more inputs which, together with its present state, determine which state it will assume next. Also, most flip-flops have two outputs, the stored bit and its complement, both of which are usually accessible to any connected external circuit.

Flip-flops are normally fabricated in integrated-circuit form as complete units. However, for illustrative purposes we describe them in this chapter as if they were made from discrete logic gates. In this way we may analyze the flip-flop circuits and obtain a better understanding of their behavior.

7.1 The SR Flip-Flop

Since it is one of the simplest to analyze, we begin with the *set–reset*, or *SR*, flip-flop, the symbol for which is shown in Fig. 7.1. The *set* input is S, the *reset* input is R, and the output, or *state*, of the flip-flop is Q. The complement $\bar{Q}$ of the state is also available as an output. The *SR* flip-flop retains a value of $Q = 0$ or $Q = 1$, even after the input that led to these values is removed. The reader may also see the *SR* flip-flop referred to in the literature as an *SR latch*. For the *SR* flip-flop, the terms *flip-flop* and *latch* are used interchangeably, but in general this is not the case.

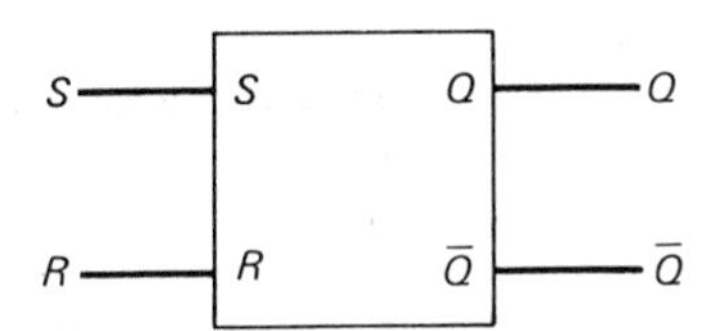

Figure 7.1 *SR flip-flop symbol*

In Fig. 7.1, if $S = 1$ and $R = 0$, the flip-flop is *set*, in which case we have $Q = 1$ ($\bar{Q} = 0$). Thus a 1 is *stored* in the flip-flop. We may then make $S = 0$ if we wish, keeping $R = 0$, and the flip-flop retains, or *remembers*, the output $Q = 1$. In other words, a 1 is still stored even after the set input ($S = 1$) is removed. If $Q = 1$ before the flip-flop is set, then Q remains 1 after it is set. That is, if the flip-flop is set ($S = 1$, $R = 0$), then if Q were previously 0, it becomes 1, and if it were previously 1, it remains 1. The flip-flop is *reset*, or *erased*, or *cleared* (a 0 is stored) by making the R input *active* (that is, making it 1). In other words, if $S = 0$ and $R = 1$, then Q becomes 0 ($\bar{Q} = 1$). If Q were previously 0, it remains 0, and if it were previously 1, it becomes 0.

From the set and reset operations we see that if $S = R = 0$, there may be either a 0 or a 1 stored in the flip-flop (that is, $Q = 0$ or $Q = 1$). If $S = 1$ and $R = 0$, then $Q = 1$, and if $S = 0$ and $R = 1$, then $Q = 0$. The only other possible input combination, $S = R = 1$, results in an indeterminate case, as we will see, and therefore is not allowed.

Fig. 7.2 is an implementation of an *SR* flip-flop using two NOR gates with

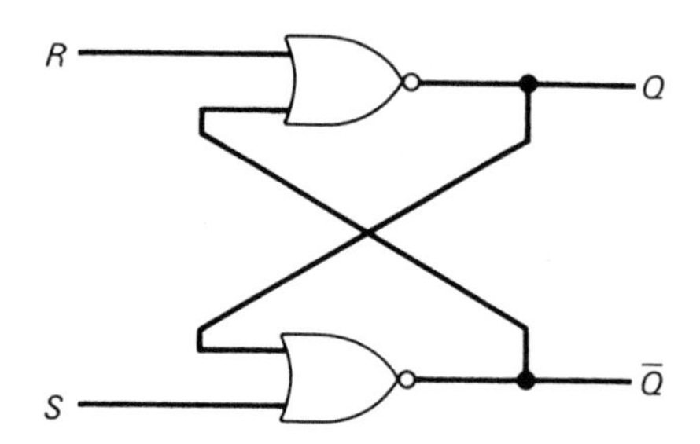

Figure 7.2 *NOR gate implementation of an SR flip-flop*

feedback. (The outputs Q and $\bar{Q}$ are *fed back* to the inputs of the NOR gates.) To see that the configuration is indeed an *SR* flip-flop we may consider the outputs

$$\begin{aligned} Q &= \overline{R + \bar{Q}} = \bar{R}Q \\ \bar{Q} &= \overline{S + Q} = \bar{S}\bar{Q} \end{aligned} \tag{7.1}$$

obtained from the figure, and the eight possible states given in Tab. 7.1.

Table 7.1 *Possible states of Fig. 7.2*

State	S	R	Q
1	0	0	0
2	0	0	1
3	0	1	0
4	0	1	1
5	1	0	0
6	1	0	1
7	1	1	0
8	1	1	1

We consider a state *stable* if for the given values of S, R, and Q no output signals (Q or $\bar{Q}$) are tending to change as long as S and R are fixed. For example, in the case of state 1 ($S = R = Q = 0$), we have from Eqs. (7.1)

$$Q = 0, \qquad \bar{Q} = 1$$

which is consistent, and therefore the state is stable. If $S = R = 0$ had forced the state to change from its given value of $Q = 0$ to the value $Q = 1$, this would have been an inconsistency indicating an unstable state.

In the case of states 2, 3, and 6, respectively, we have by Eqs. (7.1) and Tab. 7.1

$$Q = 1, \qquad \bar{Q} = 0$$

$$Q = 0, \qquad \bar{Q} = 1$$

and

$$Q = 1, \qquad \bar{Q} = 0$$

These are all consistent and therefore these states are stable. However, in the case of state 4 we have, by Eqs. (7.1), $Q = 0$, which requires a change from the given state $Q = 1$. State 4 is thus unstable. In the case of state 5, $\bar{Q} = 0$, requiring a change from $\bar{Q} = 1$, and thus state 5 is also unstable.

Finally, let us consider the forbidden states 7 and 8, where $S = R = 1$. In both cases we have, by Eqs. (7.1), $Q = \bar{Q} = 0$, which is the only situation where the outputs are not complements of each other. (Also, state 8 is unstable, since Q changes from 1 to 0.) However, there is an even more serious difficulty. Suppose now with $Q = \bar{Q} = 0$ we change both S and R from 1 to 0. From Fig. 7.2 we see that an indeterminate condition arises. Does the top gate act first to change Q to 1, which keeps $\bar{Q}$ at 0, or does the bottom gate act first to change $\bar{Q}$ to 1, which keeps Q at 0? The output is thus unpredictable, and so we must avoid the state $S = R = 1$. Therefore the only allowable stable states of Fig. 7.2 are states 1, 2, 3, and 6, as listed in Tab. 7.2.

Table 7.2 *Allowable stable states of Fig. 7.2*

State	S	R	Q	$\bar{Q}$
1	0	0	0	1
2	0	0	1	0
3	0	1	0	1
6	1	0	1	0

The results tabulated in Tab. 7.2 indicate that the circuit in Fig. 7.2 is an *SR* flip-flop. That is, if $S = R = 0$, then Q may be either 0 or 1, $S = 0$ and $R = 1$ (the reset condition) yield $Q = 0$, and $S = 1$ and $R = 0$ (the set condition) yield $Q = 1$.

The *SR* flip-flop is a sequential device, since its outputs are not determined solely by the inputs S and R. This is easily seen from the feedback in Fig. 7.2 or from the output relations of Eqs. (7.1). This may also be seen by observing that different sequences S, R, and Q at a given time t_n result in different outputs Q and $\bar{Q}$ at a succeeding time, symbolized by t_{n+1}. Specifically, if Q_n is the *present* state at time t_n and S and R are the inputs at time t_n, then Q_{n+1} is the *next* state, occurring (at t_{n+1}) as soon as the effects of the inputs are felt throughout the circuit. (We recall that there is a *propagation delay* time required before an input of a logic gate can affect the output.) A table showing the effects of the inputs and the present state on the next state is called the *transition* table; it is analogous to the truth table in the case of combinational circuits.

As an example, let us find the transition table for the SR flip-flop in Fig. 7.2. Since the four states of Tab. 7.2 are stable, Q_n (given as Q in the table) will be the same as the next state Q_{n+1} in each case. There are also two other transition table entries, for which (a) $S = 0, R = 1$, and $Q_n = 1$, and (b) $S = 1, R = 0$, and $Q_n = 0$. In these cases $Q_{n+1} \neq Q_n$, since Q_{n+1} is 0 in (a) and is 1 in (b), as the reader may verify from Fig. 7.2. In each of these two cases the state Q starts at one value (0 or 1) and changes to the other. The state transition table is given in Tab. 7.3, where the first three entries and the sixth entry are those of Table 7.2, the fourth and fifth entries are cases (a) and (b) above, and the last two entries, as noted, are the cases $S = R = 1$, which are the unpredictable, or indeterminate, cases.

Table 7.3 *Transition table for the SR flip-flop of Fig. 7.2*

Input at Time t_n		*Present State*	*Next State*
S	R	Q_n	Q_{n+1}
0	0	0	0
0	0	1	1
0	1	0	0
0	1	1	0
1	0	0	1
1	0	1	1
1	1	0	Indeterminate
1	1	1	Indeterminate

Using Tab. 7.3, we may obtain an expression for Q_{n+1} in terms of S, R, and Q_n. Using minterms, we have the sum-of-products form

$$Q_{n+1} = \bar{S}\bar{R}Q_n + S\bar{R}\bar{Q}_n + S\bar{R}Q_n \tag{7.2}$$

with the added provision that we cannot have $S = R = 1$. This last condition is equivalent to the requirement that either S or R, or both, must be 0. That is,

$$SR = 0 \tag{7.3}$$

We may simplify Eq. (7.2) to

$$\begin{aligned} Q_{n+1} &= \bar{S}\bar{R}Q_n + S\bar{R} \\ &= \bar{R}Q_n + S\bar{R} \end{aligned}$$

Adding Eq. (7.3) to this result, we have

$$Q_{n+1} = \bar{R}Q_n + S\bar{R} + SR$$

or

$$Q_{n+1} = S + \bar{R}Q_n \tag{7.4}$$

In words, this equation says that the next state is 1 if the flip-flop is "set," or if the present state is 1 and we do not "reset" the flip-flop.

If we know the present state Q_n of the flip-flop and the state Q_{n+1} that we want it to be in next, we may find the required input, or *excitation*, S and R from Tab. 7.3. However, the pertinent information may be extracted from Tab. 7.3 and arranged in a more compact table, called the *excitation table*. This is a table that shows the excitation required at time t_n to change the present state Q_n to the next state Q_{n+1}. In the case of the SR flip-flop, the excitation table is Tab. 7.4, which, as the reader may show, follows directly from Table 7.3. The x's represent "don't cares," since either a 0 or a 1 will suffice.

Table 7.4 *Excitation table for the SR flip-flop*

Present State (*time* t_n) Q_n	*Next State* (*time* t_{n+1}) Q_{n+1}	*Input at Time* t_n	
		S	R
0	0	0	x
0	1	1	0
1	0	0	1
1	1	x	0

Another useful table, called the *state table*, represents a condensation of the information given in the transition table. In the state table all the possible inputs at t_n of the flip-flop are listed along with the next state Q_{n+1}. In the case of the SR flip-flop we see from Tab. 7.3 that when $S = R = 0$ (the first two entries), Q_{n+1} is the same as Q_n. When $S = 0$ and $R = 1$, $Q_{n+1} = 0$ in both cases, and when $S = 1$ and $R = 0$, $Q_{n+1} = 1$ in both cases. From these results we may construct the state table, which is Tab. 7.5. The state table is the one most frequently used to describe the behavior of the circuit.

Table 7.5 *State table for the SR flip-flop*

t_n		t_{n+1}	
S	R	Q_{n+1}	$\bar{Q}_{n+1}$
0	0	Q_n	$\bar{Q}_n$
0	1	0	1
1	0	1	0
1	1	Indeterminate	

EXERCISES

7.1.1 Show that the given circuit is an *SR* flip-flop by finding all its allowable stable states.

Ans:

S	R	Q	$\bar{Q}$
0	0	0	1
0	0	1	0
0	1	0	1
1	0	1	0

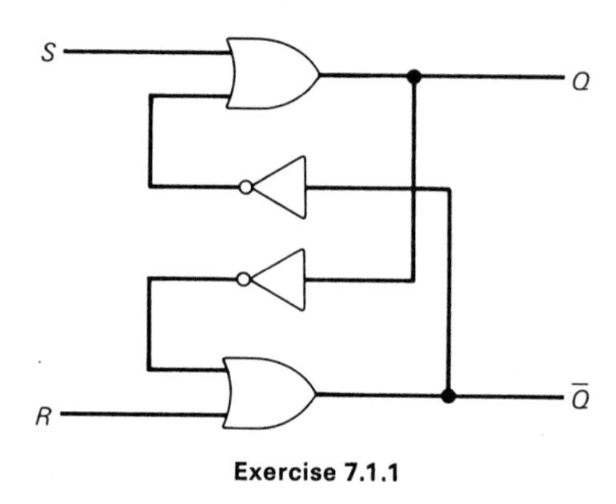

Exercise 7.1.1

7.1.2 Show that the transition table for the circuit of Ex. 7.1.1 is identical to Tab. 7.3.

7.1.3 Show that the excitation table for the circuit of Ex. 7.1.1 is identical to Tab. 7.4.

7.2 Flip-Flop Inputs

In general, there are two basic types of digital signals. A *level* signal is one such as S from t_1 to t_2 in Fig. 7.3, which remains 1 for a more or less arbitrary length of time until it is switched to 0. The other type is a *pulse* signal, which changes from

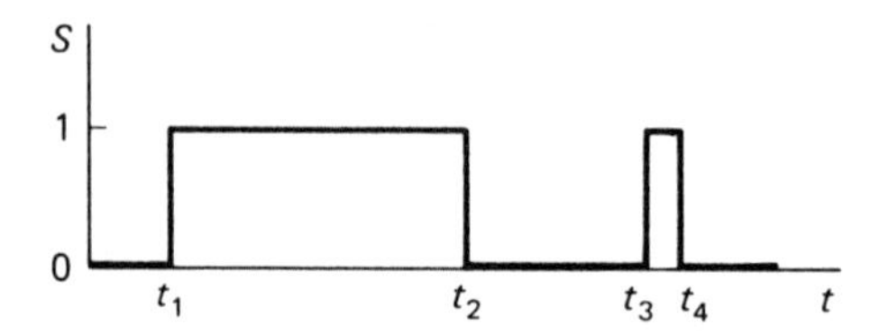

Figure 7.3 *Digital signal*

0 to 1 and then, after a fixed (usually short) time, returns to 0. Very often, but not always, the duration of a pulse is considerably shorter than that of a level. An example of a pulse might be S in Fig. 7.3 from t_3 to t_4.

The *leading*, or *positive*, edge of a pulse (or a level) is the edge at which the pulse rises from 0 to 1, and the *trailing*, or *negative*, edge is that where the pulse falls from 1 to 0. In the case of Fig. 7.3, the positive edges occur at t_1 and t_3 and the negative edges are at t_2 and t_4.

A *timing diagram* of a digital circuit is a graphical display of the sequence of input and output signals. For example, a possible timing diagram for an SR flip-flop is shown in Fig. 7.4. At time t_1 the input S is activated ($S = 1$), and after a short

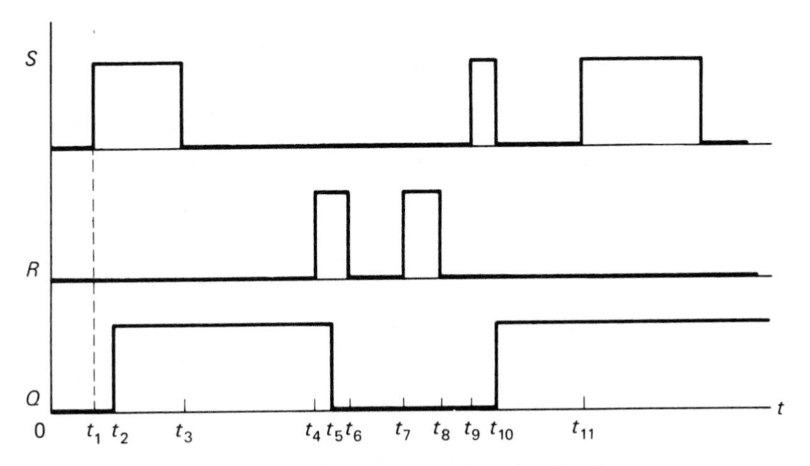

Figure 7.4 *Timing diagram for an SR flip-flop*

propagation delay the output Q becomes 1 at t_2 (the flip-flop is set). Even though S is subsequently turned off (made 0) at t_3, Q continues to be 1 until the flip-flop is reset. This occurs at t_4 when R is activated for a short time (until t_6). Again there is a delay before the circuit responds at t_5. We may note that resetting the flip-flop at t_7 has no effect since Q remains at 0. Similarly, setting the flip-flop at t_{11} after it is previously set at t_9 has no effect.

A *clock* is a device that generates a precise pattern of periodically occurring pulses, as shown in Fig. 7.5. The time for one complete cycle is called the *period*,

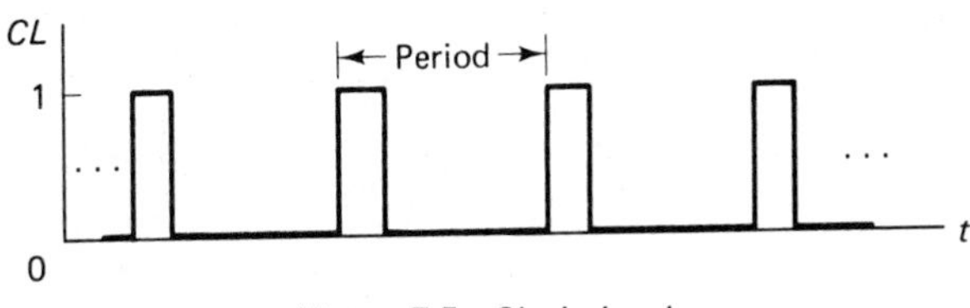

Figure 7.5 *Clock signal*

and the *frequency* of the clock is the reciprocal of the period. Frequencies of several megahertz are commonly used in generating clock signals.

Clocks are often used in sequential circuits to enable gates at periodic intervals. Thus a clock may be thought of as a *control signal* that synchronizes the sequence of operations in a sequential circuit. Circuits that are controlled in this manner are known as *synchronous sequential* circuits. Circuits that are not synchronous are called *asynchronous*. Throughout this book we will be concerned mainly with synchronous sequential circuits.

As an example of the use of a clock we note that the select signal S, considered in the previous chapter in Fig. 6.4, controls the output of the 2-to-1 multiplexer in Fig. 6.3. To illustrate further the use of a clock let us consider the circuit in Fig. 7.6, which is the SR flip-flop of Fig. 7.2 with its inputs taken as the outputs of two

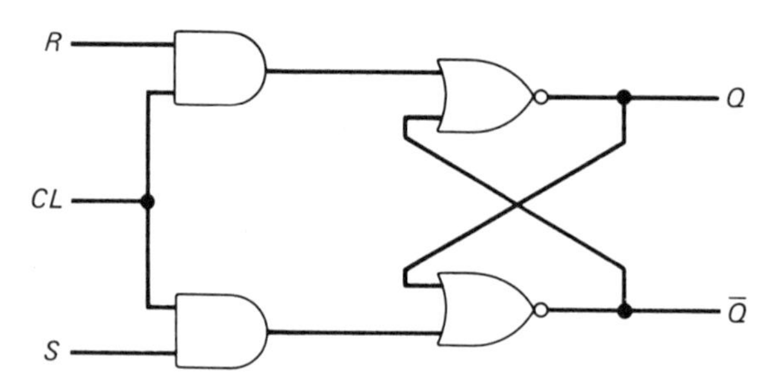

Figure 7.6 *Gated SR flip-flop*

AND gates. The inputs are S, R, and CL, the latter being the symbol for the output signal of a clock, such as that in Fig. 7.5. This configuration is sometimes called a *clocked*, or *gated*, SR flip-flop, because it functions exactly as an SR flip-flop does except that the input signals S and R can enable the flip-flop only when the clock signal CL is 1. The symbol for the gated SR flip-flop is shown in Fig. 7.7.

Some gated flip-flops are *positive-edge triggered*, which means that they respond to the positive edge of the clock pulse. The other inputs, such as S and R in the SR flip-flop, must be stable at the time the positive edge of the clock pulse occurs. Other flip-flops are *negative-edge triggered*, in which case the response takes place when the negative edge of the clock pulse occurs. As examples, Figs. 7.8(a) and (b) show, respectively, timing diagrams for positive-edge and negative-edge triggered SR flip-flops. In Fig. 7.8(a), $S = 1$, $R = 0$, and $Q = 0$ when the positive

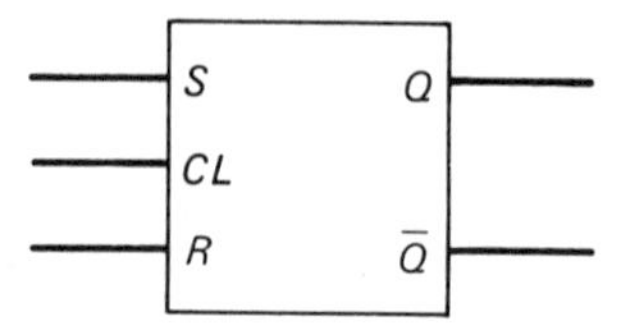

Figure 7.7 *Symbol for a gated SR flip-flop*

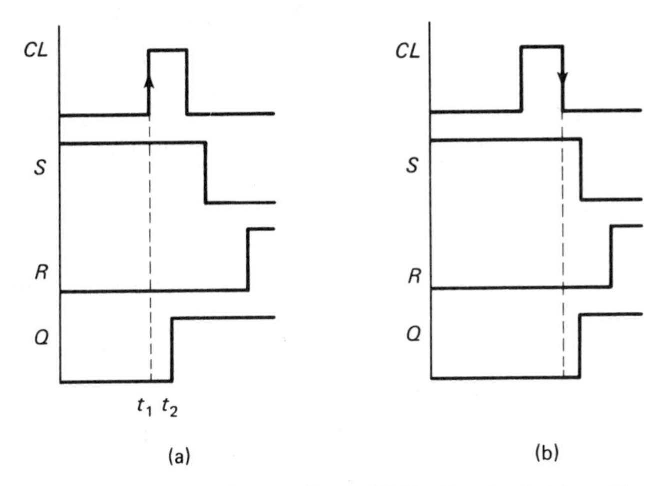

Figure 7.8 *Timing diagrams for an SR flip-flop that is (a) positive-edge triggered and (b) negative-edge triggered*

edge of the clock pulse occurs at t_1, and thus Q becomes 1 after the propagation delay at t_2. In Fig. 7.8(b) Q becomes 1 after the negative edge of the clock pulse has occurred.

A *transition* gate is a device that recognizes an edge of a signal and produces a short pulse that may be used as the activating mechanism of an edge-triggered flip-flop. Such a gate would be incorporated into the design of the flip-flop as its triggering device. An example of a transition gate is given in Ex. 7.2.5.

The sequence of inputs of a flip-flop or any other sequential circuit, together with the initial state, determines the sequence of outputs. The state table for the circuit is sufficient for determining the output sequence, since it shows what the next state Q_{n+1} will be in terms of the inputs and the present state Q_n. For example, suppose in the case of the SR flip-flop that $Q = 0$ initially and the input sequence S, R is 0, 0; 1, 0; 0, 1; 0, 1. That is, $S = 0$ and $R = 0$ is the first input, followed by $S = 1$ and $R = 0$, and so on. From Tab. 7.5 we see that the output sequence Q in this case is 0, 0, 1, 0, 0, or 00100. This is clear because the first input $S = R = 0$ results in $Q_{n+1} = Q_n$, which is the initial state 0; the second input $S = 1$, $R = 0$ with $Q_n = 0$ results in $Q_{n+1} = 1$; 0, 1 results in $Q_{n+1} = 0$; and 0, 1 again yields $Q_{n+1} = 0$.

The excitation table, such as Tab. 7.4 in the case of the SR flip-flop, can be used to determine the necessary input sequence to yield a given output sequence, provided the initial state is known. For example, suppose we wish the SR flip-flop with $Q = 1$ initially to yield the output sequence Q given by 1010100. The first input must keep the output at 1, so that by Tab. 7.4 it must be $S = \text{x}$ (don't care) and $R = 0$. The second output $Q = 0$ is a change from $Q = 1$, requiring $S = 0$ and $R = 1$. The complete sequence, obtained in this manner from Tab. 7.4, is x, 0; 0, 1; 1, 0; 0, 1; 1, 0; 0, 1; 0, x.

EXERCISES

7.2.1 Complete the timing diagram by finding Q in the gated SR flip-flop of Fig. 7.7 if S, R, and CL are as shown. Assume that the flip-flop is positive-edge triggered and that for illustrative purposes there is no propagation delay.

Ans:

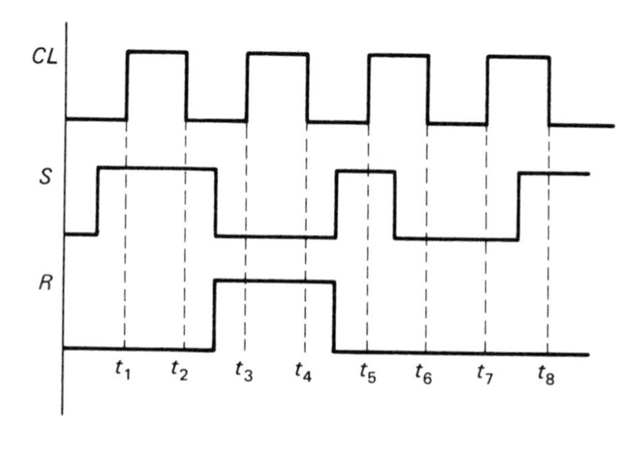

Exercise 7.2.1

7.2.2 Repeat Ex. 7.2.1 if the flip-flop is negative-edge triggered.

Ans:

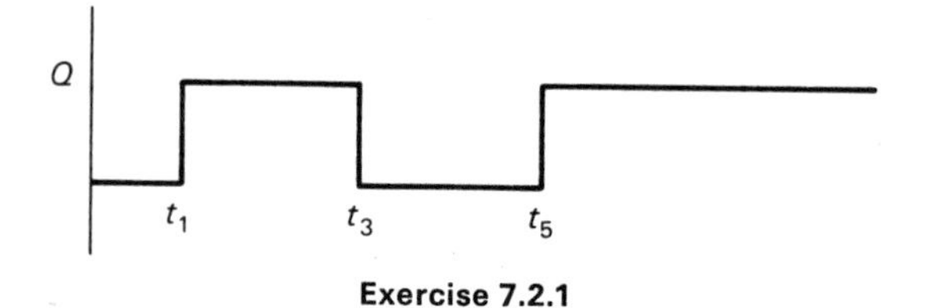

Exercise 7.2.2

7.2.3 Find the sequence of outputs Q of the SR flip-flop if initially $Q = 0$ and the input sequence S, R is 0, 0; 1, 0; 0, 0; 1, 0; 0, 1; 0, 0.

Ans: 0011100

7.2.4 Find the input sequence S, R to yield the output sequence Q given by 1100110. The initial state is $Q = 0$.

Ans: 1, 0; x, 0; 0, 1; 0, x; 1, 0; x, 0; 0, 1.

7.2.5 The given circuit is a transition gate that produces a short output pulse F in response to the positive edge of the clock signal CL. Show this by drawing the timing diagram, assuming that the positive edge of CL occurs at a time t and, for tutorial purposes, that the delay Δt_1 of the inverter is greater than the delay Δt_2 of the AND gate. (*Suggestion:* Note that because of the different delays there is a time when both CL and $\overline{CL}$ are 1.)

Ans:

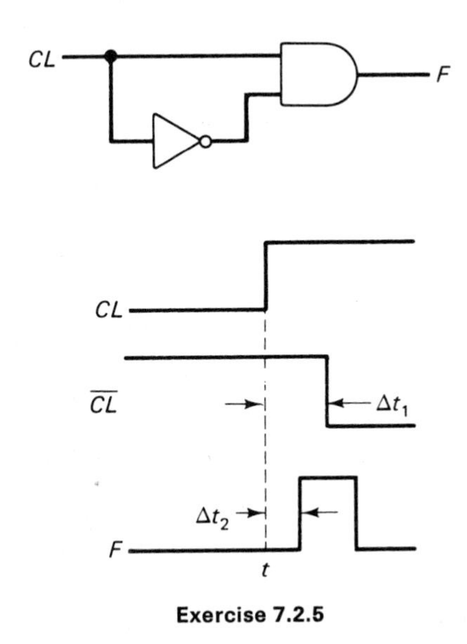

Exercise 7.2.5

7.3 The *JK* Flip-Flop

The function performed by the SR flip-flop is the basic function commonly performed by all flip-flops. However, the indeterminate state that exists when both S and R are 1 is a distinct disadvantage, which led to the development of another, much more versatile, flip-flop called the *JK flip-flop*. The symbol for the JK flip-flop, shown in Fig. 7.9, is identical to the gated SR flip-flop except for the input symbols J and K. Indeed, as we shall see, the JK flip-flop behaves exactly as the SR flip-flop does, with J and K replacing S and R, except that there is no indeterminate state.

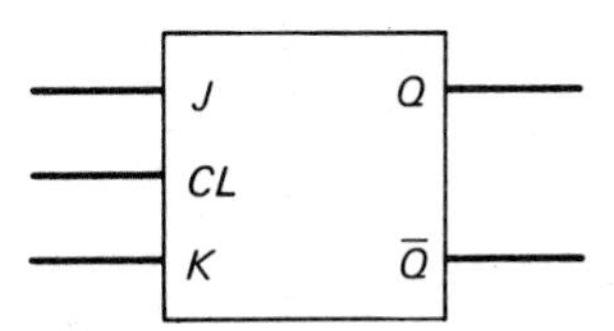

Figure 7.9 *Symbol for a JK flip-flop*

A *JK* flip-flop may be implemented with a gated *SR* flip-flop and two AND gates, as shown in Fig. 7.10. To see how the *JK* flip-flop functions, let us analyze the circuit and find its state table. To make the work easier we may use the transition table of the *SR* flip-flop given earlier as Tab. 7.3 and construct from it and Fig. 7.10 the transition table of the *JK* flip-flop.

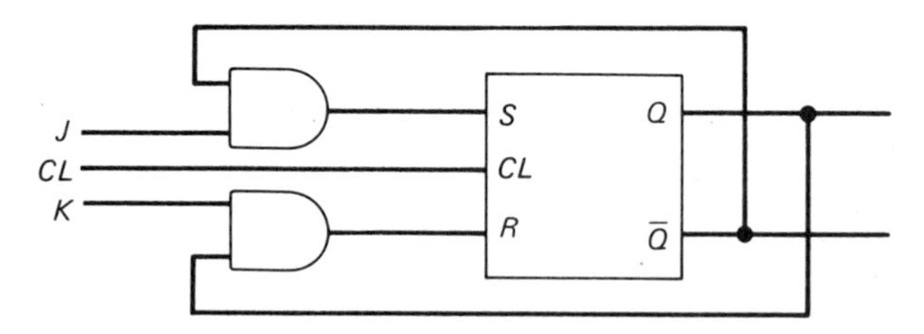

Figure 7.10 *Implementation of a JK flip-flop*

We begin by noting from Fig. 7.10 that

$$\begin{aligned} S &= J\bar{Q} \\ R &= KQ \end{aligned} \tag{7.5}$$

When $J = K = Q_n = 0$ we see by Eqs. (7.5) that $S = R = 0$ ($Q = Q_n$, the state at time t_n). Therefore, from the first row of Tab. 7.3, the conditions $S = R = Q_n = 0$ yields $Q_{n+1} = 0$ (the state at time t_{n+1}). The first row of the transition table for the *JK* flip-flop is thus $J = K = Q_n = Q_{n+1} = 0$. To obtain the second row, we let $J = K = 0$ and $Q_n = 1$. Then by Eqs. (7.5) we have $S = R = 0$. This condition ($S = R = 0$, $Q_n = 1$) yields $Q_{n+1} = 1$ from row two of Tab. 7.3. Therefore row two of the transition table of the *JK* flip-flop is $J = K = 0$, $Q_n = Q_{n+1} = 1$. Continuing in this manner, we construct the transition table of the *JK* flip-flop as shown in Tab. 7.6.

The last two rows of the transition table for the *SR* flip-flop had indeterminate

Table 7.6 *Transition table for the JK flip-flop*

Input at t_n		*Present State*	*Next State*
J	K	Q_n	Q_{n+1}
0	0	0	0
0	0	1	1
0	1	0	0
0	1	1	0
1	0	0	1
1	0	1	1
1	1	0	1
1	1	1	0

next states, but this is not the case for the *JK* flip-flop, as was noted earlier. The reader is encouraged to show this by developing the last two rows of Tab. 7.6.

The state table of the *JK* flip-flop is easily found from Tab. 7.6. In the first two rows, $J = K = 0$ and $Q_{n+1} = Q_n$. In the next two rows (rows 3 and 4), $J = 0$, $K = 1$, and $Q_{n+1} = 0$. In rows 5 and 6 we have $J = 1$, $K = 0$, and $Q_{n+1} = 1$. Finally, in the last two rows we have $J = K = 1$ and $Q_{n+1} = \bar{Q}_n$. Thus the state table is as shown in Tab. 7.7.

Table 7.7 *State table for the JK flip-flop*

t_n		t_{n+1}	
J	K	Q_{n+1}	$\bar{Q}_{n+1}$
0	0	Q_n	$\bar{Q}_n$
0	1	0	1
1	0	1	0
1	1	$\bar{Q}_n$	Q_n

Suppose we set $J = K = 1$ at time t_n. Then we see from Tab. 7.7 that at t_{n+1}, $Q_{n+1} = \bar{Q}_n$. Therefore, for this condition (J and K maintained at 1) the output Q changes, or *complements*, or *toggles*, with each clock pulse. Thus the *JK* flip-flop has two advantages over the *SR* flip-flop. It has no indeterminate states, and it can complement.

Generally, *JK* flip-flops are of two types: the *master–slave* type, which we will discuss in the next section, and the edge-triggered type. In most cases the latter type is negative-edge triggered. That is, the inputs J and K are enabled and the outputs set on the trailing edge of the clock pulse. An example of a timing diagram for this type is shown in Fig. 7.11. At t_1, after a short delay following the occurrence of the clock pulse, Q becomes 0 (since $J = 0$ and $K = 1$). The flip-flop is triggered again at t_2 with $J = K = 1$, and therefore Q toggles (goes from 0 to 1).

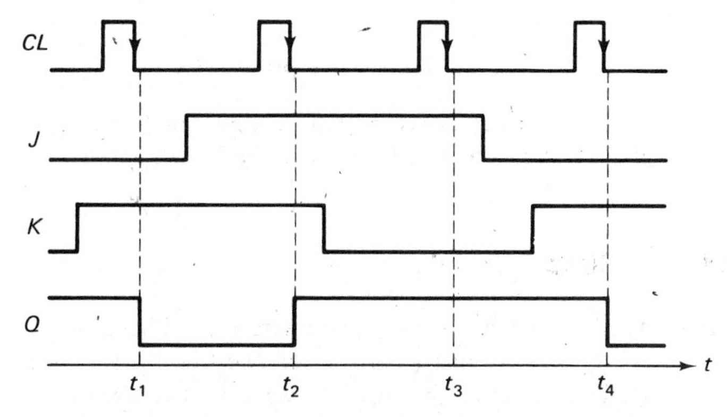

Figure 7.11 *Timing diagram for a negative-edge triggered JK flip-flop*

The clock pulse occurring at t_3 has no effect, since $J = 1$ and $K = 0$, and thus Q, which was 1, remains 1. Finally, at t_4, since $J = 0$ and $K = 1$, Q becomes 0. These statements may be easily checked by means of Tab. 7.7.

EXERCISES

7.3.1 Find the excitation table for the *JK* flip-flop.

Ans:

t_n Q_n	t_{n+1} Q_{n+1}	J	t_n K
0	0	0	x
0	1	1	x
1	0	x	1
1	1	x	0

7.3.2 Complete the timing diagram of a negative-edge triggered *JK* flip-flop if *J*, *K*, and *CL* are as shown. Assume that there is no delay and that $Q = 0$ initially.

Ans:

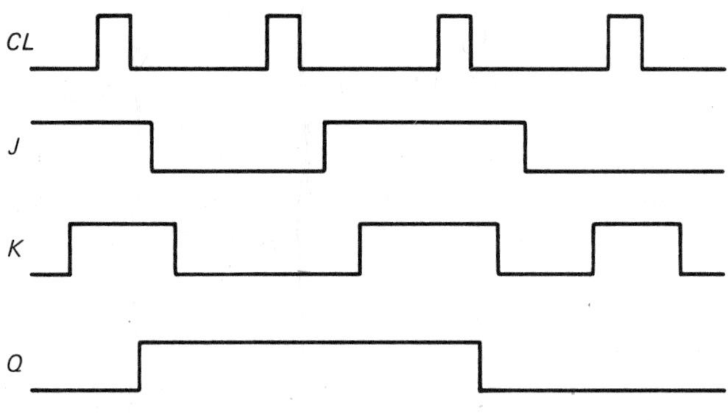

Exercise 7.3.2

7.3.3 Find the output sequence Q of a *JK* flip-flop if the initial state is $Q = 1$ and the input sequence J, K is 1, 1; 1, 0; 1, 0; 0, 0; 0, 1; 1, 1.

Ans: 1011101

7.4 The Master–Slave Principle

In many applications, such as in counters and shift registers (which we will consider in Chap. 9), the outputs Q and $\bar{Q}$ of one flip-flip are used as inputs to another flip-flop. For example, in Fig. 7.12 we have two *JK* flip-flops connected with Q and $\bar{Q}$ of the first flip-flop equal, respectively, to J and K of the second flip-flop.

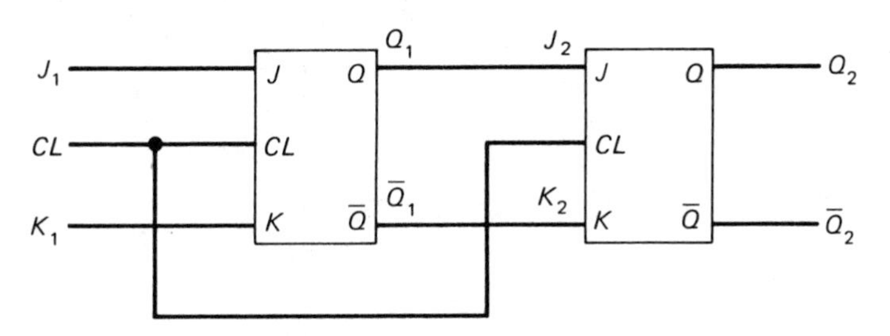

Figure 7.12 *Two interconnected flip-flops*

For clarity we have labeled the inputs and outputs of the first flip-flop as J_1, K_1, Q_1, and $\bar{Q}_1$ and those of the second flip-flop as J_2, K_2, Q_2, and $\bar{Q}_2$.

Ideally, CL enables both flip-flops simultaneously, with J_1 and K_1 determining the next states Q_{n+1} and $\bar{Q}_{n+1}$ of Q_1 and $\bar{Q}_1$, and $J_2 = Q_n$ and $K_2 = \bar{Q}_n$ (the present states of Q_1 and $\bar{Q}_1$) determining the next states of Q_2 and $\bar{Q}_2$. However, if when CL enables the first flip-flop, Q_1 changes from Q_n to Q_{n+1} *before* or *during* the enabling of the second flip-flop, then J_2 is Q_{n+1} and K_2 is $\bar{Q}_{n+1}$, rather than Q_n and $\bar{Q}_n$, when the second flip-flop is enabled. Evidently, in this case the desired results will not be achieved. The problem of the effects of the inputs *racing* through the first flip-flop before the second flip-flop is enabled, which is known as the *race* problem, must be solved if the circuit is to function properly.

The race problem is commonly solved by the use of the *master–slave* principle, as illustrated with *JK* flip-flops in Fig. 7.13. The configuration in the dashed rectangle is a version of a *JK master–slave* flip-flop, consisting of two *JK* flip-flops, a *master* and a *slave*, connected with an inverter, as indicated.

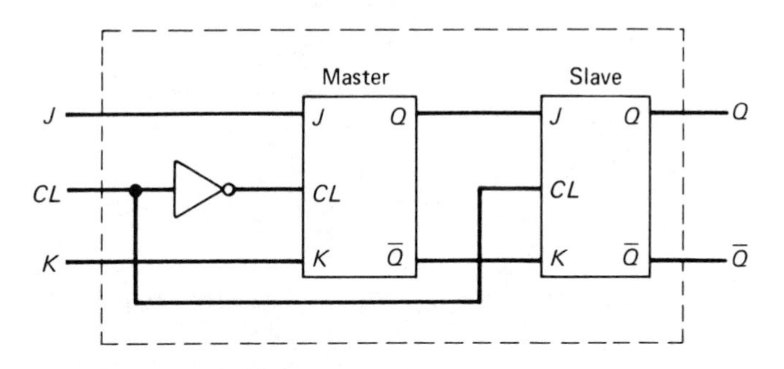

Figure 7.13 *JK master–slave flip-flop*

If the master and slave flip-flops are negative-edge triggered, then the positive edge of the CL pulse at t_1, shown in Fig. 7.14, occurs with a negative edge of the $\overline{CL}$ pulse, which enables the master. The slave is *disabled* (not enabled) at this time, because it is not activated by the positive edge of CL. At t_2 the negative edge of CL occurs, and this enables the slave, while at the corresponding positive edge of $\overline{CL}$ the master is disabled. Information on the J and K inputs is thus stored in the

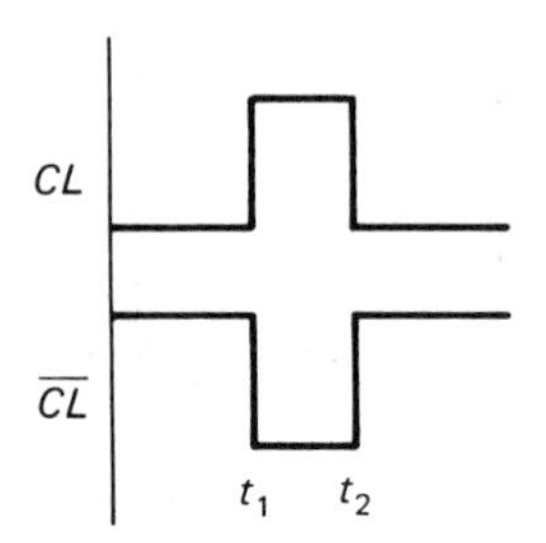

Figure 7.14 *Clock pulse and its complement*

master flip-flop on the positive edge of the clock pulse and transferred to the slave flip-flop on the negative edge of the clock pulse. The J and K inputs are isolated from the Q and $\bar{Q}$ outputs, since the slave is disabled when the master is enabled and vice versa.

To see that the race problem is solved by the master–slave principle, let us consider Fig. 7.15, which consists of two master–slave flip-flops, each of which is like the arrangement in Fig. 7.13, connected as in Fig. 7.12. The inputs Q_1 and $\bar{Q}_1$ to the second flip-flop cannot change before the second master is enabled, because the first slave is not enabled before then. (The two masters are first enabled and then disabled while the two slaves are enabled.) Thus there is no race.

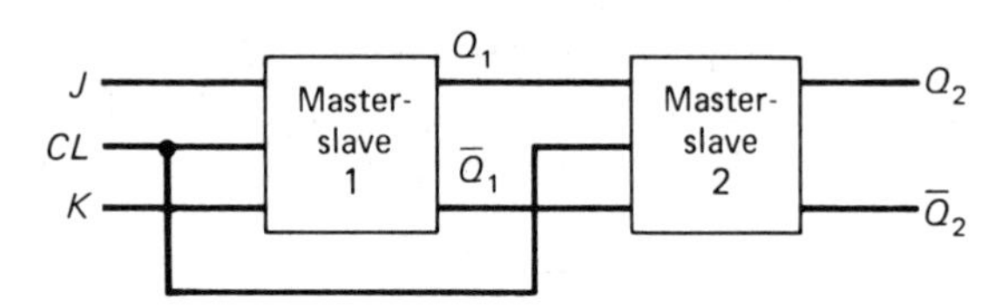

Figure 7.15 *Two interconnected master–slave flip-flops*

EXERCISE

7.4.1 If the clock signal CL in Fig. 7.15 is defined by

$$
\begin{aligned}
CL &= 0, \quad t_0 < t < t_1 \\
&= 1, \quad t_1 < t < t_2 \\
&= 0, \quad t_2 < t < t_3 \\
&= 1, \quad t_3 < t < t_4 \\
&= 0, \quad t_4 < t < t_5 \\
&= 1, \quad t_5 < t < t_6 \\
&= 0, \quad t_6 < t < t_7
\end{aligned}
$$

the outputs of both masters and both slaves are 0 at t_0, and $J = K = 1$ for $t > t_0$, sketch the outputs Q_1 and Q_2 for $t_0 < t < t_7$.

Ans:

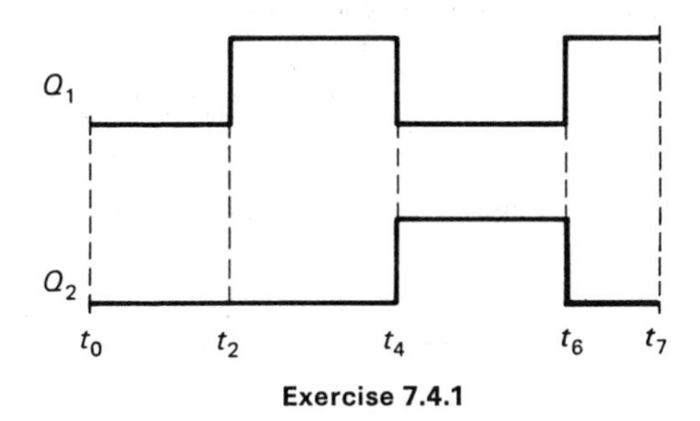

Exercise 7.4.1

7.5 The *D* Flip-Flop

Another important flip-flop is the *D flip-flop*, or *DATA flip-flop*. Its symbol is shown in Fig. 7.16, where we see that there is a D (data) input, a CL (clock) input, and outputs Q and $\bar{Q}$. The function performed by the D flip-flop is defined by the

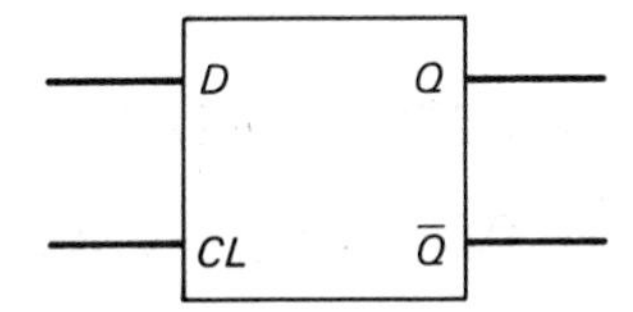

Figure 7.16 *Symbol for a D flip-flop*

state table given in Tab. 7.8. Evidently $Q_{n+1} = D$, which says that the data at D is simply transferred to the output when the input is enabled. Thus the D flip-flop is useful in applications where a single bit is to be stored. The value of the input D is stored and is available as the output Q.

Like JK flip-flops, D flip-flops are generally of the edge-triggered or the master–slave type. However, unlike the JK flip-flop, the edge-triggered D flip-flop is normally positive-edge triggered. There is also a similar device, called a *D latch*, which behaves like the D flip-flop except that it is enabled by a voltage level rather than by a clock pulse edge. Differences in the D flip-flop and D latch are illustrated in Ex. 7.5.2.

We may construct a D flip-flop with a JK flip-flop and an inverter, as shown in Fig. 7.17. To see that this is the case we note that $J = D$ and $K = \bar{D}$. Thus, when $D = 0$ we have $J = 0$ and $K = 1$, and by Tab. 7.7, $Q_{n+1} = D = 0$. Also when $D = 1$, we have $J = 1$ and $K = 0$, and therefore $Q_{n+1} = D = 1$. These results are the same as those of Tab. 7.8, and therefore we have a D flip-flop.

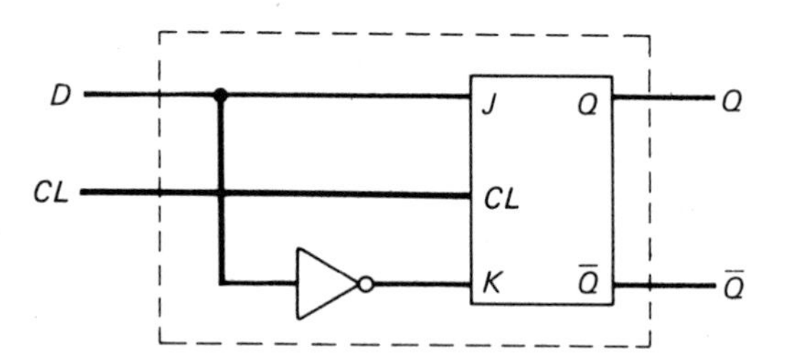

Figure 7.17 *D flip-flop constructed with a JK flip-flop*

Table 7.8 *State table for the D flip-flop*

t_n D	Q_{n+1}	t_{n+1} $\bar{Q}_{n+1}$
0	0	1
1	1	0

We may also construct a *JK* flip-flop with a *D* flip-flop and other logic gates. This may be seen from

$$D = Q_{n+1} \tag{7.6}$$

and from Tab. 7.6, by which we may write

$$Q_{n+1} = \bar{J}\bar{K}Q_n + J\bar{K}\bar{Q}_n + J\bar{K}Q_n + JK\bar{Q}_n \tag{7.7}$$

Simplifying Eq. (7.7) and combining the result with Eq. (7.6), we have

$$D = \bar{K}Q_n + J\bar{Q}_n \tag{7.8}$$

which may be implemented as shown in Fig. 7.18. Thus we have a *JK* flip-flop composed of a *D* flip-flop and four other logic gates.

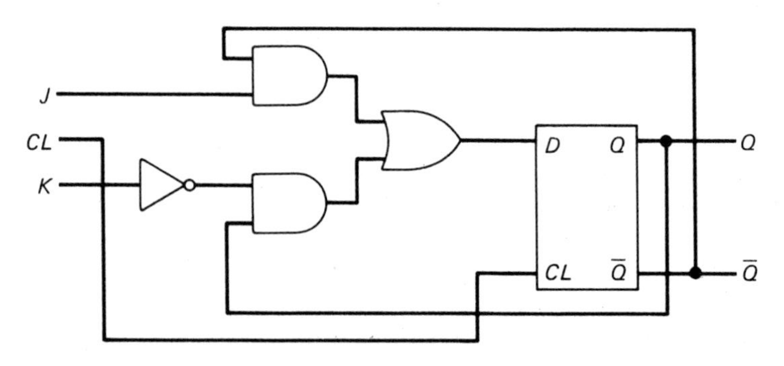

Figure 7.18 *JK flip-flop implemented with a D flip-flop and four gates*

EXERCISES

7.5.1 Show that the given circuit using a gated SR flip-flop is a D flip-flop. (*Suggestion:* Note that the presence of the inverter prevents occurrence of the forbidden state $S = R = 1$.)

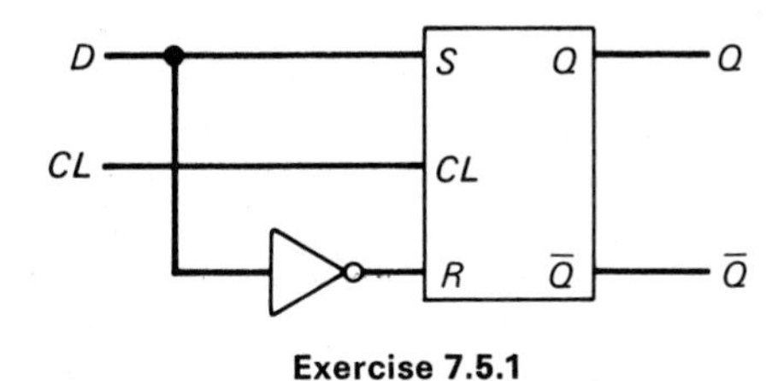

Exercise 7.5.1

7.5.2 The inputs shown are (a) those of a positive-edge triggered D flip-flop with output Q_1, or (b) those of a level-triggered D latch with output Q_2. Sketch the outputs Q_1 and Q_2, assuming no delays.

Ans:

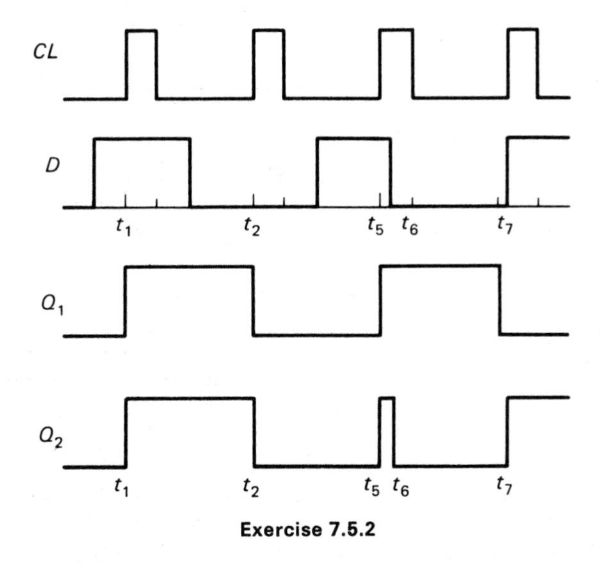

Exercise 7.5.2

7.6 The T Flip-Flop

Some applications, such as counting, require a flip-flop that will change its state, or toggle, every time an activating signal such as a clock is applied. A flip-flop that has this property, shown symbolically in Fig. 7.19, is called a T, or *toggle*, flip-flop. There is one input T and the two customary outputs Q and $\bar{Q}$.

When T changes from 0 to 1, the output Q is complemented (the flip-flop

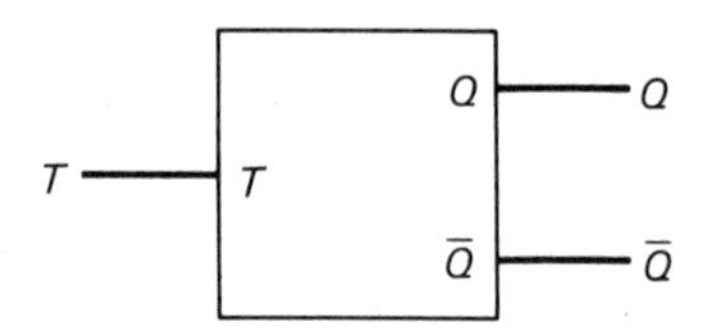

Figure 7.19 *Symbol for a T flip-flop*

toggles). However, there is no change in Q if T changes from 1 to 0, or remains 1, or remains 0. Thus each time a pulse of T occurs (T goes from 0 to 1 and back to 0), the T flip-flop toggles. Since by the state table of the JK flip-flop (Tab. 7.7), $Q_{n+1} = \bar{Q}_n$ when $J = K = 1$, we may obtain a T flip-flop from a JK flip-flop. All we have to do is make $J = K = 1$ and $CL = T$. The result is shown in Fig. 7.20.

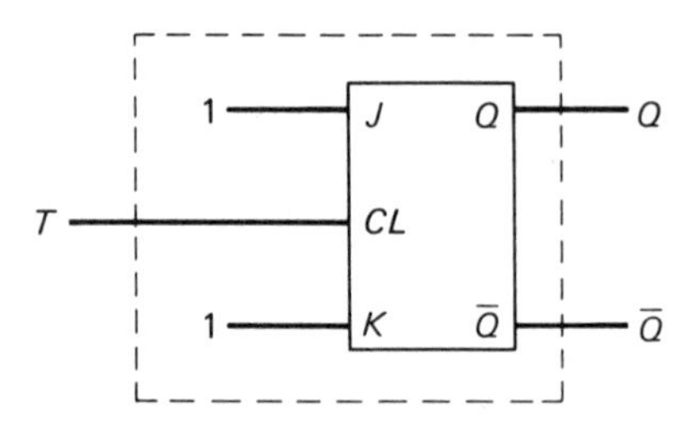

Figure 7.20 *T flip-flop from a JK flip-flop*

The state table of a T flip-flop has two entries, $T = 0$ and $T = 1$. Since for $T = 0$ the output does not change ($Q_{n+1} = Q_n$), and for $T = 1$ the output complements ($Q_{n+1} = \bar{Q}_n$), the state table is as shown in Tab. 7.9.

Table 7.9 *State table for the T flip-flop*

t_n T	t_{n+1} Q_{n+1}	 $\bar{Q}_{n+1}$
0	Q_n	$\bar{Q}_n$
1	$\bar{Q}_n$	Q_n

A clocked T flip-flop is one that is controlled by a clock signal, as symbolized in Fig. 7.21(a). A possible implementation of such a flip-flop with a T flip-flop and an AND gate is shown in Fig. 7.21(b). The signal T controls which pulse of the clock will be applied to the input terminal T, and, of course, each clock pulse that arrives at T causes the flip-flop to toggle.

Clocked T flip-flops are normally negative-edge triggered. Thus if $T = 1$ on the negative edge of the clock pulse, then the T flip-flop changes state; that is, $Q_{n+1} = \bar{Q}_n$. On the other hand, if $T = 0$, the output state remains the same, or

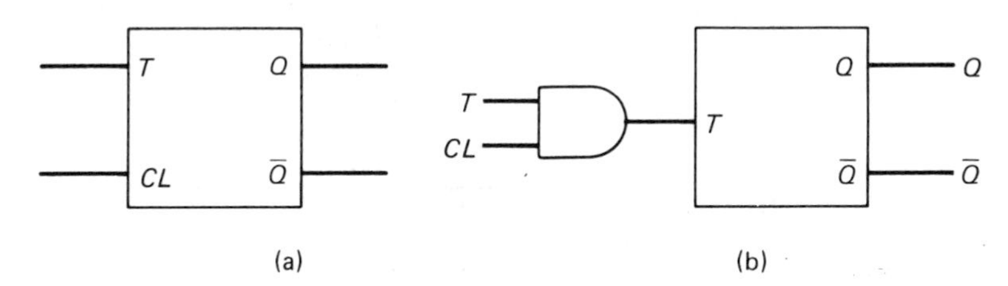

Figure 7.21 *(a) Symbol for a clocked T flip-flop, and (b) its implementation with a T flip-flop*

$Q_{n+1} = Q_n$. The behavior of a T flip-flop is illustrated by a timing diagram in Fig. 7.22. At the negative edges t_1, t_2, t_3, and t_5 of the clock pulse, $T = 1$, and thus the flip-flop changes state. However, at the edge t_4, $T = 0$, and there is no change in the state.

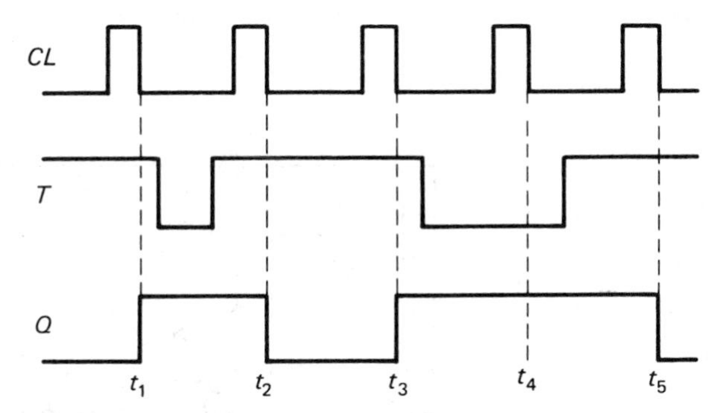

Figure 7.22 *Timing diagram for a T flip-flop*

Before leaving the subject of flip-flops we should note that many applications require that a flip-flop in a logic circuit be in a specified state at the beginning of the operation of the circuit. Therefore many flip-flops have additional inputs of PRESET, or direct SET (S_D), and/or CLEAR, or direct RESET (R_D). These inputs, which function exactly as the S and R inputs of the SR flip-flop do, are used to establish the initial state of the circuit. When S_D is made 1, Q is set to 1, and when R_D is made 1, Q is reset to 0. It is, of course, invalid as well as illogical to activate both the PRESET and CLEAR inputs simultaneously. Once the flip-flop is set or reset, the inputs S_D and R_D are returned to 0.

A JK flip-flop with S_D and R_D inputs is symbolized in Fig. 7.23. Other types of flip-flops are symbolized in an identical way.

As an example, the circuit in Fig. 7.24 with the inputs S_D and R_D removed is a T flip-flop, as the reader is asked to show in Ex. 7.6.1. With S_D and R_D present as shown, the flip-flop may be preset or cleared.

As a final example let us consider the problem of constructing a JK flip-flop with a T flip-flop and a number of logic gates, as was done in Fig. 7.18 using a D flip-flop as the memory element. To solve this problem we need a Boolean

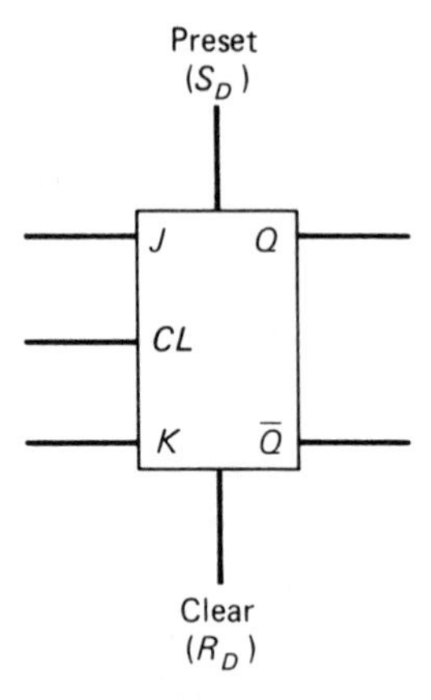

Figure 7.23 *JK flip-flop with preset and clear inputs*

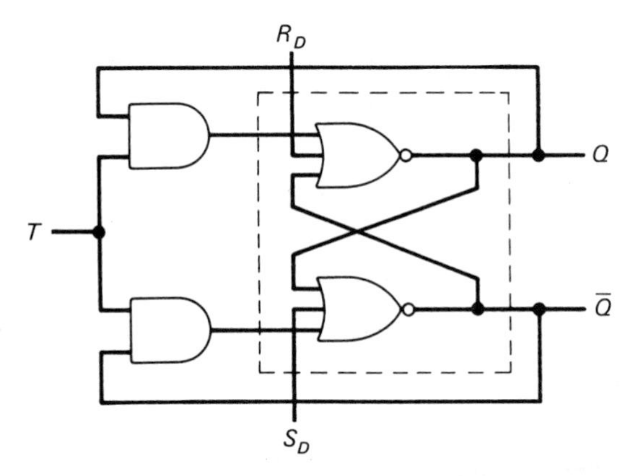

Figure 7.24 *T flip-flop with preset and clear inputs*

expression for T in terms of J, K, Q, and $\bar{Q}$, which can then be implemented with logic circuitry.

We begin by reproducing Tab. 7.6, to which we add a column for T. This column is easily determined from Table 7.9, from which the values of Q_n and Q_{n+1} may be used to determine T. In every case where $Q_{n+1} = Q_n$ we have $T = 0$, and where $Q_{n+1} = \bar{Q}_n$ we have $T = 1$. The results are given in Table 7.10.

From Table 7.10 we see that T is given in terms of J, K, Q_n, and $\bar{Q}_n$ by

$$T = \bar{J}KQ_n + J\bar{K}\bar{Q}_n + JK\bar{Q}_n + JKQ_n$$

which may be simplified to

$$T = KQ_n + J\bar{Q}_n \tag{7.9}$$

Table 7.10 *Transition table for the JK flip-flop and the corresponding required values of T*

J	K	Q_n	Q_{n+1}	T
0	0	0	0	0
0	0	1	1	0
0	1	0	0	0
0	1	1	0	1
1	0	0	1	1
1	0	1	1	0
1	1	0	1	1
1	1	1	0	1

This is the required input T to the T flip-flop, and thus the circuit for the JK flip-flop is easily obtained. The result is given in Fig. 7.25.

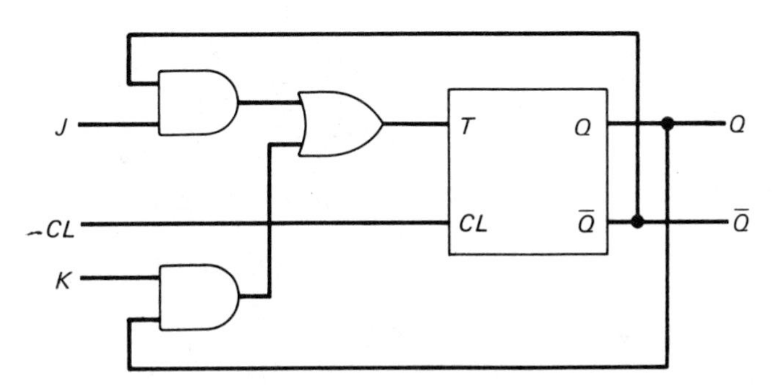

Figure 7.25 *JK flip-flop implemented with a T flip-flop and three gates*

EXERCISES

7.6.1 Show that the circuit in Fig. 7.24 with inputs S_D and R_D deleted is a T flip-flop. (*Suggestion:* Note that the circuit inside the dashed rectangle is an SR flip-flop.)

7.6.2 Show that the circuit of Fig. 7.24 with inputs S_D and R_D as shown has the properties that Q becomes 1 (set) when $S_D = 1$ and $R_D = 0$, and that Q becomes 0 (reset) when $S_D = 0$ and $R_D = 1$.

7.6.3 Show from Tabs. 7.8 and 7.9 that

$$T = D \oplus Q_n$$

and use this result to construct a D flip-flop with a T flip-flop and a number of logic gates.

Ans:

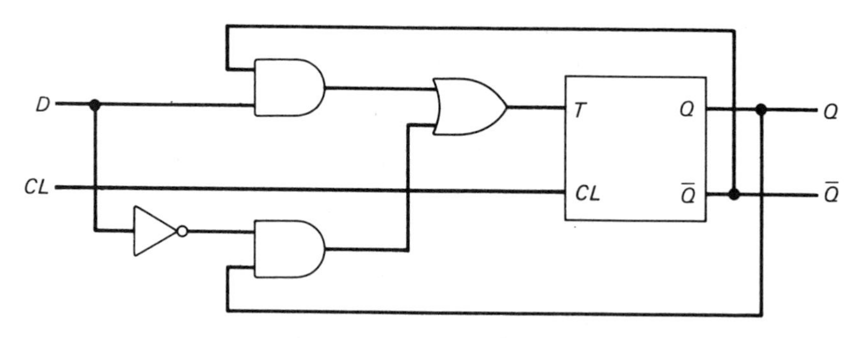

Exercise 7.6.3

7.6.4 Obtain a clocked T flip-flop from a JK flip-flop.

Ans:

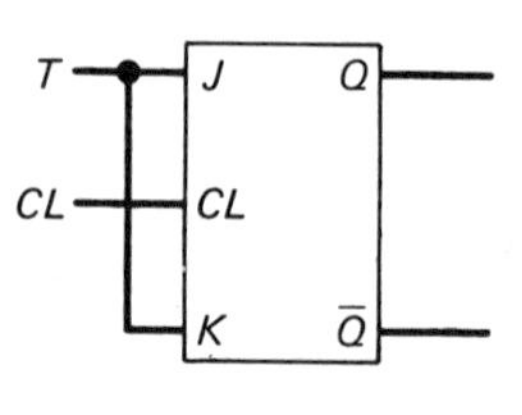

Exercise 7.6.4

7.7 Sequential Circuits

In the general case, sequential circuits are combinational circuits with one or more memory elements, such as flip-flops, embedded within them. This is illustrated by the block diagram in Fig. 7.26, where $x_1, x_2, \ldots, x_n$ are inputs, z_1,

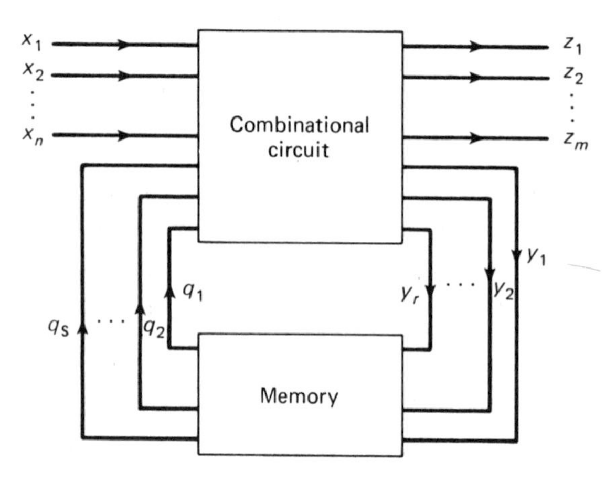

Figure 7.26 *General sequential circuit*

$z_2, \ldots, z_m$ are outputs, $y_1, y_2, \ldots, y_r$ are feedback inputs from the combinational circuitry into the memory circuitry, and $q_1, q_2, \ldots, q_s$ are inputs from memory into the combinational circuitry. The q's may be thought of as *state* variables, which describe the prevailing internal situation, or *state*, of the circuit at the time the inputs are applied. Thus the previous history of the circuit (which produced the q's) and the current inputs, the x's, determine the output z's as well as the y's. The latter are the inputs to the memory that will determine the *next* state.

As an example, let us consider the sequential circuit in Fig. 7.27, which has a single input x, a single output z, a single y, and two q's, as shown. In this case there are two state variables $q_1 = Q$ and $q_2 = \bar{Q}$, but since one is the complement of the other, we have, strictly speaking, only one state. The memory device is a D flip-flop, which is controlled by a clock.

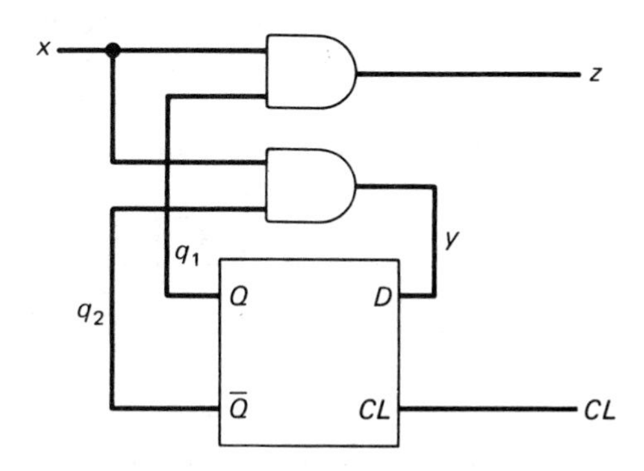

Figure 7.27 *Sequential circuit*

Let us suppose that the initial state is $q_1 = Q = 0$ and analyze the circuit in Fig. 7.27 for the input sequence $x = 101101101$. That is, let us find the output sequence z. To begin, we note from the circuit that

$$z = xQ \tag{7.10}$$

and

$$y = D = x\bar{Q} \tag{7.11}$$

For $x = 1$ and $Q = 0$ we have, at the first clock pulse,

$$z = 1 \cdot 0 = 0$$

$$y = 1 \cdot 1 = 1$$

On the next clock pulse we have $x = 0$, $Q = 1$ (the *previous* value of D, which is transferred to Q), and thus

$$z = 0, \qquad y = 0$$

The work may be arranged in the following compact form, where each calculated value of $y = D$ is the next value of Q:

x:	1 0 1 1 0 1 1 0 1
Q:	0 1 0 1 0 0 1 0 0
z:	0 0 0 1 0 0 1 0 0
$y = D$:	1 0 1 0 0 1 0 0 1
CL pulse:	1 2 3 4 5 6 7 8 9

Thus the results are $z = 000100100$ and $Q = 010100100$.

These results are also shown in the timing diagram in Fig. 7.28, where we have assumed that x changes on the clock pulse and that the flip-flop is positive-edge triggered. The values in the x, z, and Q sequences are taken as those just before the occurrence of the positive edge of the clock pulse.

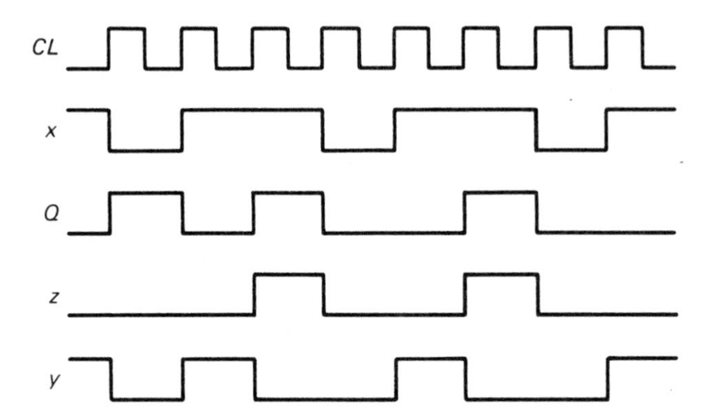

Figure 7.28 *Timing diagram for the circuit in Figure 7.27*

EXERCISES

7.7.1 If the input sequence is $x = 10111011$ and $Q = 0$ initially, find the output and state sequences z and Q in the circuit in Fig. 7.27.

Ans: 00010001, 01010101

7.7.2 For the circuit shown, the input sequence is $x = 101100100$ and the initial state is $Q = 0$. Find the output sequence z and the state sequence Q.

Ans: 011010110, 001001101

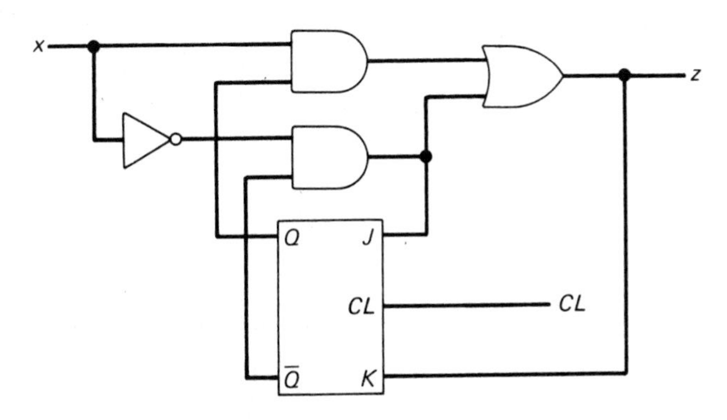

Exercise 7.7.2

PROBLEMS

7.1 Find the state table for the given circuit and show that if $X = \bar{S}$ and $Y = \bar{R}$, the result is that of Tab. 7.5 for the *SR* flip-flop. Thus the given circuit is an *SR* flip-flop for which $X = 0$ *sets* Q to 1 and $Y = 0$ *resets* Q to 0.

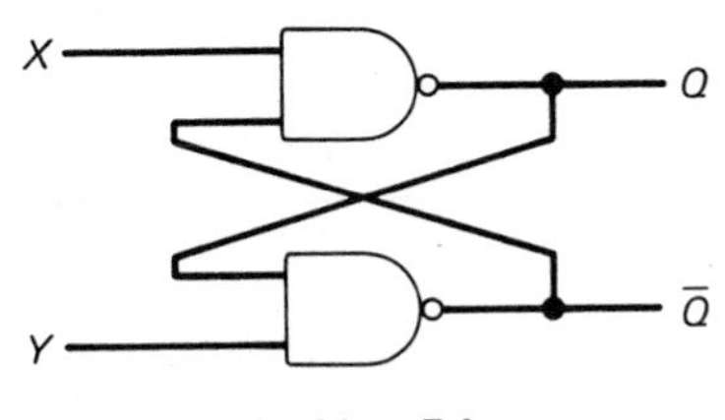

Problem 7.1

7.2 Find the excitation table for the flip-flop of Prob. 7.1.

7.3 Suppose $Q = 0$ when a sequence of inputs S, R, given by 1, 1; 0, 1; 1, 1; 1, 0; 1, 0; 0, 1 is applied at the input terminals of an *SR* flip-flop. Find the sequence of outputs Q.

7.4 Find the state table for the circuit shown.

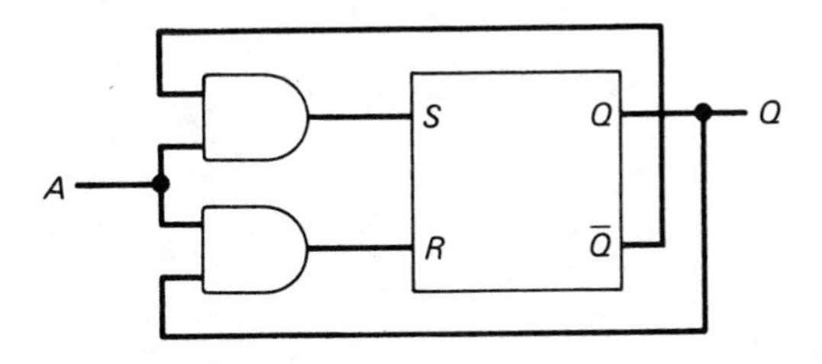

Problem 7.4

7.5 For the circuit of Prob. 7.4, find (a) the excitation table and use the result to find (b) the sequence of inputs required to yield the sequence of outputs 101110. Assume that initially $Q = 0$.

7.6 Find the sequence of outputs Q of the *JK* flip-flop in Fig. 7.10 if the sequence of clocked inputs J, K is 1, 1; 0, 1; 1, 1; 0, 0; 1, 0; 0, 1. Assume that initially $Q = 1$.

7.7 Sketch the output Q of a JK flip-flop if the negative edge of the clock occurs at t_1, t_2, t_3, and t_4 and J and K are as shown. Assume that $Q = 0$ initially and that there is no delay.

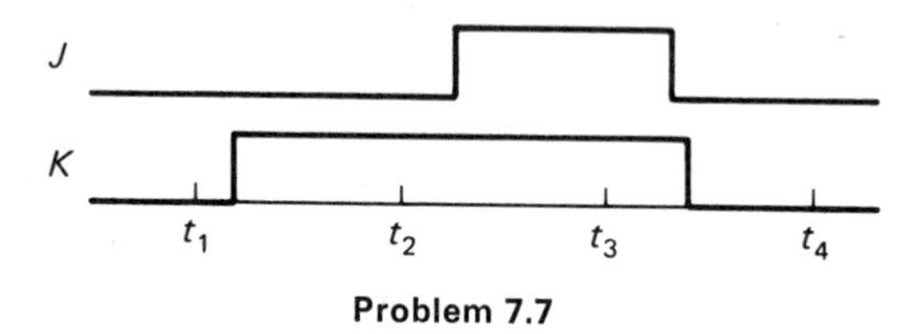

Problem 7.7

7.8 Find (a) the state table for the circuit shown and (b) use the result to find the sequence of outputs Q resulting from the sequence of clocked inputs A given by 0110110. Assume that $Q = 0$ initially.

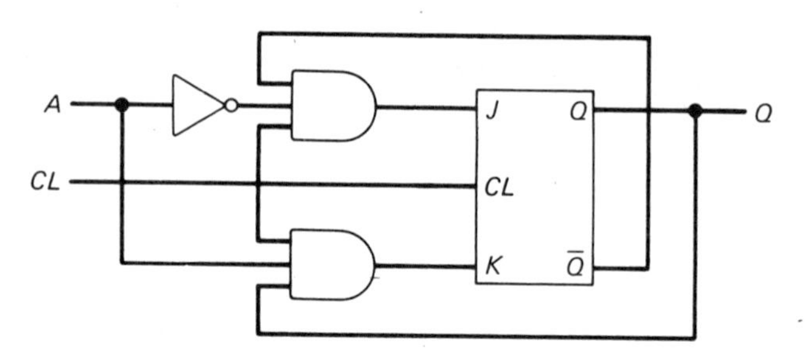

Problem 7.8

7.9 Sketch the outputs Q_1 and Q_2 for the master–slave flip-flops of Fig. 7.15 if J, K, and CL are as shown. Assume no delay and $Q_1 = Q_2 = 0$ initially.

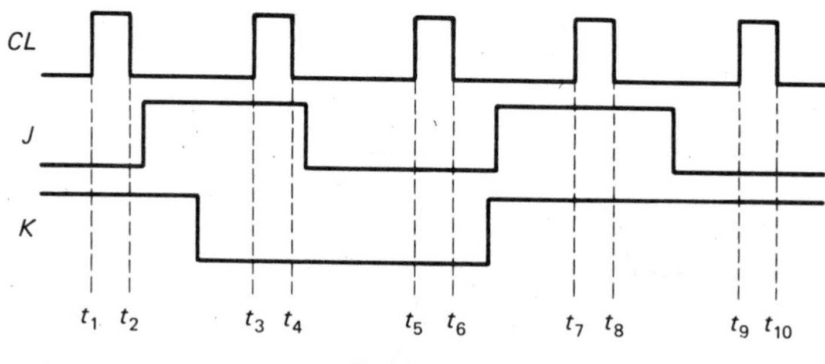

Problem 7.9

7.10 For the master–slave flip-flops of Fig. 7.15, assume that initially $Q_1 = Q_2 = 0$ and the input sequence JK is 0, 1; 0, 0; 1, 0; 1, 1; 1, 1; 0, 1. Find the output sequences Q_{m1}, Q_{m2}, Q_1, and Q_2, where Q_{m1} and Q_{m2} are the outputs of the master flip-flops of flip-flops 1 and 2, respectively.

7.11 Obtain a T flip-flop from a D flip-flop and additional gates.

7.12 Let $Q = 0$ initially in the circuit of Ex. 7.6.3 and find the output sequence if the input sequence D is 0110101.

7.13 Show that the given circuit is a D flip-flop by proving that $D = Q_{n+1}$.

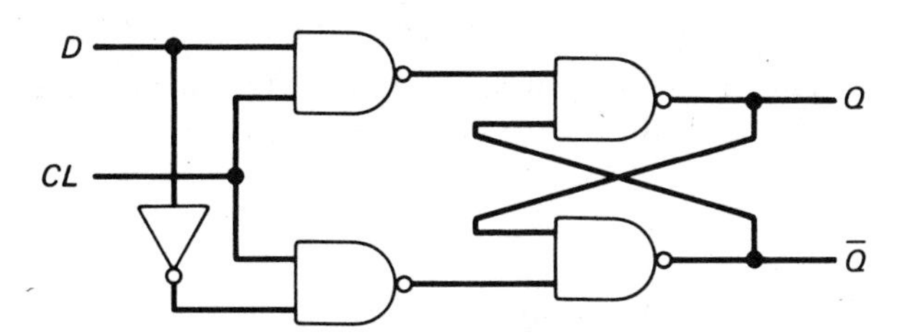

Problem 7.13

7.14 Two *JK* flip-flops are connected as shown with $J = K = 1$ in each case. If $Q_1 = Q_2 = 0$ before the first clock pulse, show that the number $(Q_2Q_1)_2$ is successively the binary equivalent of 0, 1, 2, 3; 0, 1, 2, 3; and so on. Thus the circuit is a *counter* that counts four decimal digits repeatedly.

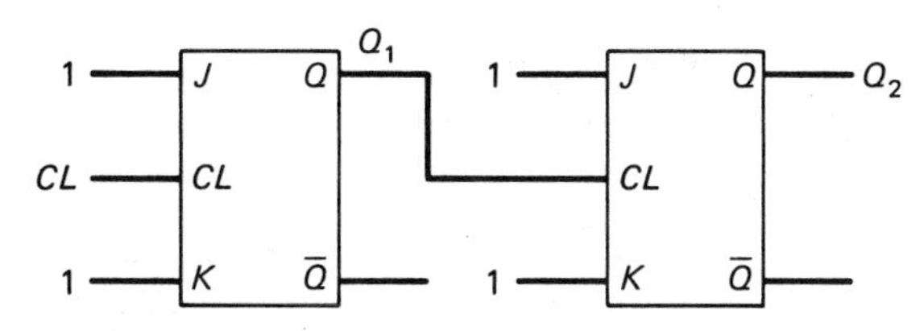

Problem 7.14

7.15 Show that in the *T* flip-flop in Fig. 7.19, if *T* is a clock signal such as *CL* in Fig. 7.22, then *Q* is a clock signal of half the frequency of *T*.

7.16 Find the output and state sequences for the circuit of Ex. 7.7.2 if the initial state is $Q = 0$ and the input sequence is $x = 1100100110$.

7.17 Find the output and state sequences if the initial state is $Q = 0$ and the input sequence is $x = 101101100$.

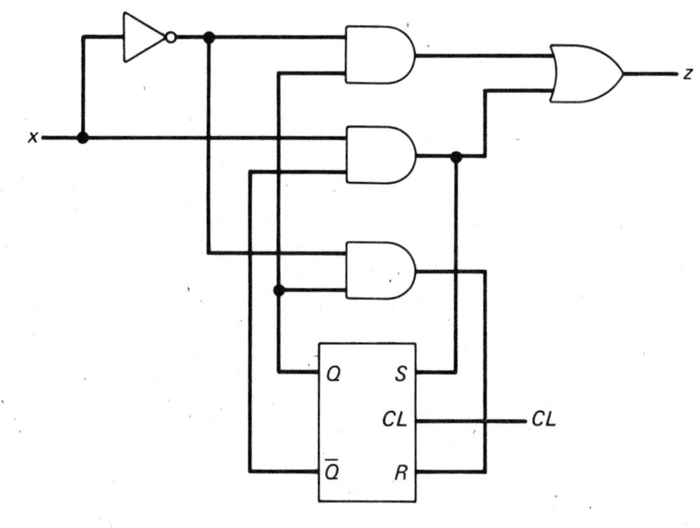

Problem 7.17

7.18 Repeat Prob. 7.17 if the input sequence is $x = 01101011$.

7.19 Find the output and state sequences if the input sequence is $x_1x_2 = 01, 10, 00, 11, 00, 11, 01, 10$, and the initial state is $Q = 0$.

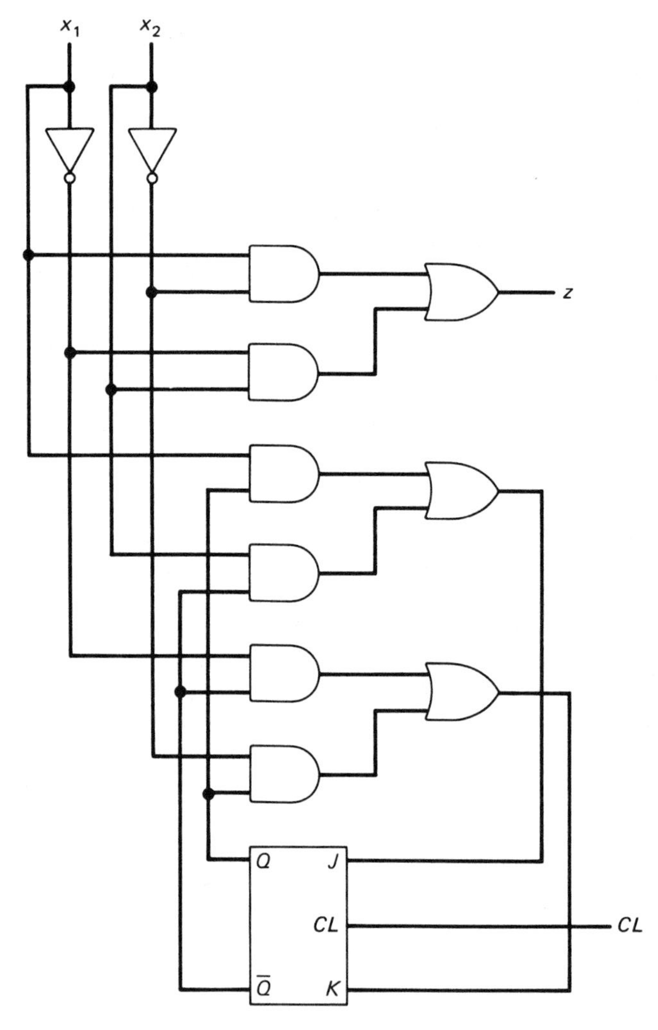

Problem 7.19

7.20 Find the output and state sequences if the input sequence is $x = 11001101$ and the initial state is $Q_1Q_2 = 01$.

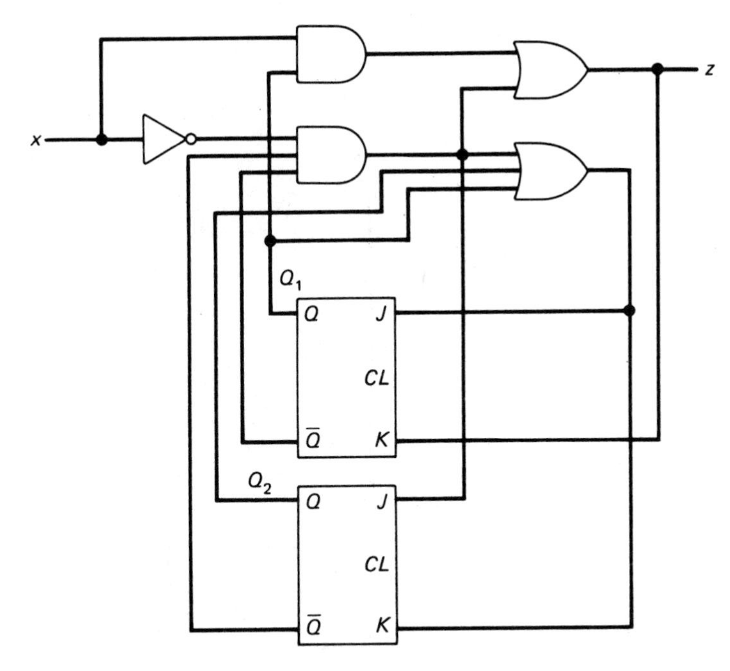

Problem 7.20

8

CODES

As we have seen, a digital circuit conveniently uses, as inputs and outputs, signals that are represented by binary numbers. However, if the circuit is to be useful, it is necessary for people to be able to interact with it. Machines may respond to binary numbers, but people still respond to decimal numbers, alphabetical characters, and so on; thus the input and output information must be put in an easily interpretable form. As we saw in Sections 4.6 and 6.1, this requirement is met by representing the input and output data by a *code*, which is an assignment of a unique combination of 0's and 1's to each number, letter, or symbol that is to be used.

In the case of decimal numbers, which users of digital circuits are most accustomed to, we have already seen that their codes may be their binary equivalents. But in many instances special binary codes for decimal numbers are preferred over their straight binary equivalents; and, of course, special binary codes must be devised for characters other than numbers.

The binary system has a special relationship to two other number systems: the octal (base 8) and hexadecimal (base 16) systems. These systems may be used to represent numbers in much more compact form than their equivalent binary numbers, and they are readily converted to or from binary. In many cases, therefore, the user of a digital circuit may prefer to record numbers in octal or hexadecimal form. Moreover, in many cases the circuit may be "instructed" to accept octal or hexadecimal numbers and convert them internally to binary for interaction with the circuit elements. As an illustration of the compactness of octal or hexadecimal numbers as compared to their binary counterparts, the binary number 110100101 is equivalent to the octal number 645, as we may see using the methods of Chap. 2.

In this chapter we give a brief discussion of octal and hexadecimal numbers and consider their arithmetic operations and conversions to and from the other bases. We also discuss many of the commonly used binary codes for numbers, letters, and other characters, such as BCD codes of various types and ASCII codes. The use of codes will be veıy important to us in Chap. 9, where we will discuss registers and counters.

8.1 Octal Number Systems

We mentioned *octal* (base 8) and *hexadecimal* (base 16) number systems in Chap. 2, but because of their special relationship to the binary system, which makes them useful for codes in digital systems, we have deferred a formal discussion of them until this chapter. Arithmetic operations and conversions to and from other bases are performed in exactly the same manner for these systems as for the decimal and binary systems. Thus the reader may wish to review Chap. 2, particularly the polynomial representation in Eq. (2.5) and the number conversion methods of Sec. 2.3, before reading these first two sections.

In the octal system there are eight digits, designated by 0, 1, 2, 3, 4, 5, 6, and 7. An octal number may be converted to decimal by means of Eq. (2.5) with $b = 8$, or by the method of Sec. 2.3. For example, the octal number $N = 257.4$ may be converted to decimal using Eq. (2.5) by noting that

$$N = 2 \times 8^2 + 5 \times 8^1 + 7 \times 8^0 + 4 \times 8^{-1}$$

which results in the decimal equivalent $N = 175.5$.

Alternatively, we may apply the method of Sec. 2.3 to the integer part 257 with the work arranged as follows:

$$\begin{array}{rrr} 2 & 5 & 7 \\ & 16 & 168 \\ \hline 2 & 21 & 175 \end{array}$$

Note that the first digit (2) is rewritten, multiplied by the base 8, and added to the next digit (5). This process of multiplying by the base and adding to the next digit is continued until the last sum is obtained (175 in this case). The last sum is the decimal equivalent, and thus $257_8 = 175_{10}$, as before. Also, since $0.4_8 = 0.5_{10}$, we have $257.4_8 = 175.5_{10}$, as in the previous case. (We could also apply the method of Sec. 2.3 to the fractional part, using base 8 instead of base 2, but in this case the conversion is simple enough to perform mentally.)

The methods of Sec. 2.3 may be used for conversion from any base b to base 10, or vice versa, by simply substituting b for 2 in the procedure. The proof was outlined in Chap. 2 (in Probs. 2.14–2.16 and 2.18–2.19).

The conversion from decimal to octal may be done also using Eq. (2.5) or the procedure of Sec. 2.3. To illustrate the latter method let us convert 175_{10} back to octal. The work is arranged as follows:

	Remainders	
$175 \div 8 = 21$	7	(LSD)
$21 \div 8 = 2$	5	
$2 \div 8 = 0$	2	(MSD)

(LSD and MSD are, respectively, the least significant digit and the most significant digit.) The decimal equivalent is therefore 257, as before. Fractional numbers are converted as in Sec. 2.3, with base 8 replacing base 2.

Because the bases 2 and 8 are related by $2^3 = 8$, binary and octal numbers are easily converted from one to the other. The exponent 3 indicates that a three-bit binary number is a one-digit octal number and vice versa. That is, by starting at the binary point of a binary number, grouping the bits in units of three in both directions, and replacing each unit by its equivalent octal digit, we obtain the equivalent octal number. For example, consider the binary number $N =$ 10110011.1111, which we group as follows:

$$N = 010\ 110\ 011.111\ 100 \tag{8.1}$$
$$2 \quad 6 \quad 3 \quad 7 \quad 4$$

Since the units are $010_2 = 2_8$, $110_2 = 6_8$, $011_2 = 3_8$, $111_2 = 7_8$, and $100_2 = 4_8$, as noted above, then the octal equivalent is

$$10110011.1111_2 = 263.74_8 \tag{8.2}$$

To see that the procedure is valid, let us consider the general binary number

$$N = (a_n a_{n-1} \cdots a_2 a_1 a_0 \,.\, a_{-1} a_{-2} \cdots a_{-m})_2$$

which, by Eq. (2.5), may be written in the polynomial form

$$\begin{aligned} N = a_0 &+ 2a_1 + 2^2a_2 + 2^3a_3 + 2^4a_4 + 2^5a_5 \\ &+ 2^6a_6 + 2^7a_7 + 2^8a_8 + \cdots + 2^{-1}a_{-1} + 2^{-2}a_{-2} + 2^{-3}a_{-3} + \cdots \end{aligned}$$

This may be written as

$$\begin{aligned} N = (a_0 + 2a_1 + 2^2a_2) &+ 2^3(a_3 + 2a_4 + 2^2a_5) \\ &+ 2^6(a_6 + 2a_7 + 2^2a_8) + \cdots \\ &+ 2^{-3}(a_{-3} + 2a_{-2} + 2^2a_{-1}) + \cdots \end{aligned}$$

or

$$N = b_0 + 8b_1 + 8^2b_2 + \cdots + 8^{-1}b_{-1} + \cdots \tag{8.3}$$

where

$$
\begin{aligned}
b_0 &= a_0 + 2a_1 + 2^2a_2 \\
b_1 &= a_3 + 2a_4 + 2^2a_5 \\
b_2 &= a_6 + 2a_7 + 2^2a_8 \\
&\cdots \\
b_{-1} &= a_{-3} + 2a_{-2} + 2^2a_{-1}
\end{aligned}
\tag{8.4}
$$

and so on. The expression in Eq. (8.3) is the polynomial form of the number N expressed in the base 8 system, with base 8 digits b_0, b_1, b_2, and so on. Therefore, by Eqs. (8.4) the octal digits in base 2 are simply $(a_2a_1a_0)_2$, $(a_5a_4a_3)_2$, $(a_8a_7a_6)_2$, and so on, formed by grouping the binary digits in units of three, as in Eq. (8.1).

The conversion from octal to binary is equally easy. For example, 346_8 is converted to binary by replacing each digit by its equivalent binary unit of three bits, using in reverse the idea illustrated in Eq. (8.1). Thus, since $3 = 011_2$, $4 = 100_2$, and $6 = 110_2$, we have

$$346_8 = 11100110_2$$

Arithmetic operations in the octal system are performed in exactly the same way as in base 2 or base 10 systems. To illustrate, let us add the octal numbers 452 and 765. The work is arranged as follows:

$$
\begin{array}{rrrrl}
1 & 1 & & & \text{Carries} \\
 & 4 & 5 & 2 & \\
+ & 7 & 6 & 5 & \\
\hline
1 & 4 & 3 & 7 & \text{Sum}
\end{array}
$$

Carries are handled exactly as in any other base, as we may see from this example. The sum $5 + 6 = 13$, or 3 with 1 to carry, and so on.

Other arithmetic examples are left to the exercises and problems.

EXERCISES

8.1.1 Convert to decimal the octal numbers

(a) 570

(b) 236.14

(c) 12.576 (*Suggestion:* Use the method of Sec. 2.3 for fractional numbers.)

Ans: (a) 376; (b) 158.1875; (c) 10.74609375

8.1.2 Convert the octal numbers of Ex. 8.1.1 to binary.

Ans: (a) 101111000; (b) 10011110.001100; (c) 1010.101111110

8.1.3 Convert to octal the decimal numbers in (a) and (b) and the binary numbers in (c) and (d):

(a) 275

(b) 31.125

(c) 11011101

(d) 101.11011101

Ans: (a) 423; (b) 37.1; (c) 335; (d) 5.672

8.1.4 Perform the following arithmetic operations in the octal system:

(a) $367 + 74$

(b) $342 - 217$

(c) 250×12

(d) $243 \div 16$

Ans: (a) 463; (b) 123; (c) 3220; (d) 13 with remainder 11

8.2 Hexadecimal Number Systems

In the base 16, or hexadecimal, system there are 16 digits, and thus we must make up new symbols for those greater than 9. The standard symbols are A, B, C, D, E, and F, which are equivalent, respectively, to the decimal numbers 10, 11, 12, 13, 14, and 15. The hexadecimal digits are listed in Tab. 8.1, where for convenience the equivalent decimal, binary, and octal numbers are also given.

Table 8.1 *Number system equivalents*

$b = 10$	$b = 2$	$b = 8$	$b = 16$
0	0	0	0
1	1	1	1
2	10	2	2
3	11	3	3
4	100	4	4
5	101	5	5
6	110	6	6
7	111	7	7
8	1000	10	8
9	1001	11	9
10	1010	12	A
11	1011	13	B
12	1100	14	C
13	1101	15	D
14	1110	16	E
15	1111	17	F

The conversion from base 16 to 10 and vice versa is accomplished in exactly the same way as that of base 8 or base 2 to base 10. For example, the hexadecimal number $N = 17AF$, by Eq. (2.5), is

$$\begin{aligned} N &= 1 \times 16^3 + 7 \times 16^2 + 10 \times 16^1 + 15 \times 10^0 \\ &= 6063_{10} \end{aligned}$$

With the method of Sec. 2.3 the work is arranged as follows:

$$\begin{array}{rrrr} 1 & 7 & 10 & 15 \\ & 16 & 368 & 6048 \\ \hline 1 & 23 & 378 & 6063 \end{array}$$

(Note that in the multiply-and-add process the multiplication is by 16 in this case.) Therefore we have

$$17AF_{16} = 6063_{10} \tag{8.5}$$

To convert from decimal to hexadecimal we may use Eq. (2.5) or the division method of Sec. 2.3. In the latter case the divisors are 16, and in the example of Eq. (8.5) the work is arranged as follows:

$$\begin{array}{rcll} & & \text{Remainders} & \\ 6063 \div 16 = & 378 & 15 = F_{16} & \text{(LSD)} \\ 378 \div 16 = & 23 & 10 = A_{16} & \\ 23 \div 16 = & 1 & 7 & \\ 1 \div 16 = & 0 & 1 & \text{(MSD)} \end{array}$$

Thus we have

$$6063_{10} = 17AF_{16}$$

which checks Eq. (8.5).

Since the bases 2 and 16 are related by $2^4 = 16$, we may use the technique previously discussed for base 8 to convert from base 2 to 16 or vice versa. The grouping will be in units of four binary bits rather than three, as was the case for base 8. For example, the binary number $N = 10110011.1111$ may be grouped as follows, with the hexadecimal equivalents of the units as shown:

$$\begin{array}{c} N = 1011 \;\; 0011 . 1111 \\ \quad\;\; B \quad\;\; 3 \;\;\; F \end{array}$$

Therefore we have

$$10110011.1111_2 = B3.F_{16} \tag{8.6}$$

The reader is asked to justify this procedure in the general case in Prob. 8.7.

As another example, let us convert $2FC3_{16}$ to binary. Since $2 = 0010_2$, F =

1111_2, $C = 1100_2$, and $3 = 0011_2$, we have

$$2FC3_{16} = 10\ 1111\ 1100\ 0011_2$$

where the units of four bits are identified by separating them by a space.

Arithmetic operations in base 16 are carried out exactly as in base 2, base 8, or base 10. For example, let us add $2F45_{16}$ to $B961_{16}$. The work is arranged as follows:

1				Carry
2	F	4	5	
B	9	6	1	
E	8	A	6	Sum

In the process we note that $9 + F = 18(1 \times 16 + 8)$, or 8 with 1 to carry. Other arithmetic examples are left to the exercises.

EXERCISES

8.2.1 Convert to decimal the hexadecimal numbers

(a) 1BF

(b) A6C.3D

Ans: (a) 447; (b) 2668.23828125

8.2.2 Convert to binary the hexadecimal numbers

(a) 2AD

(b) F3.2A

Ans: (a) 1010101101; (b) 11110011.00101010

8.2.3 Convert to hexadecimal the decimal numbers in (a) and (b) and the binary numbers in (c) and (d):

(a) 357

(b) 15.546875

(c) 1111010101

(d) 1101110.00111

Ans: (a) 165; (b) F.8C; (c) 3D5; (d) 6E.31

8.2.4 Perform the following operations in hexadecimal:

(a) $1A2 + 9BF$

(b) $34D - F6$

(c) $2BA \times 3F$

(d) $24D \div 1A$

Ans: (a) B61; (b) 257; (c) ABC6; (d) 16 with remainder 11

8.3 BCD Codes

To code the ten decimal digits, ten unique symbols consisting of the binary digits 0 and 1 are needed. A code of this type, which represents the decimal digits with binary digits, is called a *binary-coded decimal*, or *BCD*, code, and there are several such codes in use. Since there are only $2^3 = 8$ combinations of three-bit binary numbers available and there are ten decimal digits, any BCD system must consist of binary numbers having at least four bits. (It is interesting to note that there are 8008 ways to form BCD codes. This is the number of ways one can select 16 things ten at a time.)

The most common form of BCD is the one in which the ten decimal digits are simply represented by their binary equivalents. This system is sometimes called the *natural binary-coded decimal* (NBCD). In this natural case the place values, or *weights*, of the bits in the code are 2^3, 2^2, 2^1, and 2^0, or 8, 4, 2, and 1. Accordingly, the NBCD is also called the 8421 BCD, or simply BCD. It is shown in the second column of Tab. 8.2, along with other BCD codes that we will discuss later. Because the bits represent weights, the NBCD code is sometimes called a *weighted* code.

Table 8.2 *BCD codes*

Decimal	8421 *BCD*	2421 *BCD*	*Excess* 3	*Gray*
0	0000	0000	0011	0000
1	0001	0001	0100	0001
2	0010	0010	0101	0011
3	0011	0011	0110	0010
4	0100	0100	0111	0110
5	0101	1011	1000	0111
6	0110	1100	1001	0101
7	0111	1101	1010	0100
8	1000	1110	1011	1100
9	1001	1111	1100	1000

The BCD code of a decimal number of more than one digit is obtained by replacing each digit by its four-bit BCD code. For example, the BCD representation of 23 is 0010 0011, since 0010 is the BCD code for 2 and 0011 is the BCD code for 3. We note that the BCD code for a decimal number larger than 9 is not the straight binary equivalent. In the example just considered the binary equivalent of 23 is 10111, which is not its BCD code. In fact, the six binary numbers that are the equivalents of the decimal numbers 10 through 15 are not valid BCD numbers, and they are not used, because they do not represent decimal digits.

As an example, let us convert the 8421 BCD number 1000 0111 0101 to the decimal number that it represents. The first four bits are the BCD code for 8, the next four for 7, and the last four for 5. Therefore we have

$$1000\ 0111\ 0101_{BCD} = 875_{10}$$

As another example, as the reader may verify, the BCD code for 269_{10} is

$$269_{10} = 0010\ 0110\ 1001_{BCD}$$

It is possible to add two BCD numbers, but if the sum of any two representations of decimal digits exceeds the equivalent of the decimal digit 9, then it will not be the BCD code for the corresponding decimal sum. This is because of the six unused binary numbers that are not allowed in BCD. For example, the sum of 5 and 9 in BCD is the binary equivalent of 14 rather than the BCD code for 14. To get the latter we must add the binary equivalent of 6, to account for the missing binary numbers that are not allowed in the BCD code. The work may be arranged as follows:

0101	BCD 5
+ 1001	BCD 9
1110	Binary 14
+ 0110	Binary 6
0001 0100	BCD 14

As another example, let us add the BCD equivalents of the decimals 364 and 271, as follows:

0011 0110 0100	BCD 364
0010 0111 0001	BCD 271
0101 1101 0101	Binary Sum
+0110	Binary 6
0110 0011 0101	BCD 635

Note that the sum of the least significant four bits is valid, but the next most significant four bits in the sum is invalid. Therefore we have added the equivalent of 6 to the middle set of bits (corresponding to 6 + 7), whose sum exceeds the equivalent of decimal 9. The carry generated is added to the next set of bits (most significant set) as it would be in ordinary addition.

Subtraction of two BCD numbers may also result in an invalid difference, which will require correction. We will consider examples of subtraction in Ex. 8.3.4.

As we see from these examples, the arithmetic process may be complicated by the use of the BCD code. In adding two BCD numbers we first find their binary sum, and then we must decide if this sum is valid. If not, we must correct it by adding the binary equivalent of 6. However, BCD is useful in applications where information using decimal digits is required, such as in electronic calculators and instrumentation. The process of encoding BCD from decimal is easier than changing from decimal to binary, or vice versa. Thus calculators use BCD because the input and output data displayed is decimal. However, the arithmetic circuitry of calculators is more complex because of the more complicated arithmetic required.

Another weighted code is the 2421 code, shown in the third column of Tab. 8.2. In this code the weights are the same as those of the 8421 code except that the weight of the MSB position is 2 instead of 8. It is possible to represent some decimal digits two ways with the 2421 code; for example, the digit 7 may be represented by 1101 or by 0111. To remove this ambiguity we have chosen to represent the first five digits with a 0 as the MSB and the last five with a 1 as the MSB.

The major advantage of the 2421 code is its suitability for use in *complementary arithmetic*, which we will consider in Chap. 10. As we will see there, we may subtract numbers by adding certain *complements* of the numbers. An example is the *9's complement* of a decimal number, which is formed by subtracting each digit from 9. That is, the 9's complement of 743 is 256, and so on. The 1's complement of a binary number is formed by subtracting each bit from 1, so that for example, the 1's complement of 1011 is 0100. Evidently, an equivalent procedure is simply to interchange 0's with 1's and vice versa.

A *self-complementing* code may be defined as one for which the code for the 9's complement of a decimal number is the 1's complement of the code for the decimal number. According to this definition the 2421 code is self-complementing and thus is very useful in complementary arithmetic. We will show this in Chap. 10, but for the present we illustrate the self-complementary property with an example. Consider the decimal number $N = 274$ with 2421 code given by

$$274_{10} = 0010\ 1101\ 0100_{2421} \tag{8.7}$$

The 9's complement of N is 725, with 2421 code given by

$$725_{10} = 1101\ 0010\ 1011_{2421}$$

which is the 1's complement of the 2421 coded number in Eq. (8.7).

The excess 3 code, shown in the fourth column of Tab. 8.2, is an *unweighted* BCD code, obtained by adding decimal 3 (binary 0011) to the 8421 BCD code. Like the 2421 BCD code, the excess 3 code is also self-complementing. For example, the decimal number $N = 857$ in excess 3 is given by

$$857_{10} = 1011\ 1000\ 1010_{\text{EXCESS 3}}$$

and its 9's complement 142 is given by

$$142_{10} = 0100\ 0111\ 0101_{\text{EXCESS 3}}$$

Evidently one is the 1's complement of the other.

As a final example, the Gray code, shown in the fifth column of Tab. 8.2, is another unweighted BCD code. Its distinguishing feature is that only one bit changes in going from one number to the next. The representation chosen for 9 is such that this property holds also in going from 9 to 0. An example of a Gray code was considered in Ex. 6.1.1, for six decimal digits. Since only one bit at a time changes in progressing from one number to the next, the Gray code is suited for encoding a physical parameter, such as a shaft position of a motor. As the shaft turns continuously from one position to the next, only one bit in the code changes, and this minimizes the likelihood of generating an improper representation of the shaft position.

EXERCISES

8.3.1 For the decimal number 847 find

(a) the 8421 BCD code

(b) the 2421 BCD code

(c) the excess 3 code

(d) the Gray code

Ans: (a) 100001000111; (b) 111001001101; (c) 101101111010; (d) 110001100100

8.3.2 Repeat Ex. 8.3.1 for the 9's complement 152 of 847. Note that in cases (b) and (c) the answers are the 1's complements of those of Ex. 8.3.1, but this is not true in cases (a) and (d).

Ans: (a) 000101010010; (b) 000110110010; (c) 010010000101; (d) 000101110011

8.3.3 Find the indicated sums of 8421 BCD coded numbers. Check the answers by converting all the numbers to decimal.

(a) 001001010001 + 001100100111

(b) 001001010001 + 001101100111

Ans: (a) 010101111000; (b) 011000011000

8.3.4 Note that the condition for an incorrect sum in BCD addition (the two BCD equivalents add to an equivalent of a number greater than 9) is included in the condition that a carry is generated into the next more significant BCD equivalent of a decimal digit. Considering the reverse situation, it is true that the difference of two BCD decimal equivalents $x - y$ is invalid if a borrow is necessary from the next more significant decimal equivalent. To correct the result we must subtract the binary equivalent of 6 from the number $x - y$. Using these results, perform the indicated subtractions by first converting the decimals to 8421 BCD. Give the results in BCD.

(a) 235 − 122

(b) 235 − 117

(c) 235 − 176

Ans: (a) 0001 0001 0011; (b) 0001 0001 1000; (c) 0000 0101 1001

8.4 Alphanumeric Codes

The codes we have considered so far are useful only for encoding numbers. However, in the general case of data transmission a code is required which can represent numbers as well as other forms of data such as alphabetical characters (A, B, C, etc.) and symbols such as +, =, and (. Such a code is called an *alphanumeric*, or *alphameric*, code, and two well-known examples will be presented in this section.

Since there are ten decimal digits and 26 letters in the alphabet, to accomodate these characters alone would require a code with 36 representations. Thus the words in the code must have at least six bits, since five bits only allow $2^5 = 32$ representations. In addition to the numbers and letters we must include special symbols (as discussed earlier), punctuation marks, and special control characters, such as carriage return and backspace. Therefore, in many cases the code used may require seven or even eight bits per word.

The most common alphanumeric code used to represent characters is the American Standard Code for Information Interchange, abbreviated ASCII and pronounced *askey*. There is a six-bit ASCII, which excludes lowercase characters, and a seven-bit ASCII. The latter is sometimes called full ASCII, extended ASCII, or USASCII. The seven-bit ASCII is shown in Tab. 8.3 for a number of alphanumeric characters.

In the seven-bit code, an eighth bit is often used as a *parity* or *check* bit to determine whether the data has been transmitted correctly. The value of the parity bit is determined by the type of parity being used. For example, *even* parity means that the sum of all the bits in the word, including the parity bit, is even. (Or equivalently, the number of 1's is even.) As illustrations, the seven-bit ASCII code for A is 100 0001 and for C is 100 0011. In the first case the number of 1's is even (two) and thus the parity bit is 0, resulting in an even parity code for A of 0100 0001. In the case of C there are three 1's in its code (an odd number), and therefore the parity bit must be 1. The even-parity code for C is then 1100 0011. We have placed the parity bit in the MSB position, but it may be placed in another specified position, depending on the code being used.

Odd parity means that the sum of all bits (or equivalently, the number of 1's), including the parity bit, is an odd number. Thus in the examples just cited, the odd-parity code for A is 1100 0001 and for C is 0100 0011.

Other parity conventions often used are *mark* parity and *space* parity. In the former case the parity bit is always 1, and in the latter case it is always 0.

As the reader is asked to show in Ex. 8.4.3, the circuit of Fig. 8.1 may be used to check the parity of a coded character at its inputs. That is, if *ABCDEFGH* is a

Table 8.3 *Alphanumeric codes*

Character	ASCII	EBCDIC	Character	ASCII	EBCDIC
A	100 0001	1100 0001	blank	010 0000	0100 0000
B	100 0010	1100 0010	.	010 1110	0100 1011
C	100 0011	1100 0011	(	010 1000	0100 1101
D	100 0100	1100 0100	+	010 1011	0100 1110
E	100 0101	1100 0101	$	010 0100	0101 1011
F	100 0110	1100 0110	*	010 1010	0101 1100
G	100 0111	1100 0111	)	010 1001	0101 1101
H	100 1000	1100 1000	—	010 1101	0110 0000
I	100 1001	1100 1001	/	010 1111	0110 0001
J	100 1010	1101 0001	,	010 1100	0110 1011
K	100 1011	1101 0010	'	010 0111	0111 1101
L	100 1100	1101 0011	"	010 0010	0111 1111
M	100 1101	1101 0100	=	011 1101	0111 1110
N	100 1110	1101 0101			
O	100 1111	1101 0110	0	011 0000	1111 0000
P	101 0000	1101 0111	1	011 0001	1111 0001
Q	101 0001	1101 1000	2	011 0010	1111 0010
R	101 0010	1101 1001	3	011 0011	1111 0011
S	101 0011	1110 0010	4	011 0100	1111 0100
T	101 0100	1110 0011	5	011 0101	1111 0101
U	101 0101	1110 0100	6	011 0110	1111 0110
V	101 0110	1110 0101	7	011 0111	1111 0111
W	101 0111	1110 0110	8	011 1000	1111 1000
X	101 1000	1110 0111	9	011 1001	1111 1001
Y	101 1001	1110 1000			
Z	101 1010	1110 1001			

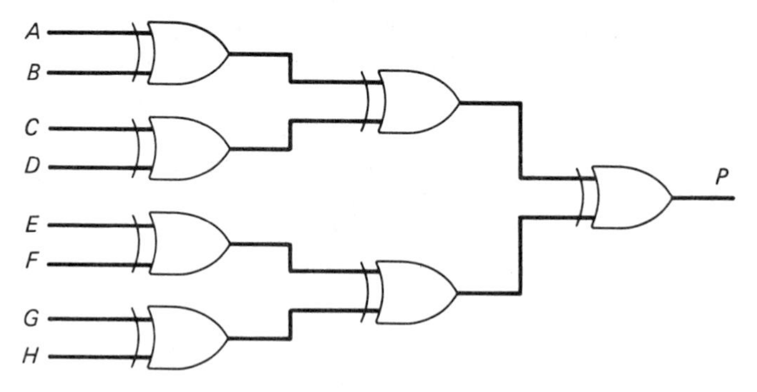

Figure 8.1 *Eight-bit even-parity checker circuit*

coded character of seven data bits and a parity bit, the output P is 0 if the number of 1 bits is even and is 1 if the number is odd. Thus if there is an error in the data resulting in an odd parity when even parity is being used, the circuit will produce a signal $P = 1$.

The ASCII code is important because it is used almost exclusively in most microcomputer systems. Another very important code is the Extended Binary Coded Decimal Interchange Code (EBCDIC), which was introduced by the IBM Corporation with its 360 line of computers. This code is also shown in Tab. 8.3. Its importance is obvious because of the dominant role of IBM in the computer industry. The EBCDIC code is used, of course, in the same way as the ASCII code.

EXERCISES

8.4.1 Encode the special message GO TO 65 in ASCII code with even parity and group the coded result into four four-bit segments.

Ans: 0100 0111 1100 1111 (GO)
1010 0000 1101 0100 (space T)
1100 1111 1010 0000 (O space)
0011 0110 0011 0101 (65)

8.4.2 Repeat Ex. 8.4.1 with odd parity.

Ans: 1100 0111 0100 1111
0010 0000 0101 0100
0100 1111 0010 0000
1011 0110 1011 0101

8.4.3 For the circuit in Fig. 8.1, find P and show that if the number of 1 bits in the binary number $(ABCDEFGH)_2$ is odd, then $P = 1$, and if even, $P = 0$. (*Suggestion:* See Sec. 4.3 and note that the circuit is equivalent to a single XOR gate with eight inputs.)

Ans: $A \oplus B \oplus C \oplus D \oplus E \oplus F \oplus G \oplus H$

PROBLEMS

8.1 Convert the following numbers to octal numbers:

(a) 11011011001_2

(b) 1011110.1101_2

(c) 3612_{10}

(d) 573.6875_{10}

8.2 Convert the numbers of Prob. 8.1 to hexadecimal numbers.

8.3 Convert to decimal the numbers

(a) 327_8

(b) 351.72_8

(c) AF19_{16}

(d) 3B7C.1A_{16}

8.4 Convert the numbers of Prob. 8.3 to binary.

8.5 Perform the following operations in the base 8 numbers shown:

(a) 237 + 175

(b) 2176 − 756

(c) 142 × 73

(d) 645 ÷ 71

8.6 Perform the following operations in the base 16 numbers shown:

(a) 2BF3 + 97A

(b) 2BC6 − 96B

(c) BC × 1A

(d) 12A ÷ B3

8.7 Express the binary number

$$N = (a_n a_{n-1} \ldots a_2 a_1 a_0 \,.\, a_{-1} a_{-2} \ldots a_{-m})_2$$

in the form

$$N = b_0 + 16b_1 + 16^2 b_2 + \cdots + 16^{-1} b_{-1} + \cdots$$

which is analogous to the octal number of Eq. (8.3). From this show that the b's are the digits in the hexadecimal equivalent of N and may be found by grouping the binary number in groups of four bits as was done to obtain Eq. (8.6).

8.8 Convert to decimal the 8421 BCD numbers

(a) 11101011001

(b) 100110111

(c) 110011000

8.9 Convert the numbers of Prob. 8.8 to

(a) 2421 BCD

(b) excess 3

(c) Gray code

8.10 Add the BCD numbers in Prob. 8.8(a) and (b) and obtain a valid BCD sum. Repeat this for the numbers of Prob. 8.9(b) and (c).

8.11 Find the following differences of 8421 BCD numbers, obtaining valid results in BCD.

(a) 11101011001 − 100110111

(b) 11101011001 − 110011000

8.12 Obtain a digital network with output f that will produce a logic 1 ($f = 1$) when a four-bit binary number $(ABCD)_2$ is not a valid 8421 BCD number.

8.13 **(a)** Find the 2421 BCD codes for the decimal 3756 and its 9's complement and show that the codes are 1's complements of each other.

(b) Repeat part (a) using the excess 3 codes.

8.14 A *binary* Gray code for the first 16 decimal numbers with their binary equivalents is as shown in the table. Prove that

$$b_3 = a_3 \oplus 0$$
$$b_2 = a_2 \oplus b_3$$
$$b_1 = a_1 \oplus b_2$$
$$b_0 = a_0 \oplus b_1$$

or, noting that $b_4 = 0$,

$$b_i = a_i \oplus b_{i+1}; \quad i = 0, 1, 2, 3$$

The number $(a_3a_2a_1a_0)$ is the Gray code equivalent of the binary number $(b_3b_2b_1b_0)_2$. (*Suggestion:* Recall that $0 \oplus 0 = 0$, $0 \oplus 1 = 1 \oplus 0 = 1$, and $1 \oplus 1 = 0$, and examine the given table.)

Decimal	*Gray Code* $a_3\ a_2\ a_1\ a_0$	*Binary* $b_3\ b_2\ b_1\ b_0$
0	0 0 0 0	0 0 0 0
1	0 0 0 1	0 0 0 1
2	0 0 1 1	0 0 1 0
3	0 0 1 0	0 0 1 1
4	0 1 1 0	0 1 0 0
5	0 1 1 1	0 1 0 1
6	0 1 0 1	0 1 1 0
7	0 1 0 0	0 1 1 1
8	1 1 0 0	1 0 0 0
9	1 1 0 1	1 0 0 1
10	1 1 1 1	1 0 1 0
11	1 1 1 0	1 0 1 1
12	1 0 1 0	1 1 0 0
13	1 0 1 1	1 1 0 1
14	1 0 0 1	1 1 1 0
15	1 0 0 0	1 1 1 1

8.15 Decode the bit configuration 0010010000110101 if it represents

(a) four decimal digits in NBCD

(b) four decimal digits in the Gray code of Tab. 8.2

(c) two characters in even parity ASCII

8.16 Code the expression $A + B = C$ in

(a) odd parity ASCII

(b) EBCDIC

8.17 Show that if $(a_3a_2a_1a_0)_{8421}$ is an 8421 BCD coded number and $(b_3b_2b_1b_0)_{2421}$ is the corresponding 2421 BCD number, then we may obtain the latter in terms of the former by the relations

$$b_3 = a_3 + a_2(a_0 + a_1)$$
$$b_2 = a_3 + a_2(\bar{a}_0 + a_1)$$
$$b_1 = a_3 + a_1\bar{a}_2 + a_0\bar{a}_1a_2$$
$$b_0 = a_0$$

(*Suggestion:* Note that the 8421 binary equivalents of 10 through 15 are don't cares.)

8.18 Using the results of Prob. 8.17, obtain an 8421 BCD to 2421 BCD decoder.

9

COUNTERS AND REGISTERS

As we saw in Chap. 7, a flip-flop is a sequential device that can store one bit. That is, its output Q is either 0 or 1. More general sequential devices that can store and process one or more bits of information include *counters* and *registers*, which, as we will see in this chapter, may be constructed by interconnecting a number of flip-flops.

A *counter* is a sequential circuit that tallies, or *counts*, the number of input pulses it receives. Basically, a counter is a memory device that stores the number of input pulses. This number, or *count*, can be determined at any time, since its bits are those stored at that time in the flip-flops that comprise the counter. Counters are used in timing circuits, signal generators, and many other digital systems. They are, of course, fundamental to digital computers.

Registers are also memory devices used for storing and manipulating data, and they are essential components of most digital systems. They are found by the thousands in digital computers. A counter may be thought of as a type of register, but in general, registers are not restricted to counting. For example, registers may store information (memory registers) or process information (shift registers).

Registers may be classified according to how their stored information is entered or removed. A *serial* register is one in which the data is entered or removed one bit at a time, and a *parallel* register accepts or transfers all bits of data simultaneously. We may also have *serial–parallel* or *parallel–serial* registers, in which the data is entered one way and removed the other.

9.1 Binary Counters

Perhaps the simplest type of counter is a *binary counter*, which stores the number of input pulses as a binary number. For example, if the number of pulses received since the counting began is 13, the number stored in the counter is 1101, the binary equivalent of 13.

A basic binary counter can be constructed using any of the flip-flops of Chap. 7, with one flip-flop required for each bit needed in the count. A four-bit binary counter using four negative-edge triggered T flip-flops is shown in Fig. 9.1(a). To see that this is the case, let us consider the timing diagram in Fig. 9.1(b), where P is an input of pulses, as shown. Let the enable input E be 1 with all the flip-flops initially cleared (for this purpose a direct RESET input, which is not shown, is used). When the first pulse of P occurs, the first flip-flop toggles, setting the D bit to 1. The other bits, C, B, and A, remain 0, since flip-flops 2, 3, and 4 all have $CL = 0$. (For flip-flop 2 to toggle requires the occurrence of a negative edge of

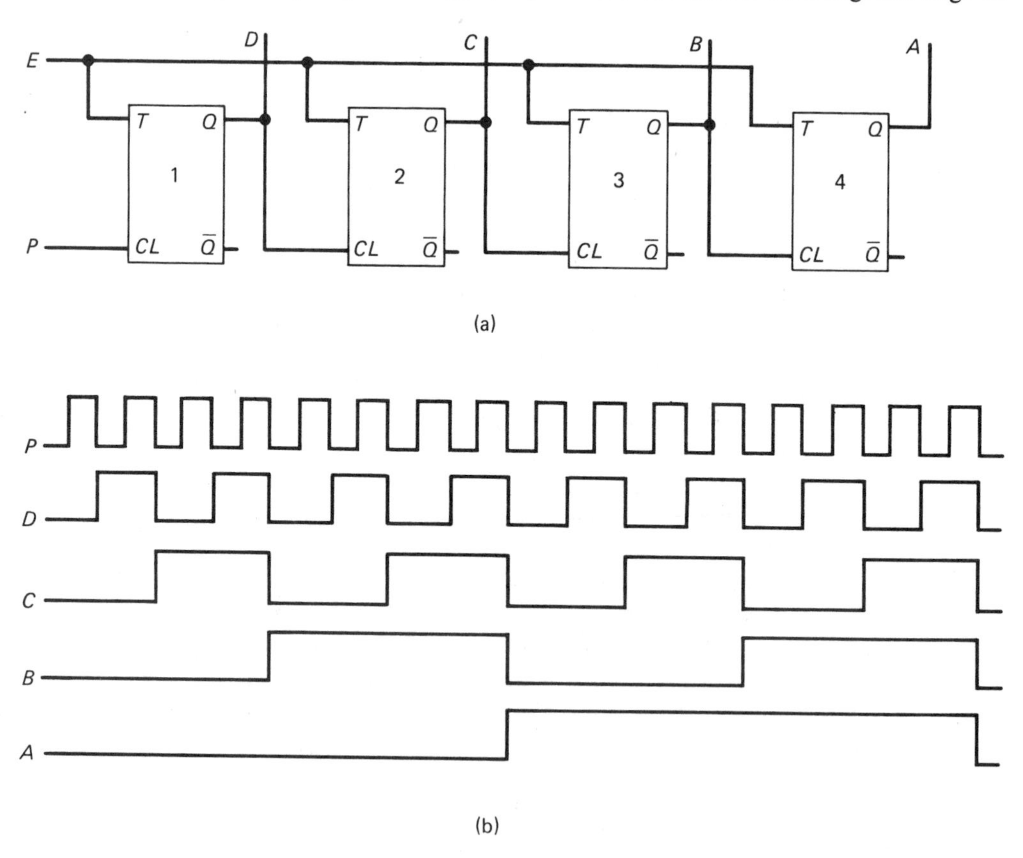

Figure 9.1 *(a) Circuit and (b) timing diagram for a four-bit binary counter*

$D = CL$.) The next pulse P resets flip-flop 1 to 0; thus D goes from 1 to 0, causing flip-flop 2 to toggle. We then have $C = 1$. The third pulse P causes flip-flop 1 to toggle, resulting in $D = 1$ once again, with C remaining 1. The fourth pulse resets D and C to 0, and the transition of C from 1 to 0 sets B to 1. The successive chain of events is easily followed in Fig. 9.1(b).

If we regard the outputs of the flip-flops as the stored binary number $(ABCD)_2$, we see from the timing diagram that its successive values as the negative edge of P occurs are 0000, 0001, 0010, 0011, 0100, 0101, 0110, . . . , 1111. These are, of course, the binary equivalents of 0, 1, 2, 3, 4, 5, 6, . . . , 15. Thus the binary counter of four flip-flops is capable of counting $2^4 = 16$ decimal integers before starting over at 0. To obtain higher counts simply requires more flip-flops.

An example of a two-bit binary counter was considered earlier, in Prob. 7.14. It was composed of two *JK* flip-flops and was capable of counting four decimal numbers.

The circuit in Fig. 9.1(a) is known as an *asynchronous* counter, because all its flip-flops are not set simultaneously by a clock. Another term used is *ripple* counter, since a certain amount of time is required as each bit changes sequentially, so that the effect seems to "ripple" through the counter.

A counter such as that in Fig. 9.1(a) is an *up* counter, since it counts *up* in the sequence 0, 1, 2, ... A *down* counter is one that counts down, as in the sequence 15, 14, 13, ... As the reader is asked in shown in Ex. 9.1.1, the circuit in Fig. 9.1(a) is a down counter if the outputs A, B, C, and D are taken at the $\bar{Q}$ terminals of the flip-flops.

A counter that can be changed from an up counter to a down counter and vice versa by appropriate input signals is called an *up-down* counter. Since the circuit in Fig. 9.1(a) is an up counter if the outputs are taken at the Q terminals and is a down counter if the outputs are taken at the $\bar{Q}$ terminals, it can be converted easily to an up–down counter by adding appropriate digital circuitry. The circuitry can be designed so that on an *up count* input signal the outputs of the counter are the Q outputs and on a *down count* input signal the outputs of the counter are the $\bar{Q}$ outputs. Such an up–down counter is discussed in Ex. 9.1.2.

Any negative-edge triggered flip-flops capable of being toggled can be used in the circuit in Fig. 9.1(a) in place of the T flip-flops. For example, *JK* flip-flops could be used if the inputs J and K are both made 1, since in this case the *JK* flip-flop toggles with each clock pulse.

The ripple counter requires a certain amount of time for each flip-flop to change state, and this causes the last flip-flop in the counter to react later than the first. A *synchronous* counter is one in which all the flip-flops are triggered at the same time by a clock. Thus synchronous flip-flops are faster and less susceptible to errors due to unwanted pulses or shortened pulses caused by the cumulative delay of the asynchronous counter.

A four-bit synchronous counter is shown in Fig. 9.2. The first flip-flop changes state D with each clock pulse, since $J = K = 1$. The other three flip-flops will have

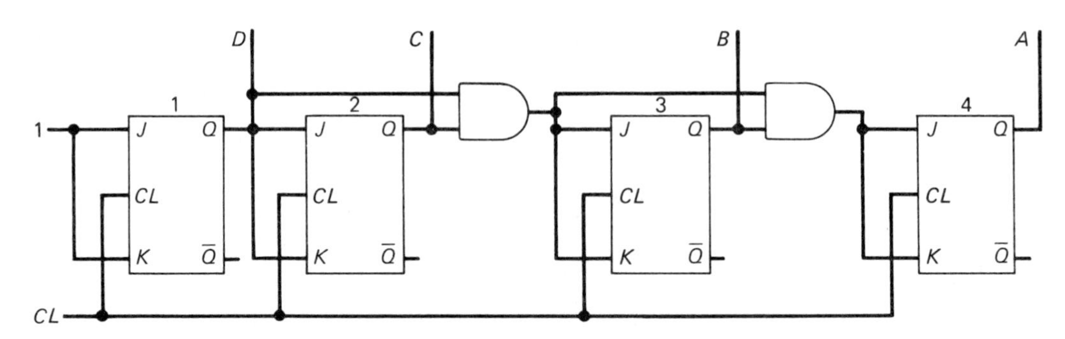

Figure 9.2 *Synchronous binary counter*

either $J = K = 0$, in which case the flip-flop remains in its present state, or $J = K = 1$, in which case it toggles. The inputs to flip-flop 2 are $J = K = D$; thus it toggles only when $D = 1$, which is every other clock pulse. The inputs to flip-flop 3 are $J = K = CD$, so that it toggles only when $C = D = 1$. Finally, the inputs to flip-flop 4 are $J = K = BCD$, and therefore it toggles only when $B = C = D = 1$. The timing diagram, which may be sketched from this information, is identical to that in Fig. 9.1(b) if $P = CL$. In the asynchronous case we have not shown the delays, and the signal P is represented as periodic, like a clock signal, although this is not necessarily true.

Fig. 9.2 is an up counter. As in the asynchronous case, we may also have synchronous down counters and up–down counters. Examples of these will be left to the exercises and problems (Ex. 9.1.3 and Prob. 9.3).

EXERCISES

9.1.1 Show that the given circuit is an asynchronous down counter by drawing the timing diagram. Assume that $E = 1$, P is as shown in Fig. 9.1(b), and $Q = 0$ initially for all the flip-flops (which are negative-edge triggered).

Ans: A, B, C, and D are complements of the values shown in Fig. 9.1(b).

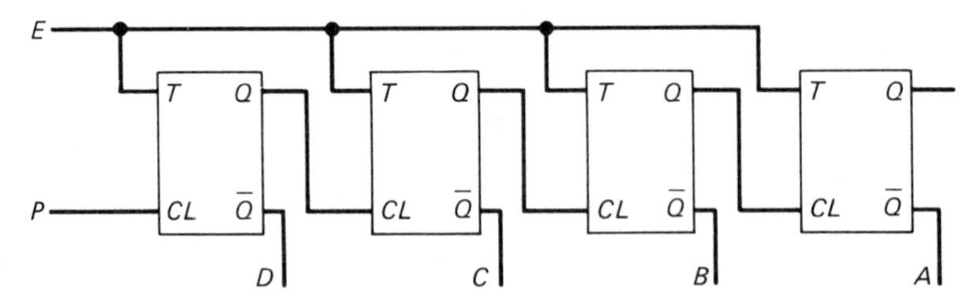

Exercise 9.1.1

9.1.2 Show that the given circuit is an up–down counter that counts up when $U/D = 1$ and down when $U/D = 0$. Assume that $E = 1$, that all the Q's are initially set to 0, and that the flip-flops are negative-edge triggered.

Ans: When $U/D = 1$, the outputs are those of Fig. 9.1(a), and when $U/D = 0$, the outputs are those of the circuit in Ex. 9.1.1.

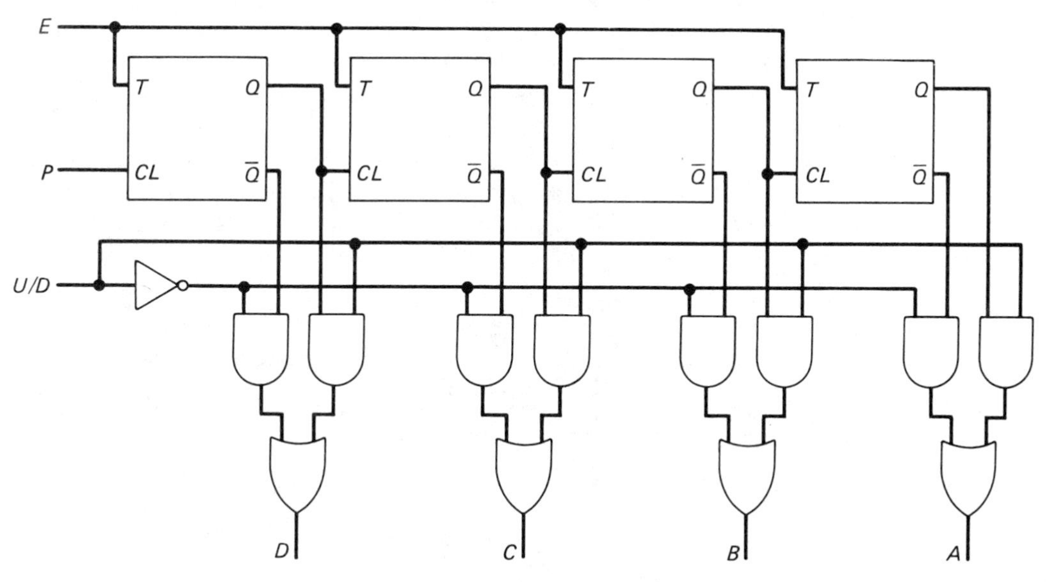

Exercise 9.1.2

9.1.3 Show that the given circuit is a synchronous down counter, assuming that the flip-flops are negative-edge triggered.

Ans: A, B, C, and D are complements of the values in Fig. 9.2.

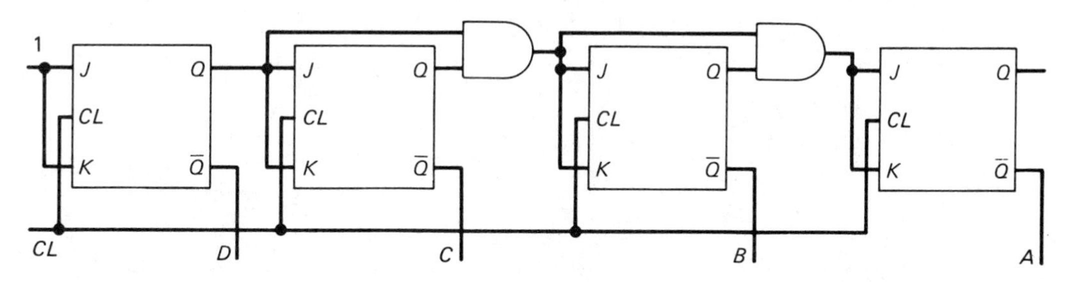

Exercise 9.1.3

9.2 BCD Counters

In some systems the ease of displaying, or *reading out*, the decimal equivalents of the numbers in a counter or register outweighs the ease of computation of binary numbers. In this case the counters or registers are designed to store the equivalent of each decimal digit in a group of four flip-flops. In other words, each group of four flip-flops contains the BCD code for a decimal digit, 0, 1, . . . , 9. If the count is, say, 16, then the group of flip-flops storing the least significant digit will contain the BCD equivalent of 6, while the next significant digit set of flip-flops will store the BCD equivalent of 1. Such counters are called BCD counters and, like binary counters, may be synchronous or asynchronous.

An asynchronous BCD counter is shown in Fig. 9.3(a), as may be seen from its timing diagram, in Fig. 9.3(b). Flip-flop 1 toggles with each negative edge of P, as shown in the timing diagram for signal D. Flip-flop 2 toggles with each negative edge of D, as long as $J = 1$. This is the case until Q of flip-flop 4 becomes 1 on counts 8 and 9 ($\bar{Q}$ becomes 0, which is J of flip-flop 2). Thus the negative edge of D, occurring at the end of count 9, does not cause flip-flop 2 to toggle, so that C remains 0 for the next count (labeled 0 again). Also, B and A remain 0 for this count, because $C = 0$ prevents flip-flops 3 and 4 from toggling, since C is the J

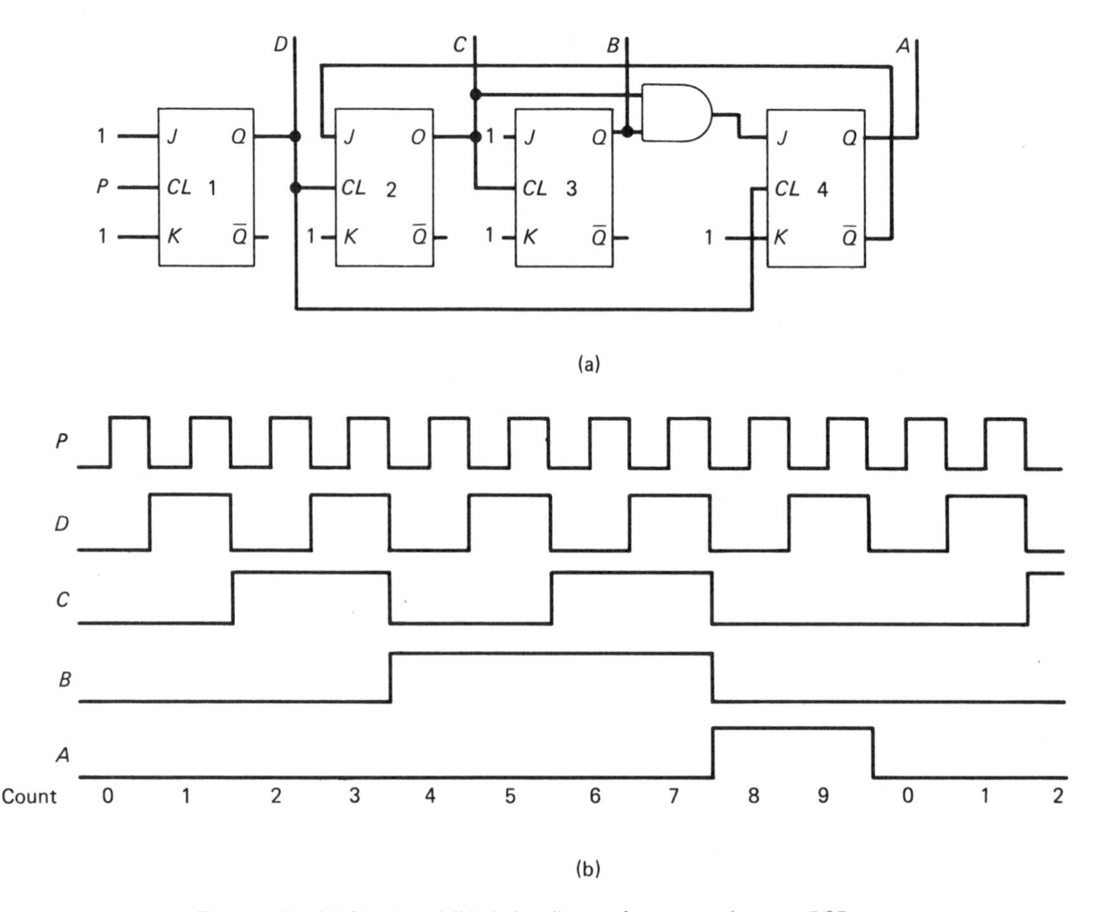

Figure 9.3 *(a) Circuit and (b) timing diagram for an asynchronous BCD counter*

input to flip-flop 3 and BC is the J input to flip-flop 4. We see, therefore, that the count proceeds as 0, 1, 2, . . . , 9, and starts over.

The BCD counter in Fig. 9.3(a) may be synthesized directly from the required timing diagram in Fig. 9.3(b). Since D goes from 0 to 1 or from 1 to 0 on each negative edge of P, we need only a JK flip-flop with $J = K = 1$, $CL = P$, and $Q = D$ to produce the output D. Thus flip-flop 1 (FF 1), which produces D, simply toggles on each negative edge of P. Flip-flop 2, needed to produce C, evidently toggles on each negative edge of D, except the one occurring on count 9 (the tenth pulse). On this count $A = 1$, in contrast to all the other counts, on which A $= 0$. Therefore, if FF 2 has $J = \bar{A}$, $K = 1$, and $CL = D$, its output Q will be the required signal C. Flip-flop 3, needed to produce B, must toggle on the negative edges of C. Therefore, for FF 3 we need $CL = C$, $J = K = 1$, and $Q = B$. Finally, FF 4, with $Q = A$ as output, toggles on the negative edge of D when $B = C = 1$. Its output goes from 1 to 0 on the negative edge of D when $B = C = 0$. Thus for FF 4 we must have $CL = D$, $J = BC$, and $K = 1$. (Recall that for $J = 0$, $K = 1$ we have $Q_{n+1} = 0$, as will be the case at count 9.) The circuit in Fig. 9.3(a) evidently satisfies the conditions required of these four flip-flops, and is thus an asynchronous BCD counter.

A synchronous BCD counter must have the same timing diagram as that shown in Fig. 9.3(b), but in contrast to the asynchronous counter, all its flip-flops must be activated by the same clock. Let us design a synchronous counter by observing from the timing diagram what the inputs and outputs of each of the constituent flip-flops must be. If $P = CL$, the timing diagram will be that shown in Fig. 9.3(b).

To begin with, we must have $P = CL$ in all four flip-flops. Flip-flop 1, which produces D, merely toggles on each negative edge of P, and therefore we must have $J = K = 1$ and $Q = D$. Flip-flop 2, which produces C, toggles on the negative edges of CL when $D = 1$, with the exception of count 9. On this count $A = 1$, whereas on the other counts $\bar{A} = 1$. Thus for FF 2 we require $J = \bar{A}D$, $K = 1$ (or equivalently in this case, $K = D$), and $Q = C$. Flip-flop 3 produces B by toggling on the negative edge of CL only when $C = D = 1$. Thus for this flip-flop we need $J = K = CD$ (or equivalently in this case, $J = CD$ and $K = 1$) and $Q = B$. Finally, FF 4 produces A, which goes from 0 to 1 on count 7 ($B = C = D = 1$) and from 1 to 0 on count 9 ($B = C = 0$, $D = 1$). One way to accomplish this is with $J = BCD$, $K = D$, and $Q = A$. The synchronous BCD counter utilizing these four flip-flops is shown in Fig. 9.4.

Clearly, the two BCD counters of this section are up counters. As in the case of binary counters, we may also have BCD down counters and up–down counters. An example of a down counter is considered in Prob. 9.7.

A *modulo-N* counter is one that counts from 0 to $N - 1$ and then repeats itself. A BCD counter is thus a modulo-10 counter, and the binary counters of Sec. 9.1 are modulo-16 counters. In general, a binary counter with n flip-flops is a modulo-2^n counter. Modulo-N counters are available as complete packages with the feature that N can be set to the desired value.

As an example, let us design a synchronous modulo-3 counter. Such a counter

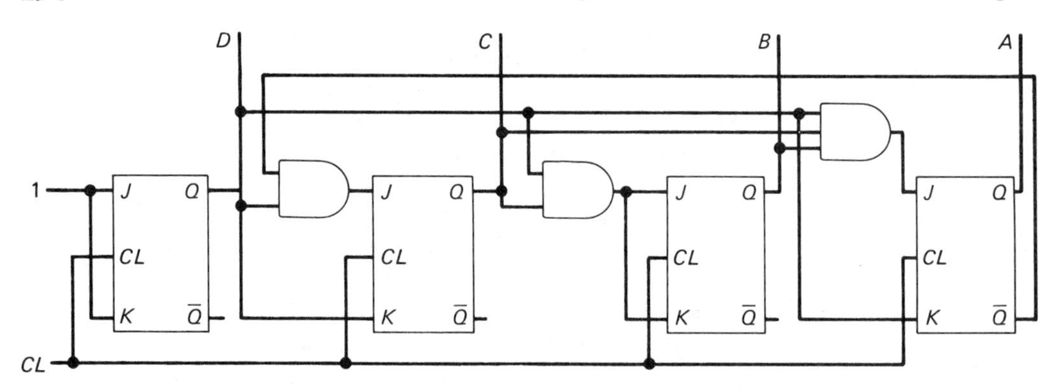

Figure 9.4 *Synchronous BCD counter*

will require two flip-flops, since the binary equivalent of 3 has two bits. The timing diagram, shown in Fig. 9.5, may be used to determine the inputs to the JK flip-flops that we will use. If FF 1 has B (the LSB) as its output, we see from the diagram that it toggles on the negative edges of CL except when $A = 1$. Thus $J = \bar{A}$ and $K = 1$ are the required inputs of FF 1. In the case of FF 2 with A (the MSB) as output, we see that on the negative edge of CL it has a transition from 0 to 1 when $B = 1$ and a transition from 1 to 0 or from 0 to 0 when $B = 0$. Therefore $J = B$ and $K = 1$ are the required inputs of FF 2. The modulo-3 counter is shown in Fig. 9.6.

As a final note on modulo-N counters, we observe from the timing diagrams

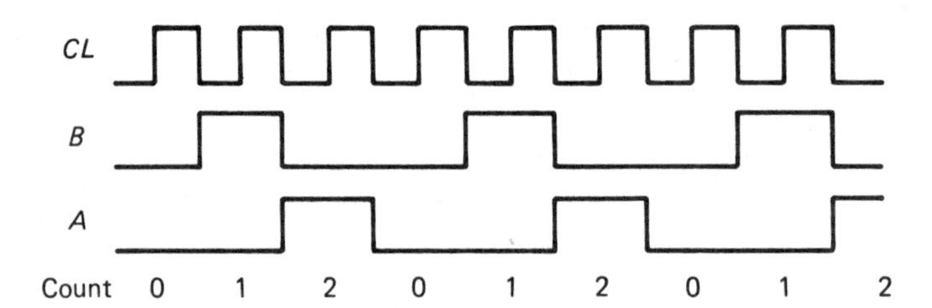

Figure 9.5 *Timing diagram for a modulo-3 counter*

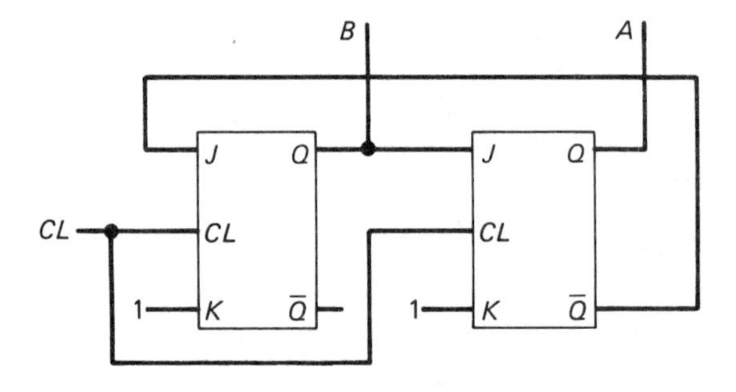

Figure 9.6 *Synchronous modulo-3 counter*

that the frequency of the MSB signal train is $1/N$ times that of the signal whose pulses are being counted. For example, in Fig. 9.1 the signal P executes 16 cycles while A executes one. Thus the frequency of A in the modulo-16 counter is $\frac{1}{16}$ that of P.

EXERCISES

9.2.1 Draw a timing diagram for a modulo-5 counter, which stores the count $(ABC)_2$. Show 10 pulses of CL.

Ans:

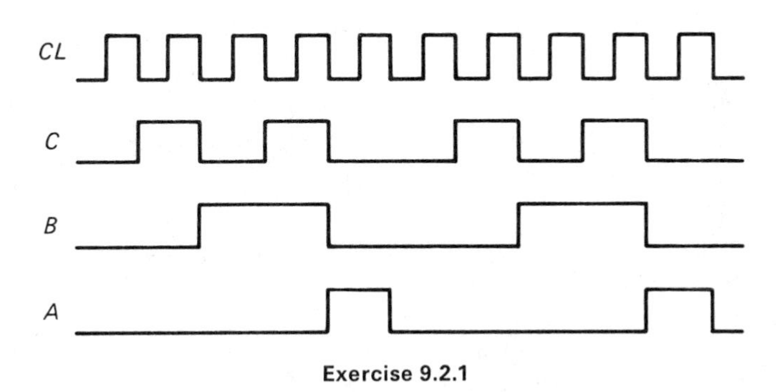

Exercise 9.2.1

9.2.2 Using the result of Ex. 9.2.1, design a synchronous modulo-5 counter using JK flip-flops.

Ans:

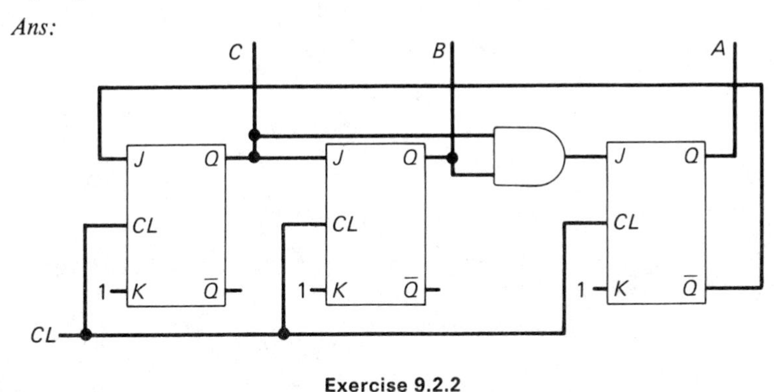

Exercise 9.2.2

9.3 Memory Registers

A *memory* register, or *storage* register, is a device capable of accepting information in the form of a binary number, holding that information after the input that provided it has been removed, and making the information available as an output.

We will use the symbol shown in Fig. 9.7 to represent a memory register. In this case the device is an eight-bit register, currently storing the binary number 10010100.

Memory registers typically provide temporary storage of data, such as the count from a counter. The device supplying the data is then free to perform other tasks while the data is preserved for future use, such as being decoded and read out or being displayed.

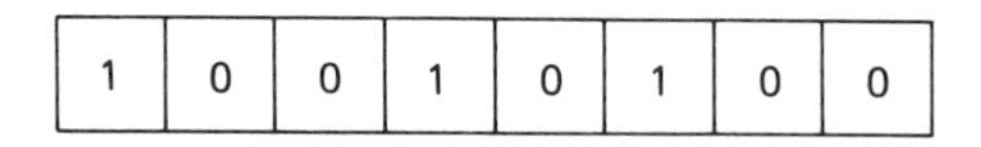

Figure 9.7 *Symbol for a memory register*

Memory registers may be constructed with flip-flops, in a manner similar to that used for counters. As in the case of counters, one flip-flop is needed for each bit in the word to be stored. For example, the circuit in Fig. 9.8, using four D flip-flops, is a four-bit memory register that stores the input $ABCD$, from a counter or from some other source, as the output $Q_1Q_2Q_3Q_4$. The clock signal CL is a *memory transfer* input that transfers A to Q_1, B to Q_2, and so on, when it is pulsed. Once $ABCD$ is stored, it will remain in the memory register until new data is stored by pulsing the memory transfer again or by clearing the register. In this case the data is said to be *latched.*

The memory register in Fig. 9.8, as is true of most memory registers, is a *parallel input–parallel output* device. That is, each flip-flop, or *stage*, of the register receives its input at the same time as every other stage, and all the outputs are read out or received by another device simultaneously. In other words, both the inputs and outputs are *parallel* data (all bits handled simultaneously), as opposed to *serial* data (one bit handled at a time).

Both Q and $\bar{Q}$ are available at each stage of a memory register, but we are concerned only with Q in Fig. 9.8. There may be occasions when we would take

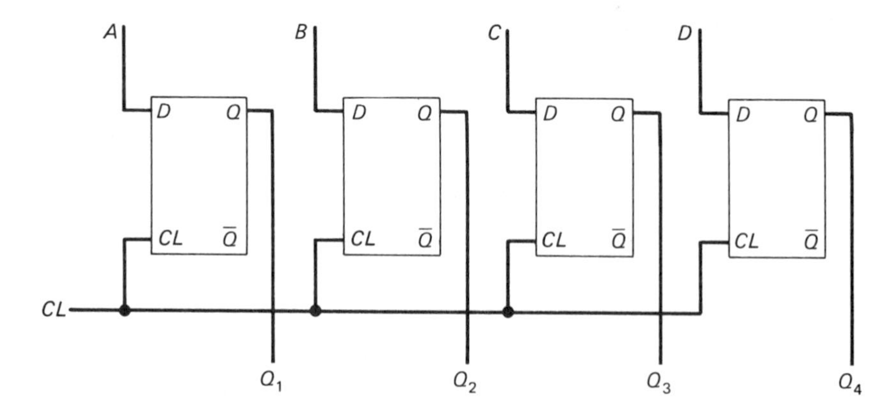

Figure 9.8 *Four-bit memory register*

the output as $\bar{Q}$ at each stage. This would be the case, for example, if we wanted the 1's complement of $ABCD$.

EXERCISES

9.3.1 Design a four-bit memory register using JK flip-flops.

Ans: The circuit is identical to Fig. 9.8 with the D flip-flops replaced by the device shown:

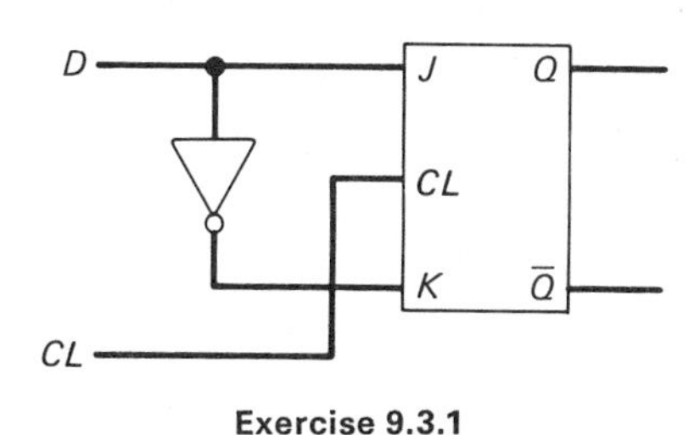

Exercise 9.3.1

9.3.2 Repeat Ex. 9.3.1 using clocked SR flip-flops.

Ans: Identical to that of Ex. 9.3.1 with clocked SR flip-flops replacing JK flip-flops.

9.4 Shift Registers

Another important type of register is the *shift* register, which can store data as a memory register does but is more often used to process, or *move*, data. Usually the movement is a shift of data from one stage of the register to an adjacent stage. The shift may be from left to right (a right-shift register), from right to left (a left-shift register), or in both directions (a left-shift, right-shift register). We will use the same symbol (that in Fig. 9.7) to represent the shift register as was used to represent the memory register.

Shift registers differ from memory registers in that adjacent stages are connected to allow shifts of data from one stage to the next. Data may be loaded in and read out of the register in parallel, as in the case of the memory register. However, shift registers may be designed for serial entry and serial readout, as well. A shift register that can accept and have available for readout both parallel and serial data is very useful in transmitting information. For example, an eight-bit ASCII code transmitted at one time would require eight separate lines, one for each bit. Using a shift register with serial and parallel capabilities, we may store the eight-bit word using parallel loading. When the word has been loaded, we may shift one bit a time from the register and transmit it serially. After the last bit has been transmitted, a new word is loaded in parallel fashion and the process repeated. At the receiving end the bits are loaded serially, and when the word is in the register a parallel readout is performed. Thus we may make one line do the work of eight—at a sacrifice, of course, of speed of transmission.

A serial-entry shift register capable of storing four-bit words is shown in Fig. 9.9. The stages are D flip-flops, but other types may be used as well. The register is serially loaded as follows. The outputs Q_1, Q_2, Q_3, and Q_4 are made 0 initially by a RESET control (not shown). The first data bit, say, D_1, is applied to the serial input terminal and on the first positive edge of CL is loaded into the first stage as $Q_1 = D_1$. Next D_2 is applied and clocked in as $Q_1 = D_2$, at which point D_1 is shifted to $Q_2 = D_1$. Continuing the process with D_3 and D_4, the next two serial entries, we have $Q_1 = D_4$, $Q_2 = D_3$, $Q_3 = D_2$, and $Q_4 = D_1$. A timing diagram is shown in Fig. 9.10 for the case $D_1 = D_4 = 1$, $D_2 = D_3 = 0$, from which we see that $Q_1 = D_1 = 1$ on the first count of the clock, $Q_1 = D_2 = 0$ and $Q_2 = D_1 = 1$ on the second count, and so on.

The output data of the register in Fig. 9.9 may be taken in either a serial or a parallel manner. That is, the outputs Q_1, Q_2, Q_3, and Q_4 may all be read simulta-

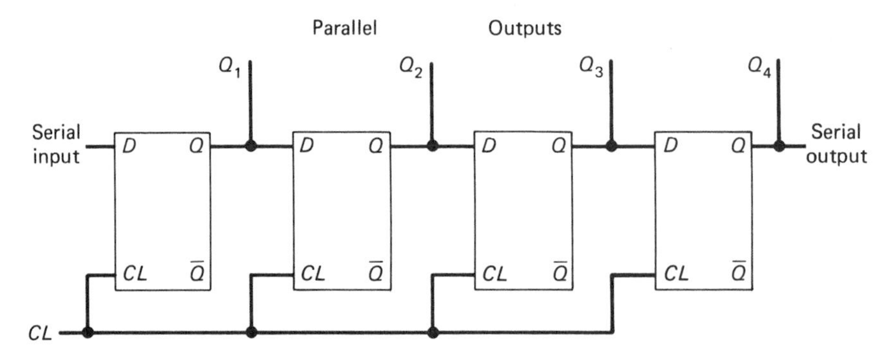

Figure 9.9 *Four-bit shift register with serial entry*

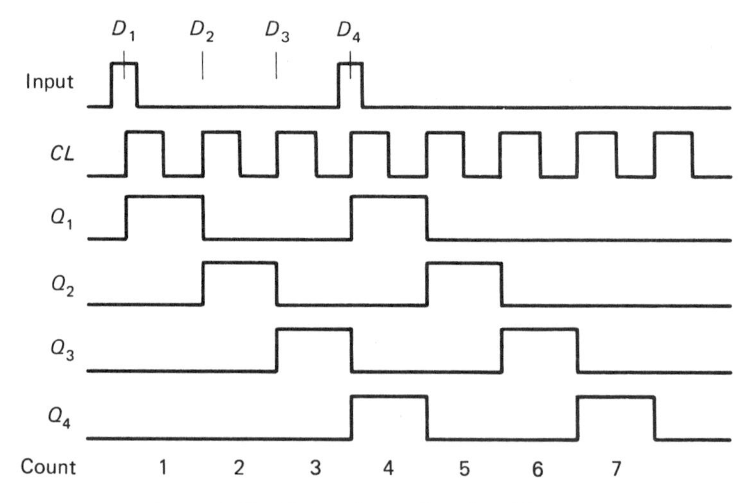

Figure 9.10 *Timing diagram for a four-bit register*

neously, or the output may be read one bit at a time at the output Q_4 of the fourth stage. In the latter case the shift register is functioning as a serial–serial register, and in the former case as a serial–parallel register. For registers with a large number of bits (and some are available with over 1000) parallel outputs are impractical because of the many connections required.

A parallel-entry shift register using *JK* flip-flops is shown in Fig. 9.11. The data inputs D_1, D_2, D_3, and D_4 may be loaded simultaneously using the parallel data lines or serially using the serial input line. In the former case the load line is activated by means of the direct set connection at the time when the inputs are to be stored. Also shown is a reset connection that can be used to clear the register before the data is entered. Following the loading operation, the data shifts one position to the right with each clock pulse. The data may also be read out in either a serial or parallel mode. Thus the device may operate as a serial–serial, parallel–parallel, serial–parallel, or parallel–serial register.

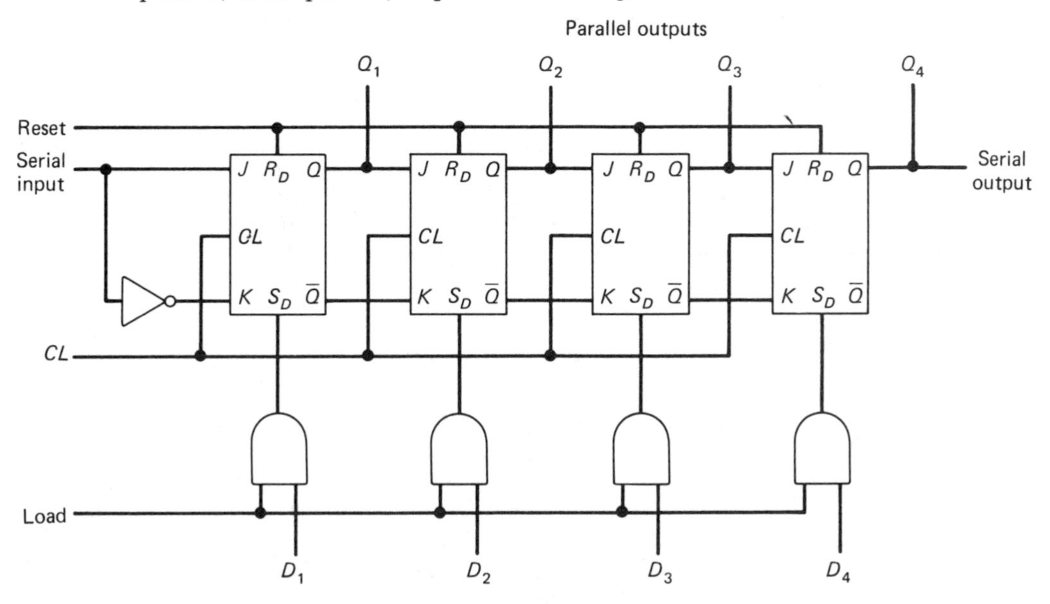

Figure 9.11 *Four-bit shift register with parallel entry*

The shift registers we have considered thus far are shift-right registers. That is, when a bit is entered at the serial input, it is stored in the first stage of the register. The contents of the first stage are shifted to the second, the contents of the second are shifted to the third, and so on, from left to right. For example, suppose an eight-bit register has the number shown in Fig. 9.12(a) stored and bits 1, 0, 1, 1 are to be loaded serially in that order. When the first bit 1 is entered, it is stored in the first stage and all the previously stored bits are shifted one stage to the right, as shown in Fig. 9.12(b), The second bit 0 is then entered, with the result shown in Fig. 9.12(c), the entry of the third bit 1 results in Fig. 9.12(d), and finally, the entry of the fourth bit 1 results in Fig. 9.12(e).

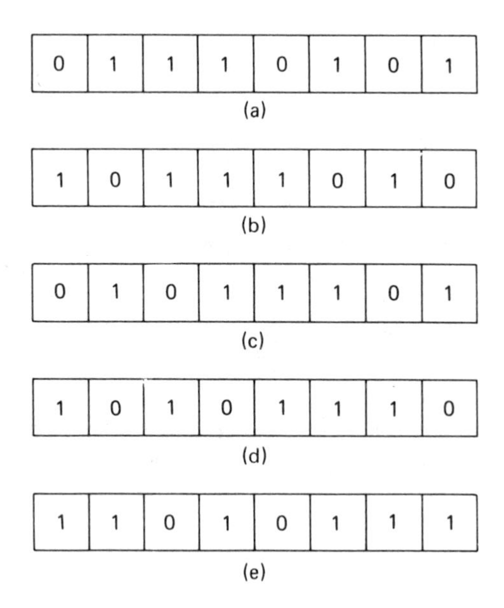

Figure 9.12 *Steps in serially loading a shift-right register*

Examples of shift-left and of shift-right, shift-left registers are considered in the exercises and problems (Ex. 9.4.2 and Prob. 9.10).

EXERCISES

9.4.1 Show that the register in Fig. 9.11 is a shift-right register, and find its stored data after four clock pulses if the sequence of serial inputs is 1101.

Ans: 1011

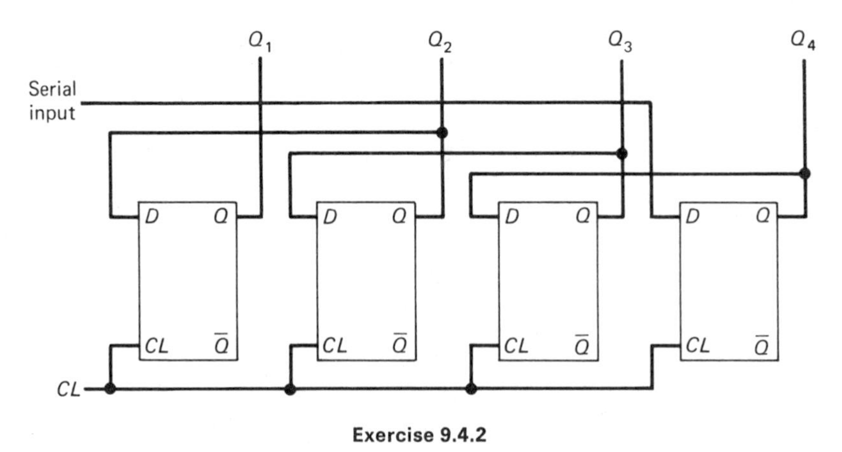

Exercise 9.4.2

9.4.2 Show that the given circuit is a shift-left register, and find its stored data after four clock pulses if the sequence of serial inputs is 1101.

Ans: 1101

9.5 Register Transfers and Buses

As we have noted, registers are used extensively in digital circuits, and in particular a digital computer contains literally thousands of registers. In general, we must be able to transfer data from any given register to almost any other register in the circuit, so that interconnecting each pair of registers by a set of lines would require a prohibitively large number of lines. In the case of a computer, the number of lines necessary would run into the millions. One way to avoid the use of large numbers of individual lines is by using *buses*, which we describe in this section.

A *bus* is a group of wires over which data is transferred one bit per wire. There are a number of types of buses, many of which will be discussed in Chap. 11 in connection with microcomputers. We will be interested here in buses that are used to communicate between two sets of interconnected registers. By the use of multiplexing, a single bus between the two sets can be used to transfer data from any register in one set to any register in the other.

To illustrate the procedure, let us consider Fig. 9.13, in which data is to be transferred from either register A or B to either register C or D. Registers A and B are multiplexed onto the single bus, as shown, and the bus is then connected to registers C and D. The inputs E_A, E_B, E_C, and E_D are enable signals used to select the two registers between which the data is to be transferred. For example if $E_A = 1$ and $E_D = 1$ while $E_B = E_C = 0$, then the contents of register A are stored in register D via the bus. The signal $E_A = 1$ selects register A to be connected to the bus, and $E_D = 1$ is the control signal that loads register D with the data connected to the bus. Of course, the pulse E_D must occur while the pulse E_A is occurring, as shown in Fig. 9.14.

If, as is usually the case, we wish to be able to transfer data from register C or D to register A or B, a dual bus system is needed. This consists of the bus arrangement shown in Fig. 9.13 with another multiplexer to load the contents of C or D, via a second bus, into register A or B. The reader is asked to design such a system in Prob. 9.11.

To transfer data between two registers in the same set, such as from register A to register B in Fig. 9.13, we may transfer the data of register A to either register C or D via the bus shown. Then we transfer the data from C or D to register B via the other bus in the dual system.

The transfer of data between register A or B and register C or D, in Fig. 9.13, may also be done in a simpler manner using gates known as *tri-state buffer* gates. The symbol for such a gate is shown in Fig. 9.15; as its name implies, it has three states. The output may be 0, it may be 1, or the output line may be opened. The truth table of the tri-state buffer gate is given as Tab. 9.1.

By means of tri-state buffer gates we may accomplish the multiplexing of Fig.

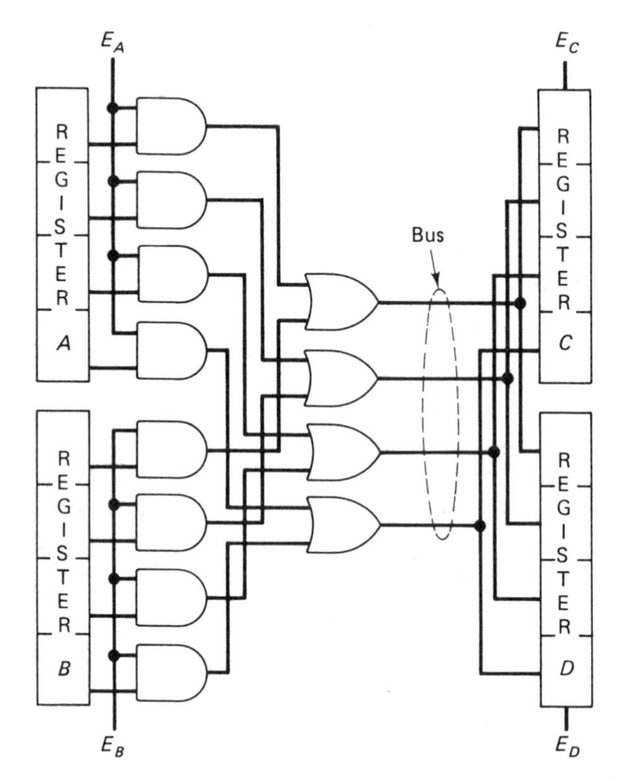

Figure 9.13 *Transfer of data between two registers via a single bus*

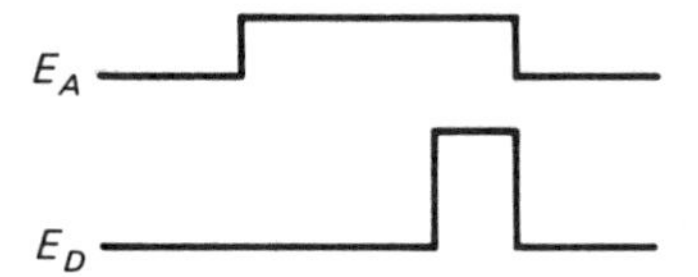

Figure 9.14 *Typical enable signals for transferring the data of register A to register D*

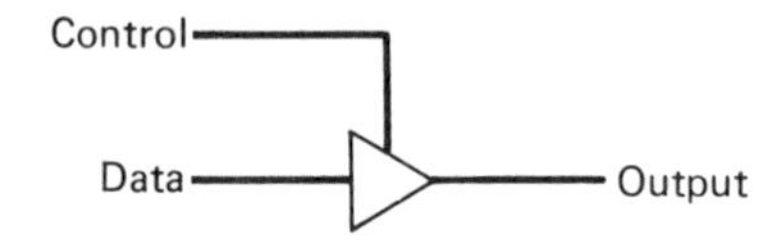

Figure 9.15 *Tri-state buffer gate*

9.13 in a much simpler way. The resulting circuit is shown in Fig. 9.16, where it may be seen that $E_A = 1$ and $E_B = 0$ connects register A to the bus and disconnects register B from the bus, and so on. For example, to transfer the data of register A to register D we make $E_A = E_D = 1$, as in Fig. 9.14, with $E_B = E_C = 0$.

Table 9.1 *Truth table for a tri-state buffer gate*

Data	*Control*	*Output*
0	0	*
1	0	*
0	1	0
1	1	1

*Output line open.

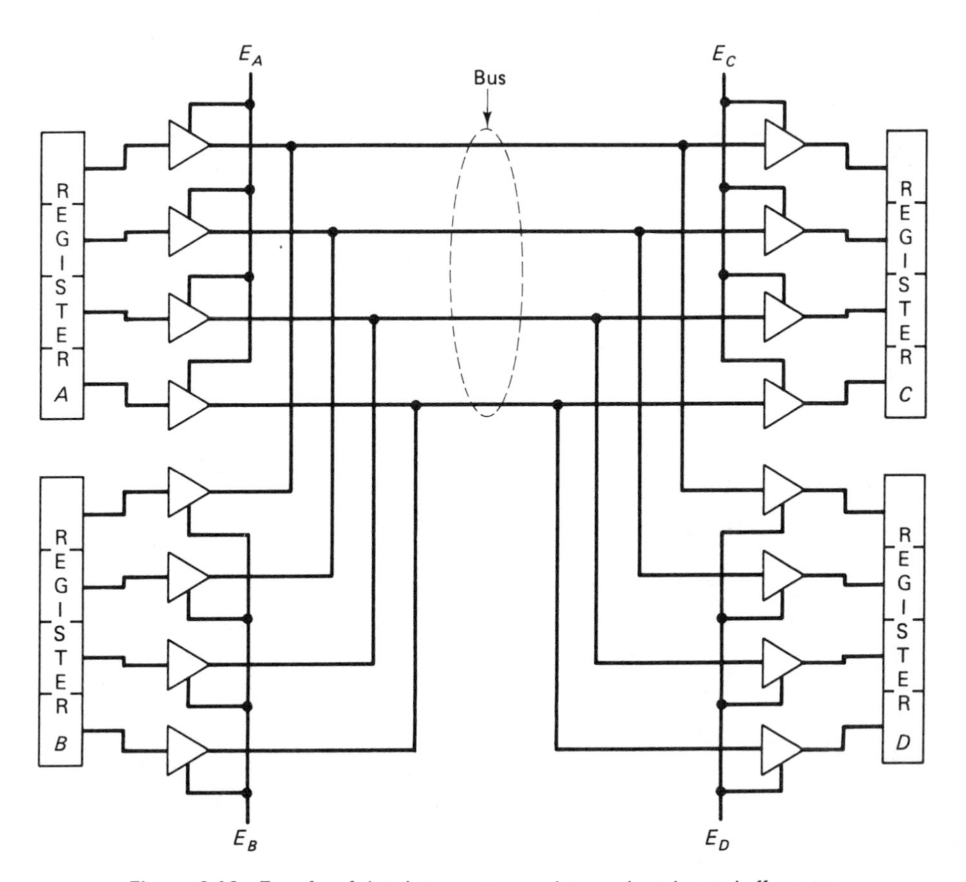

Figure 9.16 *Transfer of data between two registers using tri-state buffer gates*

EXERCISE

9.5.1 If in Fig. 9.16 the contents of register *A*, *B*, *C*, and *D* are initially 1101, 1001, 0100, and 0111, respectively, find their contents after

(a) E_A and E_C are pulsed

(b) E_A and E_D are pulsed, followed by the pulsing of E_B and E_C, and finally by the pulsing of E_A and E_C

Ans: (a) 1101, 1001, 1101, 0111; (b) 1101, 1001, 1101, 1101

PROBLEMS

9.1 Show that if the flip-flops are *positive-edge* triggered, the given circuit is a binary up counter.

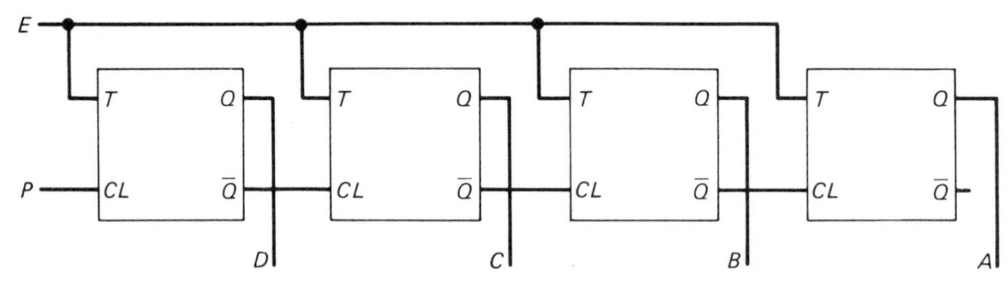

Problem 9.1

9.2 Show that if the flip-flops in Fig. 9.1(a) are positive-edge triggered, the circuit is a down counter.

9.3 Using the results of Fig. 9.2 and Exs. 9.1.2 and 9.1.3, obtain a synchronous binary up–down counter that counts up when a control signal $U/D = 1$ and counts down when $U/D = 0$.

9.4 Design a modulo-8 asynchronous counter using clocked T flip-flops. (*Suggestion:* See Fig. 9.1.)

9.5 Repeat Prob. 9.4 for a modulo-8 asynchronous down counter.

9.6 Show that the given circuit is a modulo-7 counter.

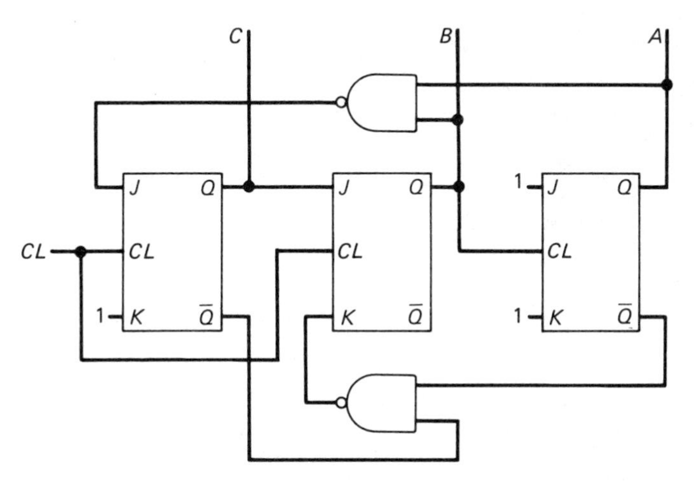

Problem 9.6

9.7 Show that the given circuit is a synchronous BCD down counter.

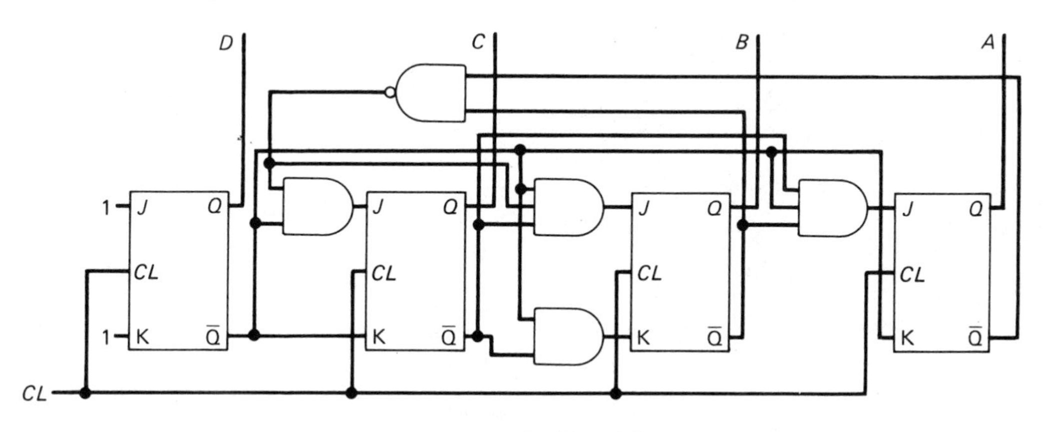

Problem 9.7

9.8 The circuit shown is a self-stopping, variable-modulus counter that can be set to count to and hold the number N equivalent in binary to $(F_3F_2F_1F_0)_2$. Illustrate this by obtaining the timing diagram for

(a) $N = 7$

(b) $N = 12$

(*Note:* N cannot exceed 15 without adding more flip-flops.)

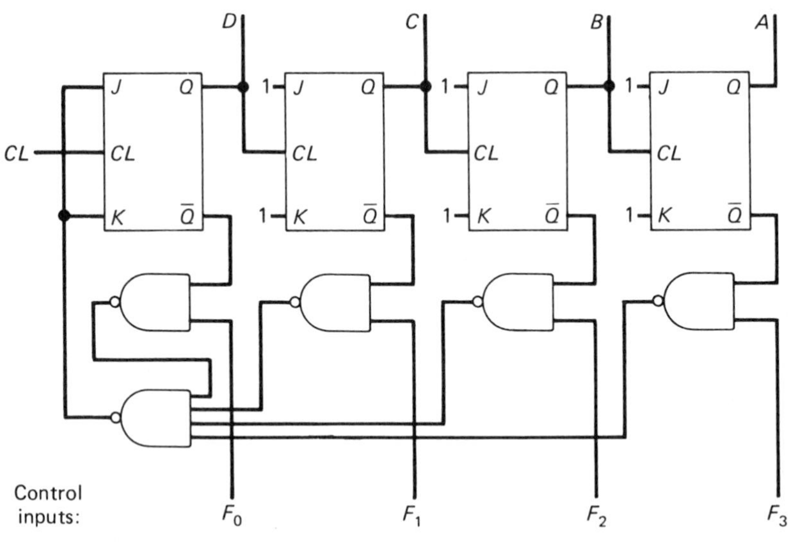

Problem 9.8

9.9 Draw the timing diagram for the four-bit register in Fig. 9.9 for eight clock pulses if the serial input sequence is 1101.

9.10 Show that the given circuit is a shift-right, shift-left register and draw the timing diagram for eight clock pulses if the serial input sequence is 1001 and the circuit is operating in

(a) the shift-left mode

(b) the shift-right mode

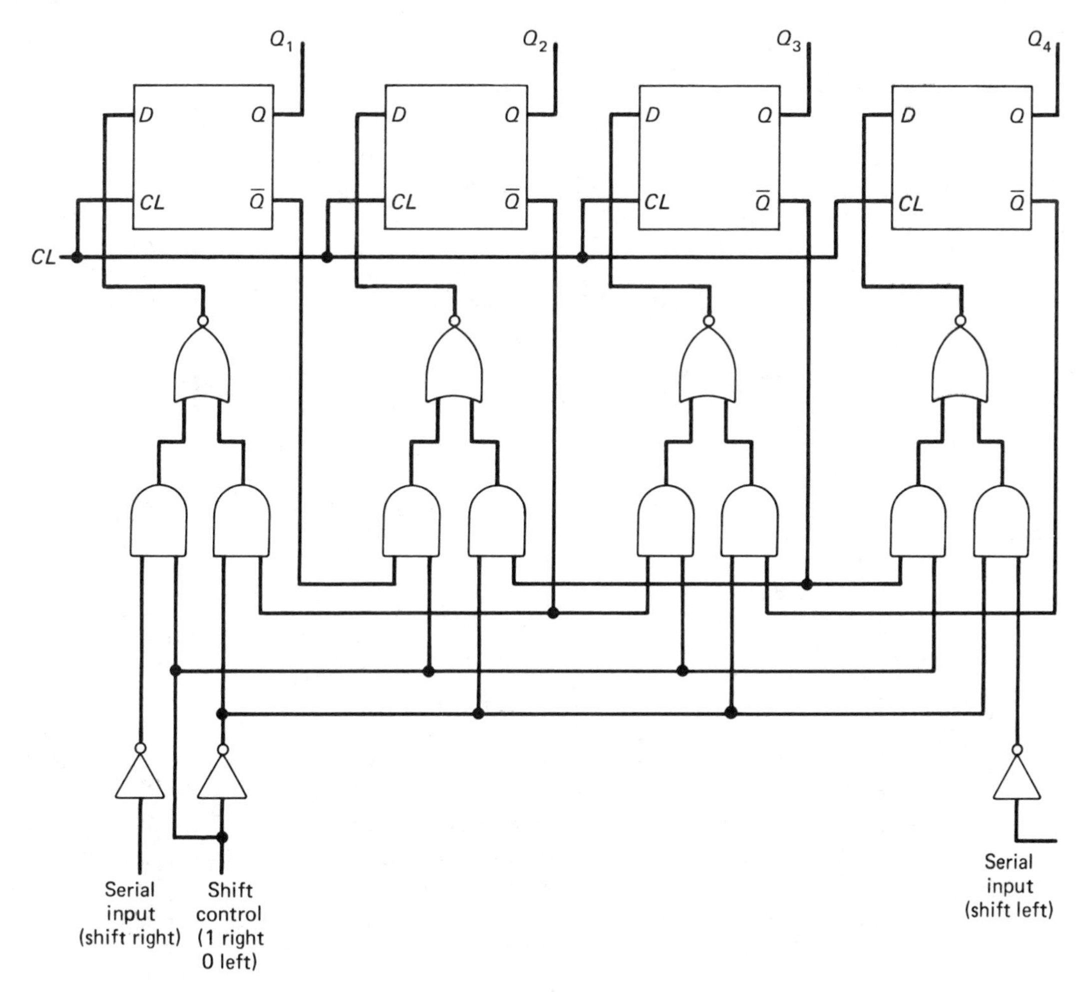

Problem 9.10

9.11 Add a multiplexing system to Fig. 9.13 so that the contents of register *C* or *D* can be transferred to register *A* or *B*.

9.12 Replace the multiplexing systems of Prob. 9.11 by systems using tri-state buffer gates.

10

SIGNED NUMBERS AND COMPLEMENTARY ARITHMETIC

In all number systems, both positive and negative numbers are used, and we must have some means of distinguishing between the two. In our everyday lives we do this by writing, say, 13 or +13 for the positive number and −13 for the negative number. If the numbers are to be processed by a digital circuit, such as a computer, we must use some code which the circuit can interpret instead of the plus and minus signs. In this chapter we consider such a coded number system, known as a *signed binary* system.

The arithmetic of signed numbers is complicated by the presence of the code representing the sign. In adding two numbers having like signs we simply add the numbers and give the sum the same sign as the two numbers. If the two numbers being added have different signs, we subtract and give the difference the sign of the larger number. Similarly, in subtracting signed numbers we change the sign of the subtrahend and add the result to the minuend. These are the ordinary rules of arithmetic of signed numbers which we apply every day. In the case of a digital circuit, applying these rules requires two different types of circuitry—one for adding and one for subtracting. We will see that *complementary arithmetic*, which we also consider in this chapter, avoids this difficulty by allowing us to subtract by adding *complements*.

10.1 Signed Numbers

The simplest signed number system is the *sign-and-magnitude* notation, often referred to as *signed binary*. In this system the most significant bit is a *sign* bit, which denotes whether the number is positive or negative. The remaining bits express the magnitude of the number. That is, if the binary number is

$$N = \pm(a_{n-1}a_{n-2} \cdots a_1a_0 . a_{-1} \cdots a_{-m})_2 \tag{10.1}$$

the sign-and-magnitude notation is

$$N = (s\ a_{n-1}a_{n-2} \cdots a_1a_0 . a_{-1} \cdots a_{-m})_{sm} \tag{10.2}$$

where the MSB is the sign bit s. If N is positive, $s = 0$, and if N is negative, $s = 1$.

We will use a comma to separate the sign bit from the magnitude bits. However, many authors omit the comma, in which case the MSB is simply read as the sign bit. For example, the number -13_{10} expressed in five-bit sign-and-magnitude notation is given by

$$-13_{10} = 1{,}1101_{sm}$$

Another example is the positive number

$$+25_{10} = 0{,}11001_{sm}$$

which requires six bits to include the sign.

Table 10.1 shows the decimal equivalents for four-bit signed binary numbers. Note that there are two representations for 0, since $-0 = +0$. Thus only 15 numbers rather than 16 can be represented in this fashion.

Table 10.1 *Decimal equivalents of four-bit signed binary numbers*

Binary	*Decimal*	*Binary*	*Decimal*
0,000	+0	1,000	−0
0,001	+1	1,001	−1
0,010	+2	1,010	−2
0,011	+3	1,011	−3
0,100	+4	1,100	−4
0,101	+5	1,101	−5
0,110	+6	1,110	−6
0,111	+7	1,111	−7

The binary sign-and-magnitude representations are special cases of the more general signed-magnitude numbers of radix, or base, r. In this more general case the sign digit is again the most significant digit and is given by $s = 0$ for positive numbers and $s = r - 1$ for negative numbers. The notation in this case is

$$N = \pm(a_{n-1} \cdots a_0 . a_{-1} \cdots a_{-m})_r$$
$$= (s, a_{n-1} \cdots a_0 . a_{-1} \cdots a_{-m})_{rsm}$$

For example, the decimal number +256 is represented by

$$+256 = 0{,}256_{10sm}$$

and −256 by

$$-256 = 9{,}256_{10sm}$$

The binary system is, of course, the one we are primarily interested in, since it is the system used in digital circuits.

EXERCISES

10.1.1 Find the decimal equivalents of the signed binary numbers

(a) 1,11010

(b) 0,1011

(c) 1,10011

Ans: (a) −26; (b) 11; (c) −19

10.1.2 Express as signed binary numbers the base r numbers

(a) -29_{10}

(b) 25_8

(c) $-A7_{16}$

Ans: (a) 1,11101; (b) 0,10101; (c) 1,10100111

10.1.3 Express in the form N_{10sm} and N_{5sm} the numbers -256_{10} and -342_5, respectively.

Ans: (a) 9,256; (b) 4,342

10.2 Complements

As we will see, we may subtract a number B from a number A by adding a certain *complement* of B to A. In this section we consider two such complements of binary numbers, namely the *1's complement* and the *2's complement*. In Sec. 10.3 we will see how these complements may be used to subtract by adding.

Let us first consider a positive integer N in base, or radix, r. The *radix complement* of N, which we denote by $[N]_r$, is defined by

$$[N]_r = r^n - N \tag{10.3}$$

where n is the number of digits in N and all three numbers are expressed in base r. For example, in the case of base 10, the radix complement is called the *10's complement* and is given by

$$[N]_{10} = 10^n - N$$

To illustrate, the 10's complement of 35 (for which $n = 2$) is given by

$$[35]_{10} = 10^2 - 35 = 65$$

The *diminished radix complement* of the positive, base r integer N, denoted by $[N]_{r-1}$, is defined to be

$$[N]_{r-1} = r^n - N - 1 \tag{10.4}$$

where again n is the number of digits in N. In the case of base 10, the diminished radix complement is called the *9's complement* and is given by

$$[N]_{10-1} = 10^n - N - 1 \tag{10.5}$$

The 9's complement of 35 is

$$\begin{aligned}[35]_{10-1} &= 10^2 - 35 - 1\\ &= 99 - 35\\ &= 64\end{aligned}$$

The 9's complement of a decimal integer is particularly easy to obtain, because it is the result of subtracting each digit in the number from 9. This is clear from the example just considered and from the general case of Eq. (10.5), since $10^n - 1$ is an n-digit number whose digits are all 9's. As another example, the 9's complement of 257 is 742, obtained from 999 − 257.

We note from Eqs. (10.3) and (10.4) that the radix complement may be obtained by adding 1 to the diminished radix complement. That is, we have

$$[N]_r = [N]_{r-1} + 1 \tag{10.6}$$

As an example, the 10's complement of 257 is 743, which is the 9's complement 742 plus 1.

In digital circuits we are interested in base 2, or binary, numbers. In this case we refer to the radix complement as the *2's complement*, given by Eq. (10.3) as

$$[N]_2 = 2^n - N \tag{10.7}$$

Also, the diminished radix complement is called the *1's complement*, which we denote by $[N]_1$. Its value, by Eq. (10.4), is

$$[N]_1 = 2^n - N - 1 \tag{10.8}$$

In both Eq. (10.7) and Eq. (10.8), n is the number of bits in N and 2^n and N are expressed in binary.

As an example, $13_{10} = 1101_2$ has a 2's complement given by

$$\begin{aligned}[1101]_2 &= (2^4)_2 - 1101 \\ &= 10000 - 1101 \\ &= 0011\end{aligned}$$

and a 1's complement given by

$$\begin{aligned}[1101]_1 &= 10000 - 1101 - 1 \\ &= 1111 - 1101 \\ &= 0010\end{aligned}$$

In this last example the 1's complement is obtained by subtracting each bit in the number from 1, which is equivalent simply to replacing 0's by 1's and 1's by 0's. This is the case in general, as may be seen in Eq. (10.8), since $2^n - 1$ in binary is a sequence of n 1's. This procedure may be easily performed with digital circuitry, since it consists in complementing each bit in a number to obtain its 1's complement. Other examples are

$$[110010]_1 = 001101$$

and

$$[0101101]_1 = 1010010$$

The 2's complement of a binary number may be obtained by adding 1 to its 1's complement, as indicated in Eq. (10.6) for the general case. Thus we may obtain the 2's complement by replacing 0's by 1's and 1's by 0's and adding 1. An easier way is to note that if the rightmost bit in N is a 1 it will change to 0 (in obtaining the 1's complement) and back to 1 when the 1 is added to obtain the 2's complement. Every other bit will complement. If the rightmost bit is a 0, it changes to 1 and back to 0 in the process, but with a 1 to carry. If the next significant bit is 1 it changes to 0 and back to 1 with the addition of the carry. All other bits complement. If, however, the next significant bit is also a 0 like the rightmost bit, it changes to 1 and back to 0 with 1 to carry. Thus the rightmost 1 and all 0's to its right do not change in going from N to its 2's complement, but all remaining bits complement. Thus we may state the following rule for obtaining the 2's complement of a number:

1. Copy the rightmost 1 and any 0's to its right.
2. Replace all other bits by their complements.

As an example let us find the 2's complement of 101100. The result may be written as

$$[101100]_2$$
$$= 010100$$

The three rightmost bits remain the same, but the others are complemented.

The procedures for finding the 1's and 2's complements are reversible. To find the number N from its 1's complement $[N]_1$ we simply replace the 0's by 1's and the 1's by 0's. That is, we complement every bit. We may subtract 1 from the 2's complement $[N]_2$, resulting in the 1's complement, and then complement every bit to obtain N. However, an easier way is to apply the two rules for finding the 2's complement. Evidently these rules also apply in changing $[N]_2$ to N. As an example, suppose we have

$$[N]_1 = 10110110$$

Then complementing each bit yields $N = 01001001$. If we have

$$[N]_2 = 10110100$$

then $N = 01001100$. These results may be checked by finding their 1's and 2's complements, respectively.

EXERCISES

10.2.1 Find the radix complement of

(a) 351_{10}

(b) 247_8

(c) 1342_5

Ans: (a) 649; (b) 531; (c) 3103

10.2.2 Find the diminished radix complements of the numbers given in Ex. 10.2.1.

Ans: (a) 648; (b) 530; (c) 3102

10.2.3 Find the 1's complement of

(a) 110110

(b) 010110111

(c) 001101110000

Ans: (a) 001001; (b) 101001000; (c) 110010001111

10.2.4 Find the 2's complements of the numbers given in Ex. 10.2.3.

Ans: (a) 001010; (b) 101001001; (c) 110010010000

10.3 Complementary Arithmetic

As was pointed out earlier, we are interested primarily in 1's and 2's complements in order to subtract by adding complements. In this way the digital circuitry may be much less complex because we will need only adders in the arithmetic hardware. In this section we see how complementary arithmetic allows us to dispense with the operation of subtraction.

Suppose we have two positive n-bit binary numbers A and B and wish to find the difference $A - B$. We may write

$$A - B = A - B + 2^n - 1 - 2^n + 1$$

or

$$A - B = A + (2^n - B - 1) - 2^n + 1 \tag{10.9}$$

where it is understood that 2^n is expressed in its binary form $100 \cdots 0$, containing n 0's. Identifying the quantity in parentheses as the 1's complement $[B]_1$ of B, we may write Eq. (10.9) in the form

$$A - B = A + [B]_1 - 2^n + 1 \tag{10.10}$$

If $A > B$, then $A - B - 1 \geq 0$, from which we may write

$$A + (2^n - B - 1) \geq 2^n$$

or

$$A + [B]_1 \geq 2^n$$

Thus in this case, adding A and $[B]_1$ always results in a carry into the 2^n position. By Eq. (10.10) we see that this carry is canceled by the term -2^n, so that A plus the 1's complement of B plus 1 is the difference $A - B$, provided the carry is ignored. The carry that is ignored may be taken as the 1 that must be added to $A + [B]_1$ to obtain the correct answer. For this reason it is sometines called an *end-around carry*. (It is brought from the most significant end around to the least significant end and added.)

As an example, let us find the difference 110111 − 010101 using 1's complement arithmetic. We perform the subtraction by adding the 1's complement of 010101, which is 101010, to 110111 and adding the end-around carry. The work may be displayed as follows:

```
  1 1 1 1 1          Carries
    1 1 0 1 1 1
+   1 0 1 0 1 0
  -------------
    1 0 0 0 0 1
            + 1      End-around carry
  -------------
    1 0 0 0 1 0
```

The difference is therefore 100010, which may be checked by conventional subtraction.

If there is no end-around carry, then $A \leq B$. In the case $A = B$, $A + [B]_1$ is the same as $A + [A]_1$, which is a sequence of n 1's. Adding the 1 indicated in Eq. (10.10) to this sequence results in the binary equivalent of 2^n, which cancels with the -2^n, leaving 0. That is, the carry generated by adding the 1 is ignored.

For $A < B$, we may write Eq. (10.10) in the form

$$A - B = -\{2^n - (A + [B]_1) - 1\}$$
$$= -[A + [B]_1]_1 \tag{10.11}$$

That is, $A - B$ is the negative of the 1's complement of $A + [B]_1$. For example, let us compute 01101 − 11100. The result is

$$-[01101 + [11100]_1]_1 = -[01101 + 00011]_1$$
$$= -[10000]_1$$
$$= -01111$$

We will obtain a better method of handling this case in the next section.

The 1's complement method of subtracting by adding is important because of the ease with which the 1's complement of a number can be obtained. However, the necessity of adding the end-around carry is a disadvantage of the method. We may avoid this by using 2's complements instead of 1's complements, as we will see.

In the case of 2's complements the $+1$ and -1 in Eq. (10.9) are not needed. Removing them, we may write Eq. (10.10) in the form

$$A - B = A + [B]_2 - 2^n \tag{10.12}$$

As in the 1's complement case, if $A > B$, a carry is generated which is canceled by the -2^n term. Thus $A - B$ is the sum of A and the 2's complement of B with the carry into the 2^n place ignored. As an example, let us calculate 110111 − 010101, which was done earlier using the 1's complement method. The work is arranged as follows, using the 2's complement 101011 of 010101:

```
 1 1 1 1 1 1     Carries
   1 1 0 1 1 1
+  1 0 1 0 1 1
  ------------
   1 0 0 0 1 0
```

The carry is ignored, and the answer, as before, is 100010.

If $A = B$, then $A + [B]_2$ results in $100 \cdots 0$, with n 0's. The carry is ignored, as before, since it is canceled by the term -2^n in Eq. (10.12). Thus the answer is $00 \cdots 0$, as it should be. As an example, let $A = B = 110111$, having 2's complement 001001. Then $A - B$ is performed as follows, ignoring the carry:

$$\begin{array}{rl} 1\ 1\ 1\ 1\ 1\ 1\ \ & \text{Carries} \\ 1\ 1\ 0\ 1\ 1\ 1 & \\ +\ \ 0\ 0\ 1\ 0\ 0\ 1 & \\ \hline 0\ 0\ 0\ 0\ 0\ 0 & \end{array}$$

If $A < B$, we may show by rearranging Eq. (10.12) that

$$A - B = -[A + [B]_2]_2 \tag{10.13}$$

That is, the difference is the negative of the 2's complement of $A + [B]_2$. An example of this type is left to the exercises. (See Ex. 10.3.2.)

EXERCISES

10.3.1 Using the 1's complement method, find the differences

(a) 110110 − 101101

(b) 11000101 − 10111001

(c) 101110 − 110001

Ans: (a) 001001; (b) 00001100; (c) −000011

10.3.2 Show that Eq. (10.13) holds and use the result to find 101111 − 110001.

Ans: 000010

10.3.3 Solve Ex. 10.3.1 using the 2's complement method.

10.4 Arithmetic Using 2's Complement Representation

The arithmetic methods of the previous section were restricted to positive numbers. That is, we were finding $A - B$ where A and B were positive. Moreover, there was no provision for including a sign bit in the calculations, so that if $B > A$, we had to attach a minus sign to the result $A - B$. A method that can handle both positive and negative numbers with sign bits, using a so-called *2's complement representation*, will be considered in this section. This method is the one employed by most digital computers. (A similar procedure based on 1's complements could also be developed, but because of the disadvantage of the end-around carry we will not consider it.)

We *represent*, or code, a binary integer N as follows. If N is positive, its representation will be simply its binary equivalent with sign bit 0. Thus in a five-digit representation the decimal number 13 will be represented by 01101. (In this section we are omitting the comma separating the sign and magnitude bits.) If N is negative, its representation is the 2's complement of $-N$. Thus, for example, the representation of the decimal -13 is the 2's complement of the binary equivalent 01101 of $+13$. This is, of course, 10011. Note that the sign bit is 1, which occurs automatically in the complementing process and indicates a negative number. This representation, which we call the *2's complement representation*, is shown in Tab. 10.2 for the decimal integers from -15 to $+15$. The code chosen is a five-bit code, since four bits are necessary for the magnitudes.

Table 10.2 *Five-bit 2's complement representations*

Decimal	*2's complement*	*Decimal*	*2's complement*
0	00000		
1	00001	−1	11111
2	00010	−2	11110
3	00011	−3	11101
4	00100	−4	11100
5	00101	−5	11011
6	00110	−6	11010
7	00111	−7	11001
8	01000	−8	11000
9	01001	−9	10111
10	01010	−10	10110
11	01011	−11	10101
12	01100	−12	10100
13	01101	−13	10011
14	01110	−14	10010
15	01111	−15	10001

The addition of more 0's in front of a binary number does not change its value, as we know. For example, $01101 = 0001101$. In like manner, adding additional 1's in front of the 2's complement representation does not affect it. That is, 10111 and 1110111 both represent -9, since the 1 in the MSB indicates a negative number and the binary numbers they represent (in magnitude) are 01001 and 0001001, both of which are 9 in decimal. (The binary number N represented in Tab. 10.2 is simply the binary number if $N \geq 0$ and is the 2's complement of $-N$ if $N < 0$. In these cases the sign bit is 0 or 1, respectively, in the table.)

To illustrate the use of 2's complement representations, we note that $A - B$ may be performed by finding $A + (-B)$. If we add the 2's complement representations of A and $-B$, the result is the 2's complement representation of $A - B$, which may then be converted to decimal or to binary. If $A > B$, the 2's complement representation *is* the binary result, which will be evident because the sign bit will be 0. If the sign bit in the sum is 1, the result is negative and the decimal or binary equivalents must be decoded. In the computer the result is held in 2's complement representation and decoded when it is read out.

The procedure works because the arithmetic is simply 2's complement arithmetic. In finding $A - B$ for both A and B positive, the representation of A is its binary equivalent and that of $-B$ is the 2's complement of B, as was the case in the previous section. In this case, for $A > B$, we must ignore the carry as before.

To illustrate the procedure, let us calculate 15 − 12, which is done by adding the 2's complement representations of 15 and −12, as follows:

$$\begin{array}{rl} & C_{s+1},\ C_s \\ 1\ 1\ 1\ 0\ 0 & \text{carries} \\ 0\ 1\ 1\ 1\ 1 & \\ +\quad 1\ 0\ 1\ 0\ 0 & \\ \hline 0\ 0\ 0\ 1\ 1 & \end{array} \tag{10.14}$$

The result 00011 is the 2's complement representation of +3, which is the answer. Note that the sign bit is 0, indicating a positive number.

In Eq. (10.14) we have identified a *sign bit carry* C_s and a *carry-out of the sign bit* C_{s+1}, both of which are 1. As in 2's complement arithmetic, we have ignored the carry-out bit C_{s+1}.

If arithmetic using 2's complement representations is to be valid in every case and not just in the case of the subtraction of one positive number from another, then we must modify the disposal of the carry bit C_{s+1}. For example, let us perform the addition 5 + 3 of two positive decimal numbers. Adding their 2's complement representations, we have

$$\begin{array}{rl} 0\ 0\ 1\ 1\ 1 & \text{Carries} \\ 0\ 0\ 1\ 0\ 1 & \\ +\quad 0\ 0\ 0\ 1\ 1 & \\ \hline 0\ 1\ 0\ 0\ 0 & \end{array} \tag{10.15}$$

The result 01000 is the 2's representation of 8. However, there was no carry of 1 to be ignored. The answer is obviously correct, since in the addition of positive numbers we are merely adding binary equivalents, which are *defined* to be 2's complement representations. Note that in this example $C_s = C_{s+1} = 0$, and this, as we will see, is the key to the correctness of the result.

As another example, let us find 15 + 12 by adding their representations as follows:

$$\begin{array}{rl} 0\ 1\ 1\ 0\ 0 & \text{Carries} \\ 0\ 1\ 1\ 1\ 1 & \\ +\quad 0\ 1\ 1\ 0\ 0 & \\ \hline 0\ 1\ 1\ 0\ 1\ 1 & \end{array} \tag{10.16}$$

Here, $C_{s+1} = 0$ and $C_s = 1$, and we have not ignored C_{s+1} but have added it in as the MSB. This is equivalent to ignoring it as far as the magnitude of the answer is concerned, but we notice that ignoring C_{s+1} altogether leaves a 1 in the MSB or sign bit, indicating a *negative* answer for the sum of two positive numbers. Moreover, note that the answer 011011, the binary equivalent of 27, is correct, but that it has five magnitude bits instead of four. Thus an *overflow* has occurred. We cannot write 27 as a four-bit number. Obviously, we are merely adding binary numbers, and both the overflow and the use of the carry bit are correct.

As still another example, let us find $-12 - 14$, which is performed, using 2's complement representations, as follows:

$$\begin{array}{rl} 1\,0\,0\,0\,0 & \text{Carries} \\ 1\,0\,1\,0\,0 & \\ +\;\; 1\,0\,0\,1\,0 & \\ \hline 1\,0\,0\,1\,1\,0 & \end{array} \tag{10.17}$$

Again, we have used $C_{s+1} = 1$ as the sign bit (otherwise, adding two negative numbers would yield a positive result). Also, there is an overflow (five digits are required to determine the magnitude). The result 100110 is the 2's complement representation of -26, which is correct.

In all these examples we have ignored the carry C_{s+1} if $C_{s+1} = C_s$, as in Eqs. (10.14) and (10.15), and we have used C_{s+1} as the MSB, or sign bit, if $C_{s+1} \neq C_s$, as in Eqs. (10.16) and (10.17). In the latter case an overflow occurs. These are the rules in general, as we may see by considering the ways in which overflows may occur.

Obviously, we cannot have an overflow if a positive and a negative number are added. In the case of two positive numbers their MSB's are both 0 and we may have either $C_s = 0$, in which case $C_{s+1} = 0$, or $C_s = 1$, in which case $C_{s+1} = 0$ as well. In the first case, $C_s = C_{s+1}$, the sign bit is 0 as it should be, and no overflow has occurred. Thus we ignore the carry. In the second case, $C_s \neq C_{s+1}$, and an overflow has occurred, for otherwise the sign bit is 1, which is impossible for the sum of two positive numbers. Thus the carry bit is not ignored but must be taken as the sign bit 0. The remaining bits are one more in number, since they now include the old sign bit position. This is, of course, the nature of the overflow. A similar analysis can be carried out for the sum of two negative numbers.

As a final example, let us find $7 - (-5)$, which in 2's complement representation is $00111 - 11011$. Since we subtract by adding complements, we write $00111 + [11011]_2$, which we arrange as follows:

$$\begin{array}{rl} 0\,0\,1\,1\,1 & \text{Carries} \\ 0\,0\,1\,1\,1 & \\ +\;\; 0\,0\,1\,0\,1 & \\ \hline 0\,1\,1\,0\,0 & \end{array}$$

Since $C_{s+1} = C_s = 0$, we ignore the carry and leave the answer as 01100, the 2's complement representation of +12.

EXERCISES

10.4.1 Using 2's complement representation arithmetic, perform the following operations:

(a) 13 + 2

(b) 13 − 2

(c) 12 + 13

(d) 12 − 13

(e) −11 − 8

(f) 12 − (−3)

(g) 14 − (−6)

Ans: (a) 01111; (b) 01011; (c) 011001; (d) 11111; (e) 101101; (f) 01111; (g) 010100

10.4.2 Find the cases in Ex. 10.4.1 where overflows have occurred and note that in every case $C_s \neq C_{s+1}$.

Ans: (c), (e), (g)

10.4.3 Find the equivalent decimal readouts of the following numbers stored in the memory of a digital circuit:

(a) 0110101

(b) 1011101

(c) 1111011010

Ans: (a) 53; (b) −35; (c) −37

PROBLEMS

10.1 Perform the following arithmetic operations on the binary numbers given in the sign-and-magnitude notation:

(a) 0,1101 + 1,1001

(b) 0,1011 − 1,0110

(c) 0,1001 + 0,0111

(d) 1,1000 − 1,1101

10.2 Express as signed binary numbers the base *r* numbers

(a) -35_{10}

(b) 47_{10}

(c) -236_8

(d) $-3AF_{16}$

10.3 Find the radix complement of

(a) 2751_{10}

(b) 357_8

(c) 1253_6

10.4 Find the diminished radix complements of the numbers in Prob. 10.4.

10.5 Find the 1's complement of

(a) 011101

(b) 10111011

(c) 011011101111

10.6 Find the 2's complements of the numbers in Prob. 10.5.

10.7 Find the binary numbers whose 2's complements are

(a) 10111011

(b) 0110111101

(c) 0011010000

10.8 Perform the following operations using 1's complement arithmetic:

(a) 110111 − 101110

(b) 1111001 − 0110110

(c) 11100011 − 11110100

10.9 Solve Prob. 10.8 using 2's complement arithmetic.

10.10 Perform the following operations using 2's complement representation arithmetic. (The numbers are decimals.)

(a) $17 - 14$

(b) $17 + 14$

(c) $2 - 9$

(d) $-6 - 15$

(e) $-6 - 5$

(f) $12 - (-2)$

(g) $15 - (-9)$

(h) $-11 - (-4)$

11

MICROPROCESSORS

In the previous chapters we were primarily concerned with the basic logic elements used in the construction of digital networks, or *machines*. One of the most important of these in use today is the digital computer. Digital computers employ, in some form, virtually all the elements that we have discussed so far. In fact, technological developments have combined the CPU and memory into a single integrated-circuit package, and these devices are referred to as microcomputers.

The microprocessor is one of the most important technological achievements since the development of the transistor. It was first introduced into the marketplace by Intel Corporation in 1971. The microprocessor is a *central processing unit* (CPU), which is usually implemented in a single integrated-circuit package. When the microprocessor is combined with *memory* and *input/output* (I/O) devices, a microcomputer is formed.

Like all computers, microcomputers, manipulate binary information, or *data*, represented by groups of bits, or words. The number of bits making up a word varies among the different microprocessors; words usually consist of either four, eight, twelve, or sixteen bits. Two commonly used binary definitions are the *nibble* and the *byte*, which represent binary words of four and eight bits, respectively.

Microprocessors are having a dramatic impact on the design of nearly all digital systems. We have already studied methods for interconnecting logic elements to perform specified digital operations. Networks such as these are often referred to as *random logic* networks. Many random logic networks are now being replaced by microcomputers that offer the designer an added dimension of flexibility known as *software*. Software consists of an organized sequence of steps, called a *program*, that the microcomputer executes. Each step in the sequence is called an *instruction*. Software allows the designer to modify the task a microcomputer network performs by merely changing the computer program but not the physical parts, or *hardware*, making up the network. In contrast, changing the task of a random logic network requires modifying its hardware.

Building microcomputers requires both hardware and software design. We will examine the hardware makeup, called *architecture*, of a typical microcomputer in this chapter. We will then consider the software in Chap. 12.

11.1 Microcomputer Architecture

All computers, including microcomputers, are digital machines that satisfy five criteria first set forth by Charles Babbage in 1830. Babbage described a machine which he called the Analytical Engine as a device having the following features:

1. An *input* medium, by which data and instructions can be entered.
2. A *memory*, from which data and instructions can be obtained and in which results can be stored in any desired order.
3. A *calculating* section, which is capable of performing arithmetic and logical operations on any data.
4. A *decision* capability, by which it can select alternative courses of action on the basis of computed results.
5. An *output* medium, by which results can be delivered to the user.

Machines that satisfy these five properties are known as *Harvard* class computers. In addition to these properties, many machines also have the feature that

6. Data and instructions are stored in the same form and in the same memory, each being equally accessible to the calculating section of the machine. In this way instructions can be treated as data, allowing the machine to modify its own instructions.

Computers that have this additional feature are known as *von Neumann* or *Princeton* class machines. Microcomputers are available in both Harvard and von Neumann classes.

The design of most microcomputers is based on four basic building blocks. These are input devices (feature 1), memory (feature 2), a microprocessor (features 3, 4, and perhaps 6), and output devices (feature 5), as shown by the block diagram in Fig. 11.1.

The input devices convert input signals into the proper binary form for the microprocessor. Some typical input devices are keyboards, teletypewriters, analog-

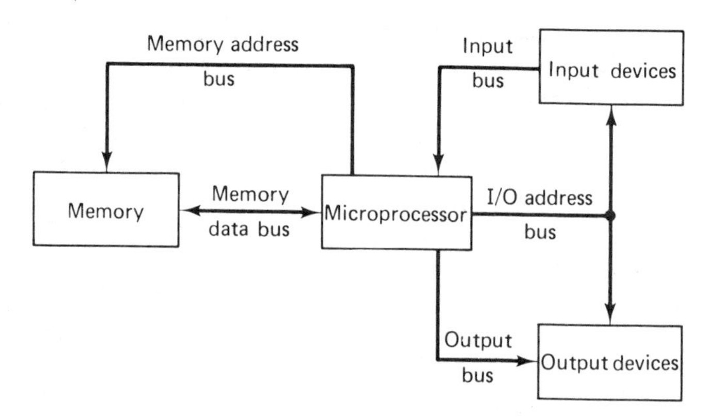

Figure 11.1 *Block diagram of a basic microcomputer*

to-digital (A/D) converters, and cassette tape decks. An *interface* network that transforms the input data into a compatible digital form for the microprocessor is usually necessary. This data is passed to the microprocessor by way of an input bus. The input device is selected by the microprocessor by sending out an address on the I/O address bus.

Memory has the ability to store binary numbers that describe in detail the instructions the computer is to execute. It also stores data (in binary form) that is to be operated on by the computer and eventually output to the users. In performing a memory operation, which is often referred to as *accessing memory*, the microprocessor requests a memory location by sending out an address on the memory address bus. The memory, in turn, either accepts data from the memory data bus into this location (*write operation*) or it sends data from this location to the microprocessor over the memory data bus (*read operation*). The instruction being executed by the microprocessor determines whether a memory access is a read or write operation.

The microprocessor contains a CPU that consists of the circuitry required to access the appropriate locations in memory and interpret resulting instructions. The execution of these instructions also takes place in this unit. The CPU contains an *arithmetic/logic unit* (ALU), a combinational network that performs arithmetic and logical operations on the data, a *control section*, which controls the operations of the computer, and various data registers for temporary storage and manipulation of data and instructions.

Output devices convert the binary output data into a useful form. Examples of these devices include printers, tape punches, cathode ray tube (CRT) displays, and digital-to-analog (D/A) converters. Data is transmitted from the microprocessor to the output devices via an output bus. As in the case of input device selection, the output device required is selected by the microprocessor via the I/O address bus.

In many microcomputer systems, the input and output devices, as well as the memory, share one or more buses, as shown in Fig. 11.2. In this configuration

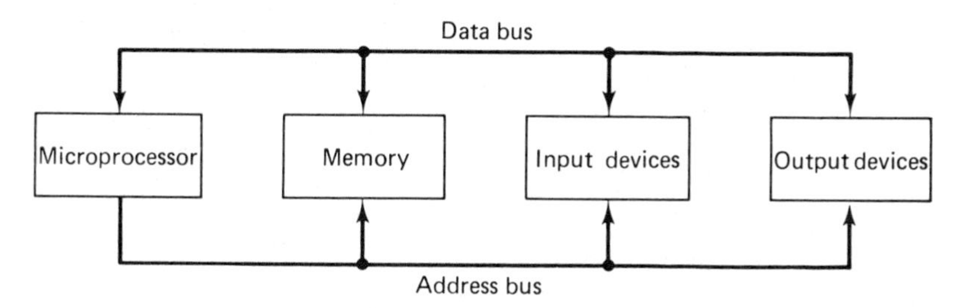

Figure 11.2 *Block diagram of a microcomputer system sharing data and address buses*

the memory data bus, the input bus, and the output bus share the single data bus. Similarly, the memory address bus and the I/O address bus are implemented into a common address bus.

The microcomputer operates in synchronism with a clock. A number of clock cycles are required to accomplish the tasks specified by one instruction. The execution of one instruction is called an *instruction cycle*, and an instruction consists of one or more *machine cycles*. During a machine cycle, the following subcycles are performed:

1. Fetch Cycle.
   (a) The CPU provides the address of an instruction via the address bus.
   (b) The address is decoded by the memory, and the instruction is read from memory into the CPU via the memory data bus.
2. Execute Cycle.
   The instruction is decoded by the CPU, and the requested operation is performed.

A typical timing sequence for these cycles is shown in Fig. 11.3.

During an instruction cycle, two types of words are processed. These are *instruction words* and *data words*. Let us first consider the operations that take place on instruction words. The flow of an instruction word is shown in Fig. 11.4. In this

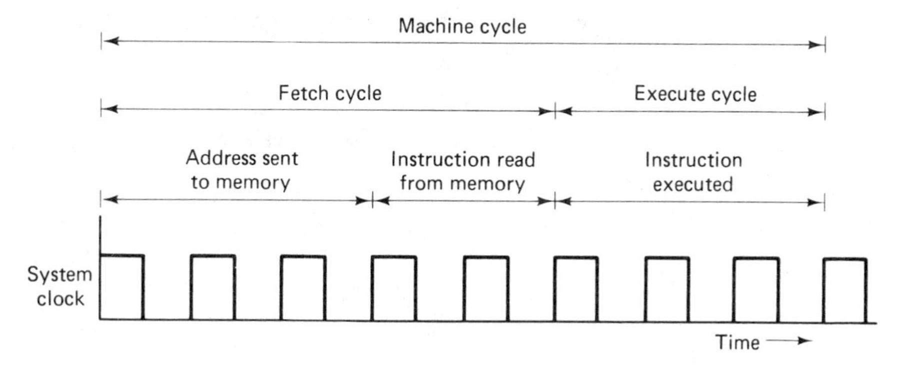

Figure 11.3 *Typical machine cycle*

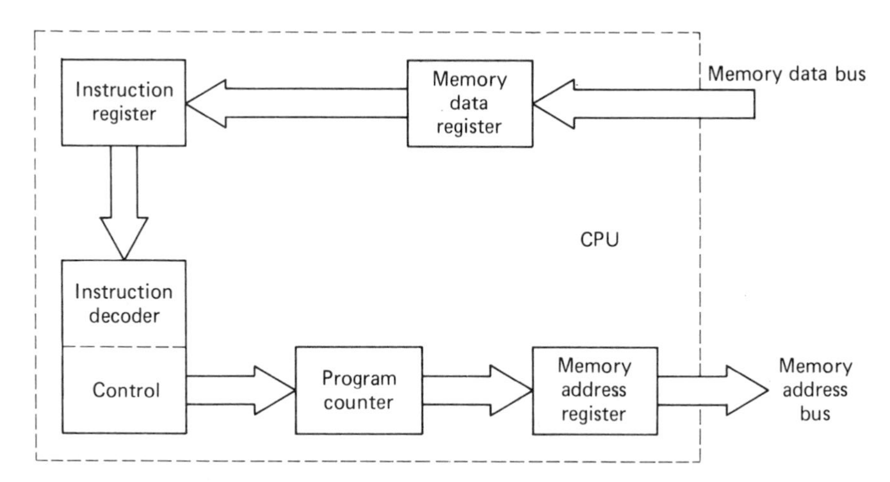

Figure 11.4 *Flow of an instruction word*

figure are shown several registers that are required to process an instruction word. The memory data register provides temporary storage for data received from memory via the memory data bus. The instruction register stores instructions while they are decoded by the instruction decoder and executed by the control section of the CPU. The program counter contains the memory address of the next instruction to be executed by the microprocessor, and the memory address register provides temporary storage for memory addresses that are passed to the memory via the memory address bus.

During a machine cycle the following operations take place for an instruction word.

1. At the beginning of a cycle, the content of the program counter is placed in the memory address register.
2. The content of the memory address register is transferred to the memory and decoded to select the instruction.
3. The instruction is read from memory via the memory data bus to the memory data register.
4. The instruction is placed in the instruction register.
5. The instruction is decoded by the instruction decoder.
6. The instruction is executed.
7. The program counter is incremented or reset, according to the instruction executed, so that it contains the address of the next instruction to be executed.

Steps 1–5 and 6–7 represent the fetch and execute cycles, respectively, of the machine cycle.

The execution of an instruction often requires an operation on data. The flow of a data word is shown in Fig. 11.5. Data is input via the data bus from either the memory or an input device. In many microprocessors the data must enter by way of a register called the *accumulator*. The accumulator also functions as the destination of all data operated on by the ALU. After the operations are completed, the data words are output either to memory via the data bus or to an output device via the output bus. As mentioned previously, the data bus and output bus may share a common bus, as shown in Fig. 11.2. All operations are controlled by the control section in accordance with the instruction being executed. All operations on data words take place during an execute cycle.

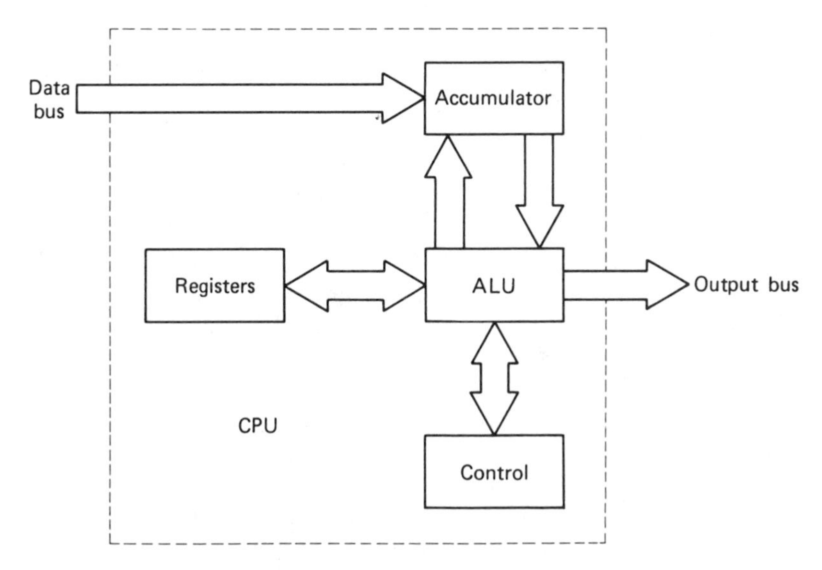

Figure 11.5 *Flow of a data word*

The structure described above does not necessarily conform to any specific microcomputer, but it has been simplified for tutorial purposes. Each microcomputer has its own unique architecture, which combines or expands certain features that have been described.

Next we describe the memory, microprocessor, and several commonly used I/O devices that constitute the microcomputer.

EXERCISE

11.1.1 The first eight-bit microprocessor (INTEL 8008) has a single eight-bit bus on which both addresses and data must be multiplexed. Sketch a block diagram similar to that in Fig. 11.2 for this type of microcomputer.

11.2 Memory

The instructions making up the computer program, as well as data, must be stored and recalled at the appropriate time in order for the computer to perform its designated function. This is the task of the memory. Memory may be broadly divided into two classes: *read-only memory* (ROM) and *read/write memory* (RWM).

Read-Only Memory (ROM)

Read-only memory, or ROM, forms an important part of the memory requirement in most microcomputer systems. The ROM, which we described in Sec. 6.4, is a memory array whose contents, once programmed, are permanently fixed and cannot be altered by the microprocessor using the memory.

The ROM, being a fixed memory, is *nonvolatile*; that is, loss of electrical power or system malfunction does not change the contents of the memory. Therefore it is very useful in situations where the memory values do not change. Microcomputers are commonly used in systems that perform dedicated tasks. In these systems, the computer program does not change and the ROM provides an ideal medium for program storage. Since a loss of power in these systems (due to power failure or an ON/OFF switch) has no effect on the contents of the ROM, the computer program memory remains intact and the system is immediately functional when power is restored.

Most ROMs have the feature of *random access*, which means that the access time for a given memory location is the same as that for all other locations. An example of a nonrandom-access memory is a cassette tape, since the time required to find a stored element depends on the tape position.

A block diagram of a typical ROM is shown in Fig. 11.6. In this figure the memory address decoder is a combinational network that directs the desired one

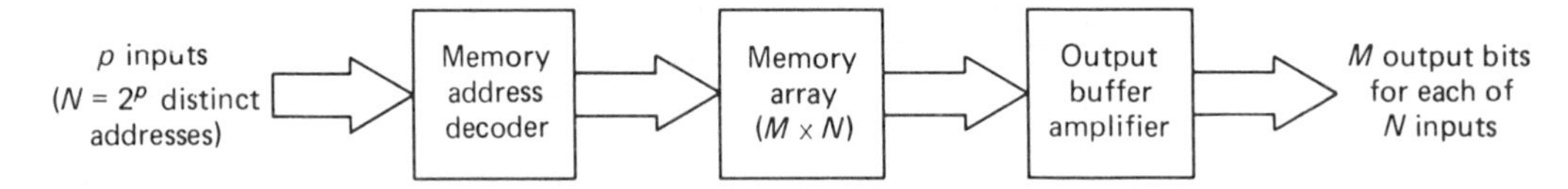

Figure 11.6 *Block diagram of a ROM*

of N possible addresses to the appropriate M bits (word) in the $N \times M$ memory array. The contents of these M bits are then transferred to the output buffer amplifier for use by the system. In order to generate the N possible input addresses, p input terminals ($N = 2^p$) are necessary at the memory address decoder. The $N \times M$ memory array contains the stored bit pattern. The output buffer amplifier provides the necessary electrical drive for the M output bits. ROMs are available in integrated-circuit form having bit storage capacities of tens of thousands of bits.

Read/Write Memory (RWM)

Read/write memory consists of elements from which data can be read and in which data can be written (stored) by the microprocessor. This memory type is used to store data that changes during the operation of the system, such as results of calculations, or for programs that are changed frequently.

Read/write memory for microcomputers is generally an array of flip-flops or similar storage devices, produced as an integrated circuit. Like the ROM, the read/write memory is a random-access memory (RAM). Nevertheless, it is commonly referred to as simply RAM. Hence the term RAM will be used to refer to read/write memory.

The two classes of RAMs are *static* and *dynamic*. A static RAM stores each bit of information in a flip-flop, and this information is retained as long as power is supplied to the circuit. On the other hand, dynamic RAMs are devices in which the information is stored in the form of an electric charge on a terminal of a metal-oxide semiconductor (MOS) transistor. This charge dissipates in a few milliseconds, and the element must be *refreshed* (charge restored) periodically. Dynamic RAMS are important because fewer transistors are required to store a bit, typically three or four, as opposed to six or eight for the static memory. They are also faster than the static RAM and consume less power in the quiescent state. However, the refreshing cycle requires additional circuitry, which is often external, and a certain number of memory elements are required before dynamic memory becomes profitable. Smaller memories are generally static, whereas larger memories are dynamic.

A diagram of an $N \times 1$ static RAM is shown in Fig. 11.7. As in the case of the ROM, p address lines are required to address N words, where $N = 2^p$. The DATA OUT is a single line to output the particular bit that is addressed onto the data bus. The DATA IN is a single line to input the particular bit to be stored in memory. The READ/WRITE is a single line on which a command to READ (output) or WRITE (input) is supplied to the circuit. The CHIP SELECT is a single line employed to disconnect the memory electrically from the output data bus and to inhibit the WRITE circuitry within the RAM. It is used to control the accessing of the chip if more than N words of memory are required.

In forming memories having M bits per word $N \times 1$ RAMs are very useful. If M of these RAMs are connected with their address lines connected in parallel, an $N \times M$ RAM is generated, as shown in Fig. 11.8 for $M = 4$. Memory can then

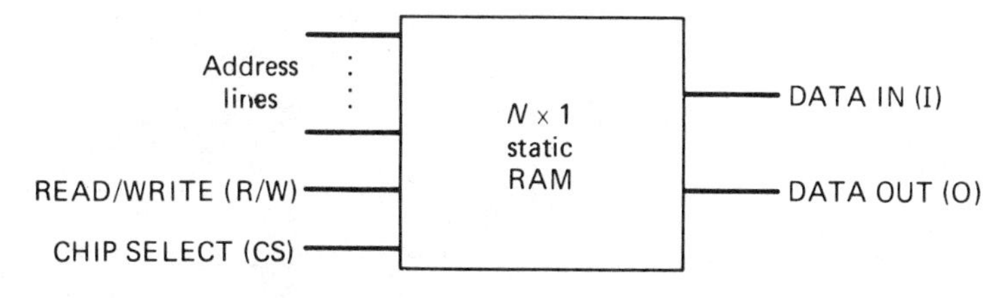

Figure 11.7 *N × 1 static RAM*

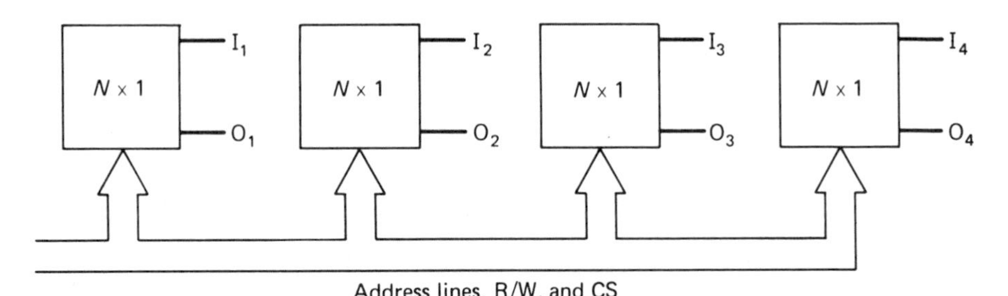

Figure 11.8 *N × 4 RAM using four N × 1 RAMs*

be expanded in $N \times M$ increments by simply adding an $N \times M$ RAM module for each increment desired. For example, three such modules would give $3N$ words of memory having M bits per word.

RAMs that have $N \times M$ bits are also fabricated. In this case, M input lines and M output lines are required on the package. Usually in these circuits, the input and output lines share the same terminals on the package and their data is interpreted by the memory as input or output data, depending on the state of the READ/WRITE line.

Dynamic RAM has the same terminals as those described for the static case, and in addition it has a PRECHARGE line. The PRECHARGE line is used in conjunction with the refresh circuitry to restore the charge in the MOS transistors.

EXERCISES

11.2.1 A 1024×1 RAM is a common memory used in microcomputer systems. How many address lines are required for this RAM?

Ans: 10

11.2.2 How many 1024×1 RAMs are required to form

(a) a 1024×4 RAM?

(b) a 1024×8 RAM?

(c) a 4096×4 RAM?

(d) a 8192×8 RAM?

Ans: (a) 4; (b) 8; (c) 16; (d) 64

11.3 Microprocessor

In our description of microcomputer architecture we saw that the microprocessor contains the CPU, which consists of an ALU, appropriate registers, and control circuitry. The execution of instructions takes place in this unit. The steps involved in executing instructions consist of transferring binary quantities from one register to another and performing arithmetic and logical operations on these quantities.

Registers

A number of important registers are normally included in the CPU. These are used for temporary storage of data and instructions by the machine. The most common registers, including several previously mentioned, are as follows:

(1) Memory Address Register (MAR)

The MAR holds the address of the word to be accessed in memory. The size of the MAR determines the number of words in memory that the microprocessor can directly address. For example, a 16-bit MAR will address $2^{16} = 65{,}536$ words directly.

(2) Memory Data Register (MDR)

The MDR receives and holds a word from memory via the data bus. The size of the MDR is determined by the number of bits in a data word (for example, a one-byte word size requires an eight-bit MDR). A word being written into memory is also held in this register until the write operation is complete. The MDR may be thought of as a buffer in the microprocessor.

(3) Accumulator (AC)

The results of arithmetic and logical operations in the ALU are typically stored in the AC. The AC also serves as one input to the ALU. It accumulates the results of ALU operations, as shown in Fig. 11.9. For example, suppose the AC is initially *cleared* (contains all zeros). If register B contains 0101 initially, then an arithmetic add, which we might denote by the symbol ADD B, in the ALU will result in 0101 being stored in the AC ($0000 + 0101 = 0101$). If 0110 is then placed in register B, an ADD B instruction will cause 1011 to be stored in the AC ($0110 + 0101 = 1011$). In this case, the sum of the numbers placed in register B is accumulated in the AC.

Frequently in microprocessors all input and output data must pass through the AC. The size of the register is equal to the size of a data word.

(4) Program Counter (PC)

The program counter contains the address in memory of the instruction being processed. The set of instructions (or program steps) is normally stored sequentially in memory. For example, the following steps might be desired:

(a) Read a data word from the I/O device into the accumulator (call this instruction I_1).

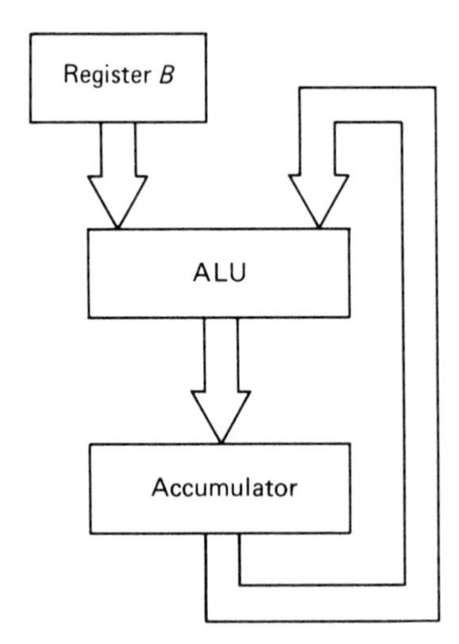

Figure 11.9 *Function of the accumulator*

(b) Add a number to the accumulator (I_2).

(c) Write the result into a certain memory location (I_3), and so on.

The location of these instructions in memory is illustrated in Fig. 11.10. In the normal step-by-step set of instructions, the computer accesses sequential locations in memory. For this reason the program counter is normally incremented during execution of an instruction.

It may be desirable, however, to change the order of the instructions, depending on the data received and the result of operations on this data. This may be accomplished by branching to a nonsequential location in memory (for example, the instruction I_4 in Fig. 11.10). Such a step requires that the program counter be directly set with the location of I_4 rather than being incremented. Much of the power of the computer is derived from this branching capability.

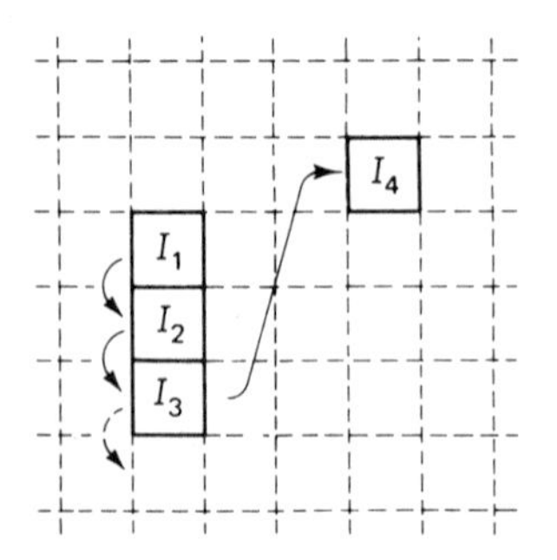

Figure 11.10 *Location of instructions in memory*

(5) Stack

The stack is an array of registers that allows words or addresses to be accessed from the top of this array on a *last-in, first-out* basis. Thus a stack is often termed an *LIFO* array, or it may be referred to as a *pushdown* (or *push–pop*) array. When a word is placed in the stack, all words previously in the stack are moved down one location; this is referred to as a *push* operation. When a word is retrieved from the stack, all words are moved up one location; this is referred to as a *pop* operation. The procedure is illustrated in Fig. 11.11. In this example the stack consists of seven registers. As a word, say, A_5, is pushed onto the stack, it is placed in the top register, and A_1 through A_4 are moved down one register. A pop operation will then recall A_5 and shift A_1 through A_4 up one register in the stack.

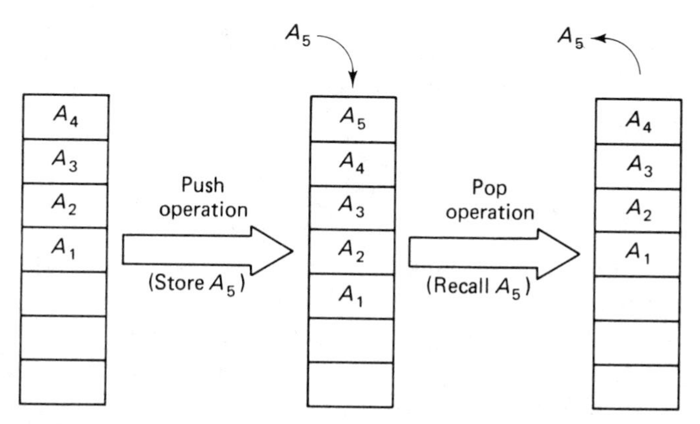

Figure 11.11 *Operation of a stack, illustrating the storage and recall of a word*

Note that it is not possible to recall A_4 before A_5. The elements are automatically accessed on a last-in, first-out basis. The stack is typically used in microprocessors to store return addresses for subroutines and to store the condition of internal registers during the processing of interrupts. Some microprocessors allow only return addresses to be stored in the stack. The stack may be implemented in memory, but this requires a memory cycle in order to access the stack. The operation is made much faster by providing a special set of registers within the CPU for this purpose. An important parameter in this case is the number of registers included in the stack.

Many push–pop arrays are implemented in a manner reminiscent of a stack of papers. Each input is placed on top of the pile. This requires a register, called the *stack pointer* (SP), to hold the address of the top element of the stack. Figure 11.12 illustrates this type of operation. The stack pointer is represented as a three-bit register with the binary representation shown. The stack pointer initially holds 011 (binary 3). This means that the top of the stack is a register 3. A push opera-

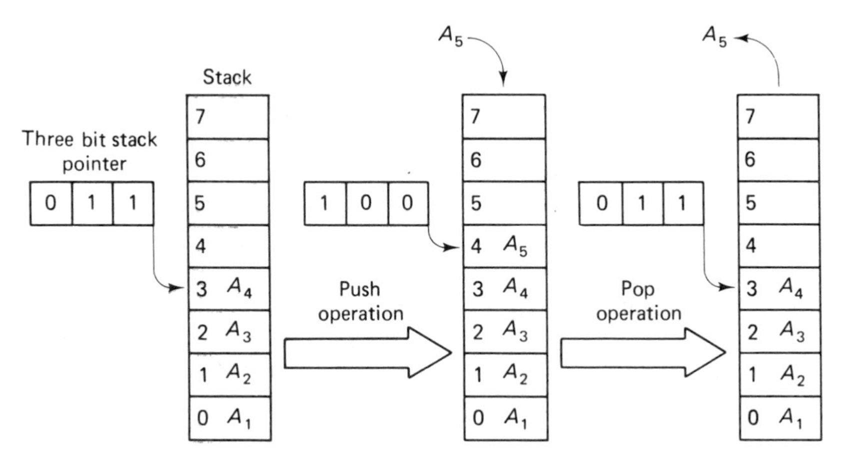

Figure 11.12 *Operation of the stack pointer (SP) for addressing a stack element*

tion loads A_5 into register 4 and changes the stack pointer to indicate that the top of the stack is now located at register 4. A pop operation reverses the procedure.

(6) Scrach-Pad Memory (General-Purpose Registers)

In most microprocessors, a number of registers are provided in the CPU for temporary storage of data and addresses. The number of these registers and the flexibility of accessing them varies greatly from computer to computer. Manufacturer's literature sometimes provides suggestions on their usage that are valuable for the novice.

(7) Instruction Register (IR)

The IR contains the instruction being decoded and executed. Input to the IR is typically from memory as the program steps are read sequentially.

(8) Status Register

A status register consisting of one or more flip-flops, often called *flags*, is used to provide indication of overflow from operations, presence of 0's in the accumulator, sign of a number in the accumulator, and carry (or link) from the accumulator. This information is vital in many arithmetic operations and is often used as a basis for deciding which program step is to be executed.

Instructions

All operations to be performed by the computer must be broken down into a series of individual tasks. The power of the computer results not from the complexity of the individual steps but from the rapidity with which it performs these steps. These individual steps, which we have previously defined, are called *instructions.* The details and number of these instructions vary greatly from computer to computer. The instructions must be encoded into binary because they are stored in memory. Theoretically, any program may be written using the instruction set of any computer, but the length of the program and the time of execution may differ greatly with different instruction sets.

Computer instructions may be divided into five functional categories: (1) transfer of data, (2) control and branch, (3) subroutine linking, (4) operation, and (5) input/output. The instructions may reference data from memory, CPU registers, or simply control the operation of the machine. Those which reference memory require an extra memory cycle to obtain the data and thus require more time to execute. The functions of these categories are as follows:

(1) Transfer of Data

This category includes the moving of data between memory locations or between register and memory. The simple transfer of information is often referred to as MOVE, LOAD, STORE, or EXCHANGE. As the data is transferred, logical or arithmetic operations may be performed on it by the ALU. Operations such as ADD, SUBTRACT, AND, OR, EXCLUSIVE-OR, and COMPARE are common capabilities. Figure 11.13 illustrates the results of an AND operation on the contents of register *B*. Most machines provide the capability of including the link or carry flag in the addition or subtraction, which facilitates multiple-precision arithmetic.

The capability of moving data to and from the stack may be provided. This instruction (PUSH or POP) makes it possible to restore the status of the computer after processing an interrupt.

(2) Control and Branch

These instructions can be classified as *conditional* or *unconditional.* The condition of a status flip-flop (CARRY, ZERO, SIGN, etc.) determines whether or not a conditional instruction will be executed. Some common instructions in this category are

(a) *HALT.* This instruction directs the control section to suspend execution of the program. External action is required for program execution to resume.

(b) *JUMP.* This instruction causes the program counter to be set directly to a nonsequential location in memory for the next instruction.

(c) *CONDITIONAL JUMP.* This instruction is conditional. If the condition is met, the content of the program counter is replaced with the address supplied in the instruction. If the condition is not met, the program counter is incremented and the next instruction is executed. For example, suppose

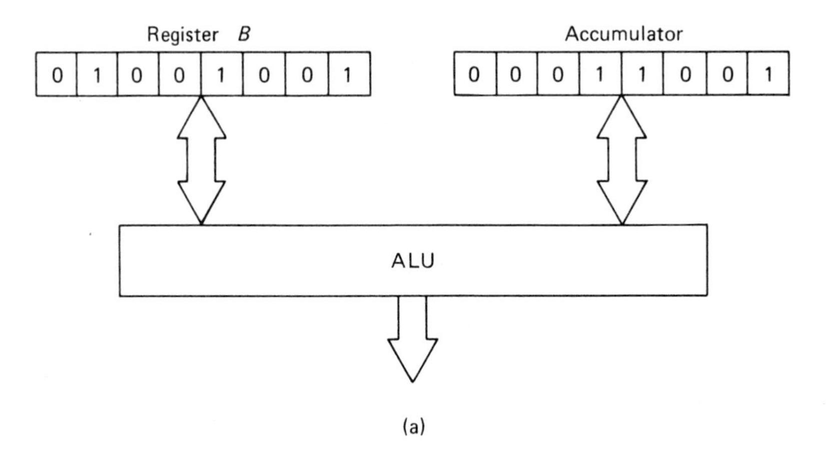

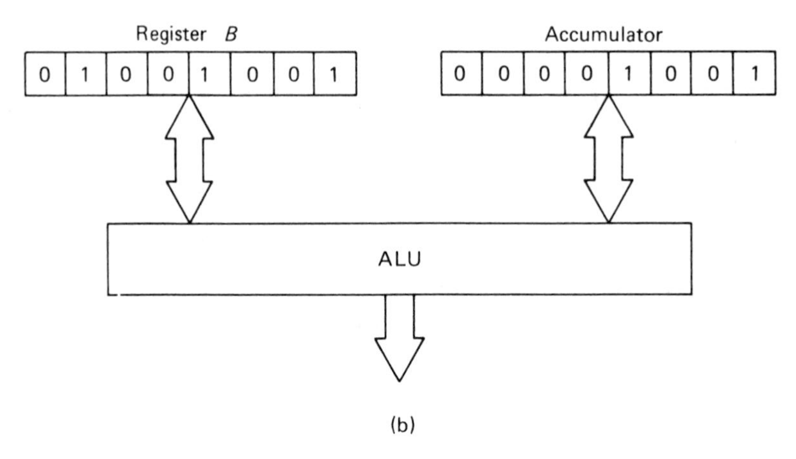

Figure 11.13 *AND operation illustrating the condition of register B and the accumulator (a) prior to the AND operation, and (b) after the AND operation*

we wish to branch to address $4C97_{16}$ if the carry flag is set. The result of this operation is shown in Fig. 11.14. We see that the PC is set to $4C97_{16}$ if the carry flag is set and to the address of the next instruction ($2A36_{16}$) otherwise.

(3) Subroutine Linking

These instructions may also be conditional or unconditional. Some typical members of this category are

(a) *CALL Subroutine.* This instruction causes the program counter to be incremented and the new address to be stored in the stack. This address is the return address for the subroutine when execution is completed. The

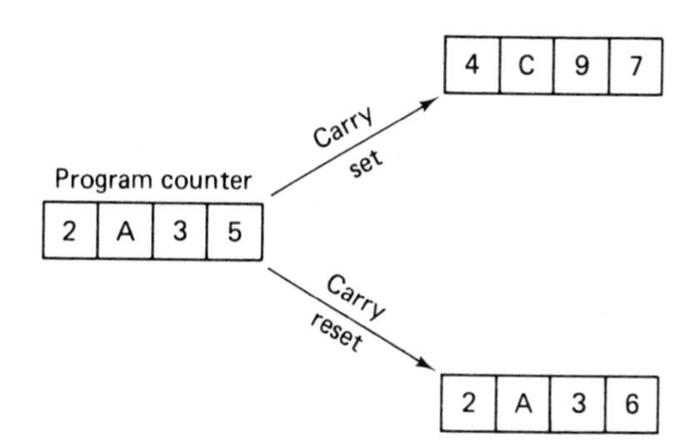

Figure 11.14 *Result of a JUMP IF CARRY SET on program counter*

program counter is then loaded with the address of the subroutine, and execution proceeds. In Fig. 11.15 the program is executing an instruction at location 2600_{16}. The stack contains a return address of $1AA0_{16}$. The instruction calls a subroutine beginning at location $00BE_{16}$. When the call is executed, the program counter is incremented to 2601_{16} and pushed onto the return address stack. Execution of a conditional CALL is dependent on the status of the appropriate conditional flag.

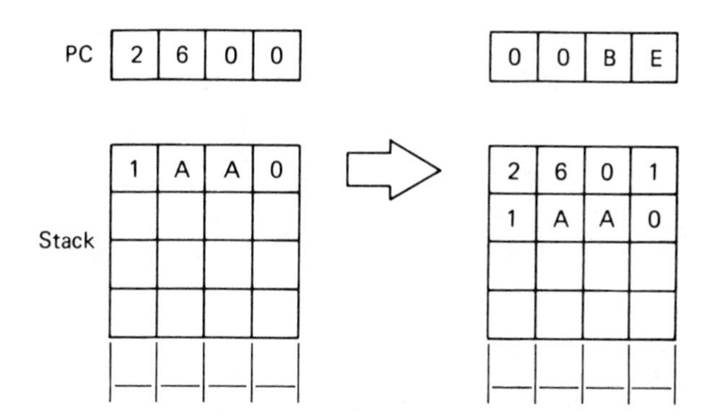

Figure 11.15 *Result of a CALL SUBROUTINE AT ADDRESS 00BE (hexadecimal) on the program counter and the stack. Note: The stack contains the return address*

(b) *RETURN.* RETURN causes the return address in the stack to pop. This address is loaded into the program counter, and program execution proceeds. In Fig. 11.16 a RETURN is executed. Before the execution of the RETURN, the stack holds the return addresses illustrated in Fig. 11.15. The execution of RETURN pops the first number in the stack (2601_{16}) into the program counter. The computer will then execute the instruction at 2601_{16}. Like a conditional CALL, a conditional RETURN is dependent on the status of the appropriate conditional flag.

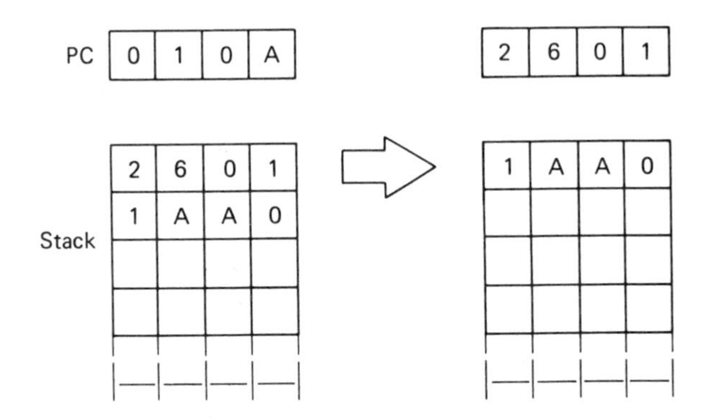

Figure 11.16 *Result of a RETURN instruction. The top address in the stack is transferred to the program counter*

(4) Operation

This category of instructions includes those that perform operations on a particular register or status flip-flop. No data transfer takes place. Typical members of this class are:

(a) *CLEAR.* Zeros are stored in the appropriate register (e.g., the accumulator).

(b) *INCREMENT.* The value stored in the register is increased by 1.

(c) *DECREMENT.* The value stored in the register is decreased by 1.

(d) *COMPLEMENT.* This instruction results in each bit in the appropriate register being complemented (1's complementation).

(e) *ROTATE.* The contents of the accumulator are rotated one bit to the right or left.

For a right rotation, the least significant bit is rotated into the position of the most significant bit, and every other bit is moved one place to the right. Figure 11.17 illustrates the result of a right rotation of the accumulator.

The ROTATE instruction may include the CARRY (or LINK) flip-flop in the rotation. Figure 11.18 illustrates this for a left rotation.

Since the conditional instructions depend on the status of the flags, the

Figure 11.17 *Right rotation of contents of the accumulator*

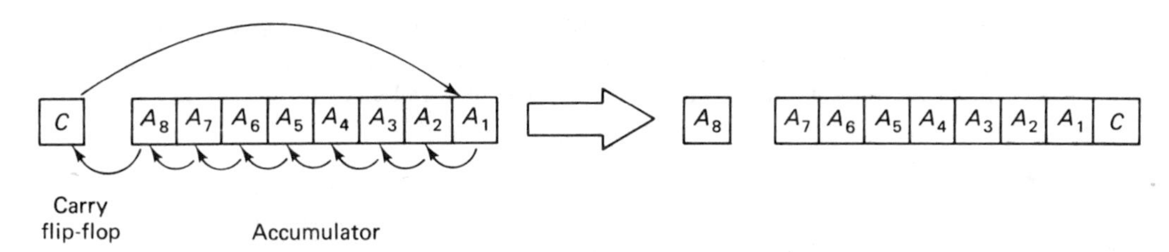

Figure 11.18 *Left rotation through the carry flip-flop*

ROTATE instruction enables a sequence of instructions to be dependent on an individual bit in a data word. For example, before a data word may be read from an I/O device, the computer must determine whether a word is ready to be read. This is done by reading a *status word* from the I/O device. The status word will contain a certain pattern if the data word is ready. Suppose a 1 in the LSB is used to indicate this status. A right rotation may now be performed to shift this bit to the carry flip-flop, and then a conditional branch to an input instruction may be performed based on the value of the carry flip-flop.

(f) *RESET CONDITION FLIP-FLOP.* This instruction clears the appropriate flag.

(g) *DECIMAL ADJUST ACCUMULATOR.* The addition and subtraction instructions are frequently implemented using 2's complement arithmetic. If the data being processed is in BCD, then the resulting addition will be erroneous if the sum of two digits is greater than 9. The instruction logically examines the result for this condition (> 9) and performs a correction by adding 6 (see Sec. 8.3). Inclusion of this instruction in the repertoire of a computer is valuable when the input/output data is in BCD.

(5) Input/Output

This category contains two fundamental instructions:

(a) *INPUT.* This instruction causes the contents of the I/O bus to be loaded into the accumulator.

(b) *OUTPUT.* This instruction causes the contents of the accumulator to be placed on the I/O bus.

Machines with large instruction sets frequently combine two or more of these operations into one instruction. The utility and effectiveness of these instructions are strongly dependent on the particular application.

Addressing Modes

An instruction word must convey the operation to be performed (*operation code*) and the address of the memory location or registers containing the data on which the operation is to be performed (*operand*). An n-bit instruction may be divided into three basic parts: (1) an operation code, (2) an address mode, and (3) an operand address, as shown in Fig. 11.19. The number of bits in each of these parts varies from microprocessor to microprocessor.

Operation code (m bits)	Address mode (n bits)	Address (p bits)

Figure 11.19 *Instruction format*

The instruction length depends on the machine and the operation being performed. An eight-bit instruction format would allow only $2^8 = 256$ possible combinations of operations and addresses. This is obviously inadequate if a reasonable-sized memory is to be accessed. For this reason two-and three-byte instructions are frequently used for memory access. Such an instruction is 16 or 24 bits long. In most cases, one byte is used to represent the operation code and address mode portions of an instruction. The number of bits used for each of these and their relative locations within the byte vary from processor to processor.

The address mode and operand part of the instruction combine to indicate the location in which the operand is stored. There are numerous modes of addressing the operand. The most important for microprocessors include direct, indirect, relative, indexed, and immediate addressing. The address mode portion of the instruction specifies how the address is to be interpreted. These addressing modes are defined as follows:

(a) *Direct Addressing.* With direct addressing, the address of the operand is specified directly in the instruction. For example, consider the instruction "add the contents of memory at address 032 to the accumulator." Such an instruction could be symbolized by

ADD 032

The operand address (032) is directly specified in this instruction (Fig. 11.20). This is a common form of addressing used in microcomputers. Direct addressing usually requires multiword instructions in four- or eight-bit microprocessors.

(b) *Indirect Addressing.* In this mode the instruction provides the address at which the address of the operand is to be found. Figure 11.21 illustrates this type of addressing. The data to be operated on is 301, located in memory location 543. Indirect addressing allows a large block of memory to be addressed by a single-word instruction. In microprocessors a form of addressing called *register indirect addressing* is commonly used. The address is stored in one or more registers within the CPU. In most cases, this architecture allows any location in memory to be addressed with a single-word instruction.

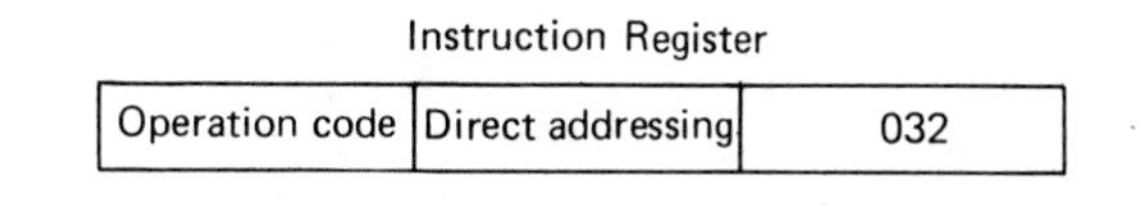

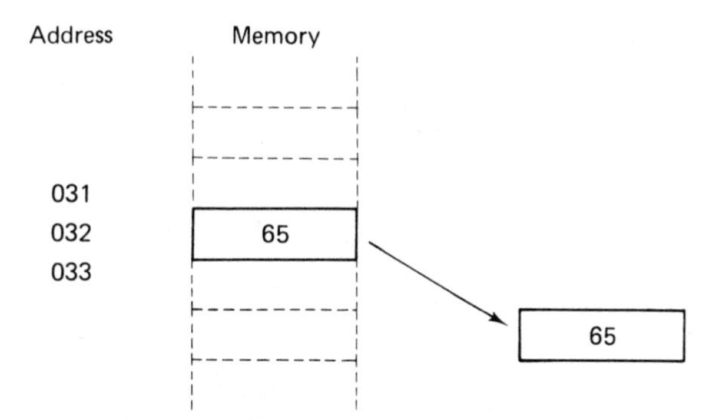

Figure 11.20 *Direct addressing*

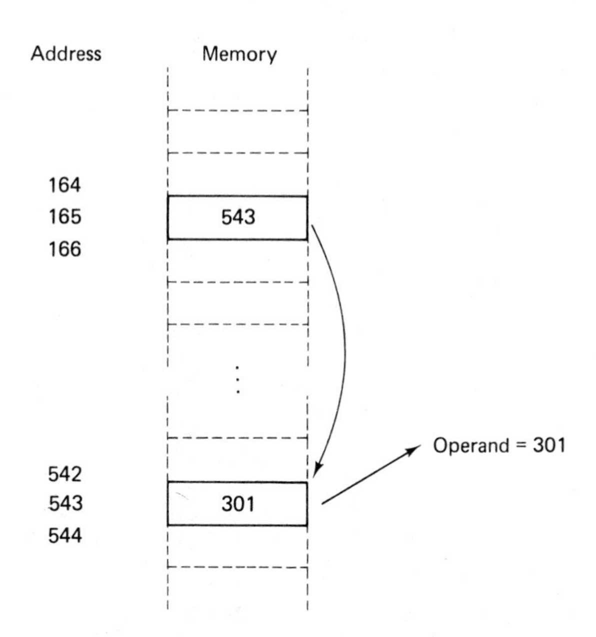

Figure 11.21 *Indirect addressing*

(c) *Relative Addressing.* In relative addressing the address is specified by its relation to the program counter. In this mode the address specified in the instruction is added to the number in the program counter to obtain the address of the operand. For example, if the address in the instruction is 11 and the program counter contains 124, then the address of the operand will be 11 + 124 = 135 (Fig. 11.22). The use of relative addressing simplifies the transfer of programs to different areas of memory.

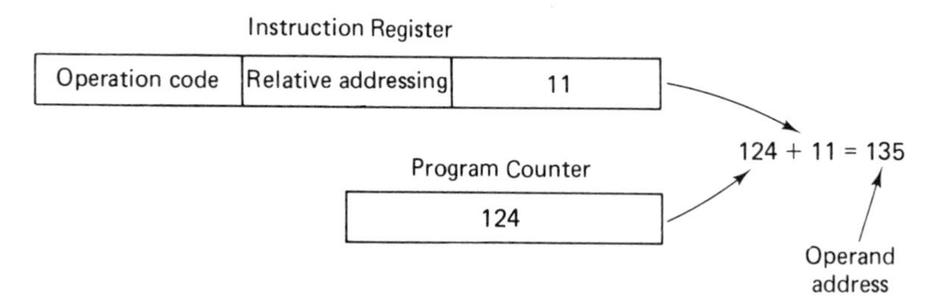

Figure 11.22 *Relative addressing*

Microcomputer memory is frequently structured into *pages*. A page may consist of 256 words of memory and is sometimes located on a single IC. A page structure divides the memory into small blocks. The use of paging reduces the necessity for multiword memory reference instructions. In conjunction with a memory page structure, a form of relative addressing called *page relative addressing* is frequently used. In page relative addressing, an operand address given in the instruction is interpreted as a location on the same page of memory addressed by the program counter. In *page-0 relative addressing*, the operand address refers to a location on page 0 of the memory, regardless of the program counter contents.

(d) *Indexed Addressing.* This mode is similar to relative addressing. However, the address specified in the instruction, is relative to a prespecified register other than the program counter. This register is called the *index*. The address given in the instruction is added to the contents of the index register to determine the address of the operand. In Fig. 11.23 register B is the index register. Since the address mode is indexed, the number stored in register B (341) is added to the operand address (32) to obtain the location of the operand (373).

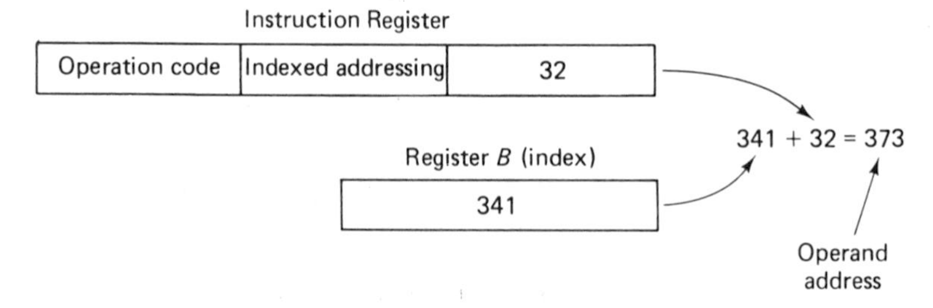

Figure 11.23 *Indexed addressing*

Indexed addressing is valuable in programs involving tables or arrays of numbers. The address of the first element of the table may be stored in the index register, and all other elements in the table may be addressed in relation to the first element.

(e) *Immediate Addressing.* In this mode the data is given directly in the instruction. For example, consider the instruction "immediately move to the accumulator the number 7." Such an instruction could be symbolized by

MVI A, 7

The operand (7) is specified directly in the instruction. The operand (7) in this instruction is located in the memory byte directly following the memory location containing the MVI instruction.

EXERCISES

11.3.1 In Fig. 11.13 the AC contains 75_{16} and register B contains 31_{16}. Find the contents of the AC following

(a) an OR operation

(b) an Exclusive-OR operation

(c) an ADD operation

Ans: (a) 75_{16}; (b) 44_{16}; (c) $A6_{16}$

11.3.2 In Fig. 11.18 the AC contains 87_{16} and the CARRY is reset. What is the content of the AC and the CARRY following four ROTATE LEFT THROUGH CARRY instructions?

Ans: 74_{16}; 0

11.3.3 In Fig. 11.12 the SP contains 010. What is the SP and the stack content following

(a) two successive CALL subroutines

(b) two successive RETURNs

(c) a CALL subroutine and two successive RETURNs

Ans: (a) 100, A_5; (b) 000, A_1; (c) 001, A_2

11.4 Input and Output Devices

Input and output devices are often referred to as *peripherals*. Generally these devices are relatively small in size and can be placed on a table top or rack-mounted in a cabinet. In contrast to peripherals employed in large computer systems, microcomputer peripherals are often at the lower end of the performance scale. Peripheral selection is important because in many instances, the cost of these devices exceeds that of the microcomputer system. In this section we describe several peripherals that are in common use.

Paper Tape Readers/Punches

Paper tape is often used in microcomputer systems as an economical method of recording programs and data. Readers are available in low-speed to high-speed versions that read from 10 to 1000 characters per second. Punches, on the other hand, punch from 10 to 150 characters per second.

Paper tape varies in width from $\frac{1}{2}$ to 1 inch, having five- to eight-bit characters, respectively. The bits are represented by perforations in the tape at each bit site. Usually a hole represents a 1, and the absence of a hole represents a 0. A commonly used paper tape is the eight-level tape shown in Fig. 11.24. It has eight lateral bit sites, which are called *characters* or *frames*. The bits within each character are called *levels* or *channels*. The row of holes positioned between levels 3 and 4 are necessary for the mechanical sprocket drive to advance the tape.

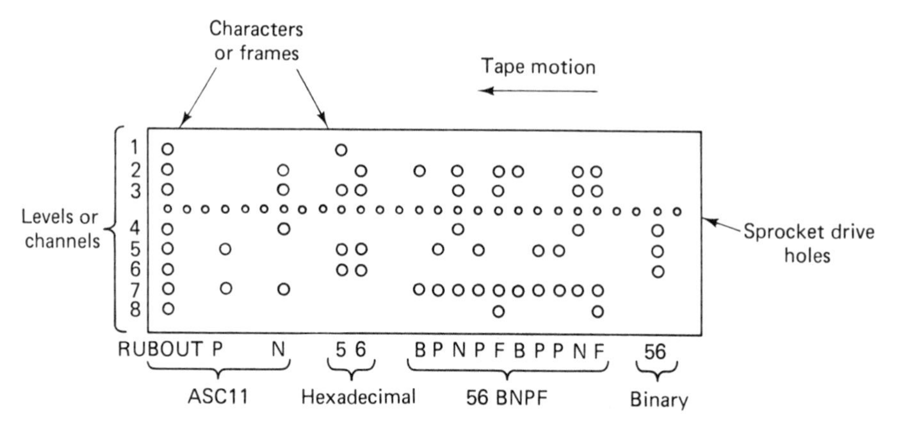

Figure 11.24 *Eight-level paper tape*

Many codes are available for representing data on the tape. In Fig. 11.24 the seven-level ASCII code (Sec. 8.4) is shown for the characters RUBOUT, P, N, 5, and 6. Note that these characters are represented by levels 1 through 7. Level 8 can be used as a parity bit (*lateral parity*) for each character. In Fig. 11.24 level 8 represents an even parity.

Numerous formats are used for representing data on paper tapes. A common one is the *BNPF* format. In this representation the ASCII character B is used to designate the beginning of a data word. ASCII characters P and N follow to represent 1's and 0's, respectively. An ASCII F denotes the end of the data word. With BNPF tapes, lateral parity can be employed. Error detection using this format is extremely good. Tapes can also be punched directly from a standard ASCII keyboard. The efficiency in terms of the number of characters required to represent a given amount of data is poor. In Fig. 11.24 we can see that at least ten ASCII characters are required to represent the decimal number 56.

Other popular representations are the *binary* and *hexadecimal* formats. In the binary case, the data is represented by eight binary characters per frame. Since level 8 is used in this representation, lateral parity is not permitted. In the hexadecimal format, the ASCII codes for the hexadecimal numbers 0 through F are employed to represent each character. The use of the ASCII code permits a lateral parity check for error detection. It also allows tapes to be generated directly from a standard ASCII keyboard, a feature not possible for a binary tape. Binary and hexadecimal formats are shown in Fig. 11.24 for the decimal number 56. Notice that the hexadecimal representation requires twice as many characters as the binary case for a given number of data words.

Tapes for machine-language programs (object tapes) are usually in binary or hexadecimal formats. They are read into the computer in discrete *records* or *blocks*. A typical block format includes a start character, record-length characters, load-address characters, data characters, and a check-sum character. The start character indicates the beginning of a block. The record-length characters are used to notify the computer (via the loader program) of the actual number of data bytes in the record. Load-address characters provide the starting memory address into which the data is to be loaded. The data characters are loaded into memory, and the check-sum character is used as a longitudinal parity check. The check-sum character contains the sum of all bytes in the record evaluated modulo 256 (that is, the carry bit in the sum is neglected). The loader program performs this sum during the loading process and compares the result to the check-sum. If the two are not equal, an error has occurred during the loading procedure. Most object tapes consist of many blocks. A typical block contains 256 data bytes.

Teletypewriters

Teletypewriters are commonly used, general-purpose I/O devices. One very popular, economical type is the ASR-33 (automatic send–receive teletypewriter set) manufactured by the Teletype Corporation. This device offers a keyboard that transmits seven-level ASCII characters, a typing unit for producing hard copy (printed output), a paper tape reader, and a paper tape punch. The device is a low-speed unit that transmits or receives data at a rate of ten characters per second.

The ASR-33 can be operated in either the LINE or the LOCAL mode. In the LOCAL mode, the unit does not communicate with the microcomputer. Characters entered from the keyboard are typed. If the paper tape punch is turned on, keyboard characters are punched on an eight-level tape in the ASCII format described previously. Parity options available include even, mark, and space parities. The contents of a tape inserted into the reader are typed and also reproduced on tape if the punch unit is activated.

In the LINE mode, the ASR-33 communicates with the computer. Keyboard characters are transmitted as serial ASCII characters. Serial ASCII characters that are received from the computer are typed or perform special functions such as line feed and carriage return. In addition, all 256 binary combinations may be punched, read, and transmitted from paper tape.

Communication between the ASR-33 and a microcomputer is performed in a serial code. The code consists of eleven elements per character in which a 0 and 1 are called a *space* and *mark*, respectively. The first element is a start bit (always a space). This is followed by eight elements (bits), which comprise the character being transmitted. The final two elements are stop bits (always marks). The start and stop bits or pulses are used for synchronization purposes. Each element or bit within the code requires 9.09 milliseconds, which results in a total of ten characters per second, or a 110 bits per second (*baud* rate). A timing diagram is shown in Fig. 11.25.

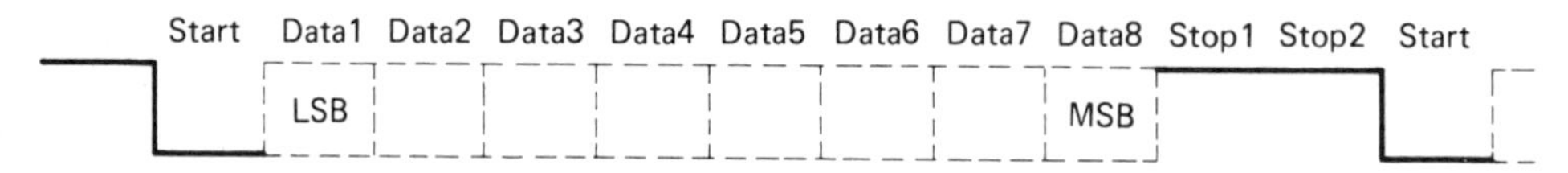

Figure 11.25 *Timing diagram of a 110-baud serial character*

CRT Terminals

The CRT (cathode ray tube) terminal is an excellent I/O device in cases where hard copy is not necessary. This device, which has a keyboard for data entry and a cathode ray tube for data display, permits a variety of rates for transmitting and receiving data. The rates are usually user-selectable and normally can be set at values from 10 to 960 characters per second, which represent baud rates of 110 to 9600.

The CRT terminal, like the ASR-33 teletypewriters, can be operated in a LINE or a LOCAL mode. In the LINE mode, the device communicates with a microcomputer using a serial ASCII code. At data rates of 10 characters per second, the serial code is identical with that described for the ASR-33. At higher rates, however, only one stop bit is normally employed, and the resulting time per bit is reduced in inverse proportion to the baud rate. For example, in transmissions occurring at 9600 baud (960 characters per second since there are 10 bits per character), the individual bit time is $\frac{1}{9600} = 0.104$ milliseconds.

Magnetic Tape Cassettes

Magnetic tape cassette units provide a substitute for paper tape readers and punches as a means of data manipulation and storage. Cassettes have the advantages of low cost, compact size, ease of loading, and superior editing capability. However, a disadvantage of cassettes is their relatively high error rate. Depending on the recording technique this rate can approach one bit per 10^6 recorded, which represents five to ten errors per cassette.

A popular cassette is the Philips cassette, which consists of two flangeless reels mounted in a plastic package having approximate dimensions of $2\frac{1}{2} \times 4 \times \frac{3}{8}$ inches. The cassette may contain 300 feet of oxide-coated Mylar tape having a width of 0.15 inches. Access to the tape is obtained through an aperture in the edge of the container. Numerous mechanisms have been designed for driving the tape. The simpler designs use friction drives, in which the tape is driven between a roller and a capstan. More sophisticated designs actually remove the portion of the tape being read from the cassette.

Cassette drives are available using either continuous recording or incremental recording techniques. Both methods usually record serially with one or two *tracks* on the tape, where a track is analogous to a channel or level of a paper tape. In continuous recorders, serial groups of eight to sixteen bits form *characters* or *words*. Groups of words (two to several thousand) form *blocks* or *records*, and groups of records (two to several hundred) constitute *files*. Records are separated by *interrecord gaps*, which are typically 0.5 to 1 inch long. Continuous recorders can record up to 700,000 characters per cassette, depending on the number of interrecord gaps and the recording density (bits per inch).

Incremental recorders read and record on the tape by starting and stopping the tape for each bit or character. This eliminates the need for large interrecord gaps, but it has the disadvantage of a lower recording density due to the mechanical inertia of the tape-drive mechanism in starting and stopping. The cassette storage capacity is typically 50,000 to 100,000 characters per cassette.

Important characteristics of cassette tapes include cassette capacity, record file length, recording technique, number of tracks, recording density, error rate and detection capability, interrecord gap length, and character transfer rates. industrial standards for cassettes have been established by the European Computer Manufacturers Association (ECMA) and the American National Standards Institute (ANSI). However, most manufacturers prefer to design units for the broadest of user markets and do not necessarily adhere to these standards.

EXERCISES

11.4.1 Sketch the tape format in Fig. 11.24 for the case of odd parity.

Ans: The eighth level of all quantities is complemented except for the binary case.

11.4.2 How many characters can be transmitted serially in 5.5 seconds at

(a) 110 baud?

(b) 300 baud?

(c) 2400 baud?

(d) 4800 baud?

Ans: (a) 55; (b) 165; (c) 1320; (d) 2640

PROBLEMS

11.1 Sketch a diagram similar to that in Fig. 11.8 for an eight-bit memory designed to store 2048 words. The RAMs to be used in implementing the memory are 1024×4 RAMs.

11.2 A microcomputer has the following data stored in its registers and memory:

	Memory Address	*Memory Content*
MAR $= 4762_{16}$	$10A1_{16}$	$B7_{16}$
MDR $= 2A_{16}$	$10A2_{16}$	$C4_{16}$
AC $= 71_{16}$	$10A3_{16}$	26_{16}
REG B $= 47_{16}$	$10A4_{16}$	33_{16}
PC $= 2B3D_{16}$	$2B3E_{16}$	91_{16}
SP $= 10A4_{16}$		
CARRY FLAG $= 1$		

(a) How many words of memory can be directly addressed by this microcomputer?

(b) What is the word size?

(c) Find the content of the AC if a ROTATE LEFT through the CARRY followed by an ADD B is executed.

(d) What is the content of the CARRY in (c)?

11.3 For the data of Prob. 11.2, a SUBROUTINE CALL to address 2744_{16} is executed. Determine

(a) The next value of the PC

(b) The next value of the SP

(c) The new value stored on the STACK

11.4 In Prob. 11.2, a RETURN instruction is executed. Find

(a) The new value of the PC

(b) The new value of the SP

(c) The new value of the STACK

11.5 Show that a ROTATE RIGHT through the CARRY instruction is equivalent to dividing the content of the AC by 2 with the remainder in the CARRY. Assume that the CARRY is initially cleared.

11.6 Show that a ROTATE LEFT through the CARRY is equivalent to multiplying the content of the AC by 2 with the carry in the CARRY. Assume the CARRY is initially cleared.

11.7 A microcomputer has the following data stored in its registers and memory:

	Memory Address	*Memory Content*
PC = 1074_{16}		
AC = $1D_{16}$	$2B47_{16}$	59_{16}
REG B = 62_{16}	$476B_{16}$	24_{16}
REG C = $2B47_{16}$	$476C_{16}$	$D1_{16}$
	⋮	
	$47CD_{16}$	BC_{16}
	⋮	
	$57DF_{16}$	$7A_{16}$
	⋮	
	$D124_{16}$	37_{16}

Determine the result in the AC of an AND $476B_{16}$ instruction if direct addressing is employed. The AND instruction performs the AND operation of the AC and the operand of the instruction.

11.8 Repeat Prob. 11.7 for indirect addressing. Assume that the indirect address points to the LSB of the data address and that the MSB of the data address is located at the next higher memory byte.

11.9 Repeat Prob. 11.7 for register indirect addressing if REG C is the register used to hold the indirect address.

11.10 Repeat Prob. 11.7 if relative addressing is employed.

11.11 Repeat Prob. 11.7 if indexed addressing is used and REG B is the index register.

11.12 Sketch the hexadecimal tape format in the form of Fig. 11.24 for the decimal number 147. Assume even parity.

11.13 Repeat Prob. 11.12 for a BNPF format. Assume that the numbers are transmitted in their decimal equivalents.

11.14 Repeat Prob. 11.12 for a binary format.

12

SOFTWARE

In Chap. 11 we discussed the component parts of a microcomputer. The hardware, of which the computer is made, serves no useful purpose without a set of instructions, or program, which causes the computer to perform a given function. We recall that a program consists of a sequence of specific instructions, often referred to as software in contrast to the hardware of Chap. 11.

The instructions must be written in binary, because the computer is only able to store and operate on binary quantities. A program in this form is said to be in *machine language*, or *machine code*. In order to simplify the development and documentation of a program, symbols are usually used to represent the binary code of the instructions and the addresses. A program expressed in this form is said to be written in *symbolic language*, or *assembly language*. There is a one-to-one correspondence between machine language and assembly language instructions. That is, one assembly language instruction translates into one machine language instruction. Translation from assembly language to machine language is generally performed by a computer program called an *assembler*.

A further improvement in documentation and ease of development is obtained through *high-level languages*. These languages are generally tailored to fit a particular class of applications. For example, FORTRAN is a scientifically oriented language, COBOL is a business-oriented language, PILOT is computer dialog language, and BASIC is a simplified, easy-to-learn, language. The common characteristic of these languages is that one statement in the language corresponds to many machine language instructions. The result is that programs written in high-level languages are shorter and easier to understand. The price paid for the simplification is that the resulting program generally requires a longer time to execute and requires more memory. Translation between a high-level language and machine code is performed by a computer. If the entire program is translated prior to execution, the task is called *compilation*. This is accomplished by means of a translating program called a *compiler*. If the translation is performed one statement at a time and executed at the same time, it is called *interpretation*. A program to accomplish this action is called an *interpreter*.

Execution time for a program using an interpreter is greater than for a program that has been compiled. However, the interpreter allows excellent operator interaction and debugging, which is desirable for *on-line* computer operation (i.e., while the computer is actually in operation).

In this chapter we provide an introduction to computer programming. We will emphasize machine language and assembly language programming because of its close association with the system hardware. The instruction set for the Intel 8080/8085 microprocessor will be used throughout for examples.

The instruction set for the 8085 is identical to that for the 8080 with the exception of a few additional instructions. The 8080 or 8085 instruction sets also contain many instructions common to that of the 8008 and Z-80 microprocessors. Thus the set of instructions used by the 8080 or the 8085 may be considered typical of eight-bit microprocessors. The 8080 and the 8085 have six internal registers (B, C, D, E, H, L) and an accumulator (the A register). A detailed explanation of the instructions used in programming the 8080/8085 is given in the Appendix. The student should make extensive use of the Appendix in studying Chap. 12.

12.1 Data Transfer Instructions

The basic data transfer instruction for the 8085 is the MOVE instruction, which is expressed by the *mnemonic*

$$\text{MOV } r_1, r_2$$

This instruction causes the contents of register r_2 to be transferred to register r_1. The machine code for the instruction is the binary word

$$01 \text{ DDD SSS}$$

The instruction word includes the operation code (01), the three-bit address of the destination register r_1 (DDD) and the three-bit address of the source register r_2 (SSS). The register addresses are

A = 111

B = 000

C = 001

D = 010

E = 011

H = 100

L = 101

M = 110

As an example, suppose we wish to transfer a word of data from the D register to the L register. The required instruction is

MOV L, D

which in machine code is

01101010

If the source or destination register is specified to be M (110), this indicates that we are addressing the memory element pointed to by the contents of the H and L register pair. This, of course, is register indirect addressing. The higher-order byte of the address is held in H and the lower-order byte in L. The instructions MOV M, *r* and MOV *r*, M allow data transfers to any location in memory from a CPU register or from any location in memory to a CPU register. For example, if the H register contains 35_{16} and the L register contains 03_{16} when a MOV M, D command is executed, the content of the D register is transferred to memory location 3503_{16}.

The seven registers of the 8080 or 8085 provide insufficient data storage for all but the most trivial problems. For this reason external RAM is provided. There are several instructions that may be used to access the data memory. The simplest form of instruction allows the address of the data to be completely specified in the instruction. The 8085 provides a 16-bit address that permits $2^{16} = 65{,}536$ memory locations to be addressed. This number is often referred to as 65K.

The STA instruction allows one to store a word from the accumulator in any of the 65K locations in memory. Conversely, the LDA instruction enables data to be transferred to the accumulator from any location in memory. For example, suppose we wish to move a byte of data from memory location $1F30_{16}$ to $3E00_{16}$. Using the STA and LDA instructions, the move may be accomplished with the instructions

Mnemonics		Machine Code (in hex)
LDA	1F30H	3A 30 1F
STA	3E00H	32 00 3E

where a number such as 1F30H represents the hexadecimal, or hex, number $1F30_{16}$. We see that the least significant byte of the address immediately follows the op code in machine code, which is typical of all 8085 instructions.

If we desire to move a 16-bit word (two bytes) from one memory location to another, we may use the LHLD and SHLD instructions. For example, LHLD 1F30H causes the data in memory locations 1F30H and 1F31H to be moved to L and H, respectively. SHLD causes the converse operation to take place. The STA, LDA, LHLD, and SHLD instructions use direct addressing.

While the MOV instruction uses the H and L registers to specify a memory address, the STAX and LDAX instructions perform the same operations using either the B and C registers (STAX B and LDAX B) or the D and E registers (STAX D and LDAX D) in pairs for holding the memory address.

Suppose, for example, that the registers contain the following numbers:

B = 2FH

C = 10H

D = 37H

E = 00H

Then the computer instructions

```
LDAX B
STAX D
```

cause the contents of memory location 2F10H to be transferred to memory location 3700H via the accumulator.

Additional data transfer instructions use immediate addressing; that is, the data is specified in the instruction. These instructions are useful in establishing constants. They are also commonly used to establish the addresses for the indirect addressing instructions.

The data transfer instructions using immediate addressing fall into two categories: those which refer to a single register, MVI r, ⟨data⟩, and those which refer to a register pair, LXI r_p, ⟨MS Byte of data⟩ ⟨LS Byte of data⟩. Suppose we wish to put 15H into the D register and 30H into the E register. This can be accomplished by either

```
MVI D, 15H
MVI E, 30H
```

or

```
LXI D, 1530H
```

There are generally several programs that will accomplish the same result. For instance, suppose we wish to move a byte of data from 537H to 105AH. One method of accomplishing this is to use the direct-addressing STA and LDA instructions, as follows:

Mnemonic	Machine Code (in hex)	
LDA 537H	3AH	
	37H	
	05H	
STA 105AH	32H	(12.1)
	5AH	
	10H	

The same result can be accomplished using the MOVE instruction, as follows:

LXI H, 537H	21H	
	37H	
	05H	
LXI D, 105A H	11H	
	5AH	(12.2)
	10H	
MOV A, M	7EH	
XCHG	EBH	
MOV M, A	77H	

where XCHG exchanges the data between the D, E and H, L register pairs

Another approach is to use the STAX and LDAX instructions, as follows:

LXI B, 537H	01H	
	37H	
	05H	
LXI D, 105AH	11H	
	5AH	(12.3)
	10H	
LDAX B	0AH	
STAX D	12H	

It is enlightening to compare the three approaches we have used. Note that the number of memory bytes required for storing programs (12.1)–(12.3) are 6, 9, and 8, respectively. The time required to execute these programs can be compared by determining the number of clock cycles required to execute the sequence of instructions. In the Appendix the number of clock cycles required for each instruction is given. The number of clock cycles required is, in the first case,

Mnemonics	Clock cycles
LDA 537H	13
STA 105AH	13
	26

in the second case,

LXI H, 537H	10
LXI D, 105AH	10
MOV A, M	7
XCHG	4
MOV M, A	7
	38

and in the third case,

LXI B, 537H	10
LXI D, 105AH	10
LDAX B	7
STAX D	7
	34

The time of execution can be calculated from the number of clock cycles by multiplying by the period of the clock. We note in the first case that the program requires less memory and executes faster than the latter two cases.

EXERCISES

12.1.1 Suppose memory location 106H contains a program instruction whose machine code is 01101000. What is the instruction?

Ans: MOV L, B

12.1.2 Suppose we desire to move the contents of memory location 20FH to register D. What program instruction steps will accomplish this? Write

(a) a program that will perform the transfer directly

(b) a program that transfers the data via the accumulator

Ans: (a) LXI H, 020FH
MOV D, M
(b) LDA 20FH
MOV D, A

12.1.3 Compare the memory requirements and time requirements for the two programs developed in Ex. 12.1.2.

Ans: (a) 4 bytes, 17 clock cycles; (b) 4 bytes, 18 clock cycles

12.2 Arithmetic and Logical Instructions

Instructions are provided to allow a full range of logical and arithmetic operations to be performed on the data contained in the microprocessor registers and memory. In order to shorten the instruction code, the accumulator, or A register, is usually specified to receive the results of all arithmetic and logical operations. It therefore serves as one of the operands in most logical and arithmetic operations.

The logical and arithmetic operations cause the condition flags (or flip-flops) in the 8085 to be set according to the result of the operation. The 8085 has five condition flags, which are

(a) *Zero Flag*: If the result of an instruction has the value 0, this flag is set; otherwise, it is reset.

(b) *Sign Flag*: If the most significant bit of A is 1 following the operation, this flag is set; otherwise, it is reset.

(c) *Parity Flag*: If the number of 1's in A is *even* following the operation, this flag is set; otherwise, it is reset.

(d) *Carry Flag*: If the instruction results in a carry (from addition) or a borrow (from subtraction or a comparison) this flag is set; otherwise, it is reset.

(e) *Auxiliary Carry Flag*: If the instruction causes a carry out of bit 3 into bit 4 of A, this flag is set; otherwise, it is reset. This flag is principally used to provide decimal correction in the case of BCD arithmetic.

The arithmetic and logical instructions for the 8085 include instructions to perform addition (ADD, ADC, ADI, DAD), subtraction (SUB, SBB, SBI), comparison (CMP, CPI), logical AND (ANA, ANI), logical OR (ORA, ORI), exclusive-OR (XRA, XRI), increment (INR, INX), decrement (DCR, DCX), and rotate accumulator (RAR, RRC, RAL, RLC) instructions. These instructions are explained in detail in the Appendix.

The instructions described provide the programmer with the capability of performing many complex logical operations. For example, suppose register C contains the ASCII code for a decimal digit. Let us find this number by performing appropriate logical operations on the word in register C. (Refer to Tab. 8.3 for the ASCII codes.) Note that the code for the digits 0 through 9 contains the binary equivalent of the digit in the four LSB's. Thus, if we can simply suppress, or *mask*, the four MSB's, we will have the required number. This can be accomplished by performing the logical AND of the word in register C with 00001111. A program to accomplish this is listed below:

Mnemonic	Machine code
MOV A, C	79H
ANI 0FH	E6H
	0FH

As another example, let us add the 16-bit number contained in registers B and C to the 16-bit number contained in registers D and E. The result of the addition should be placed in registers H and L. The assembly language instructions are

```
MOV A, C
ADD E
MOV L, A
MOV A, B
ADC D
MOV H, A
```
(12.4)

If an overflow has occurred in the addition, the carry flag will be set at the conclusion of this sequence.

Another method of accomplishing this operation is with the instructions

```
XCHG
DAD B
```

The logical and arithmetic capability of the computer allows the programmer to perform very complex logical and arithmetic operations in software. This replacement of hardware by software is an important advantage of microcomputers.

The 8085 is typical of most microcomputers in that it does not have a multiply or divide instruction. However, shifting the bits of a number left by one position effects a multiplication by 2 (see Probs. 11.5–11.6). This fact can be used to produce a multiplication algorithm consisting of shift and add operations. For example, a binary number N may be multiplied by 7 by implementing the algorithm

$$7 \times N = 4 \times N + 2 \times N + 1 \times N$$

The multiplication of a binary number by 2 can be accomplished by shifting left and multiplication by 4 by shifting left twice. Thus if $N = 5$ we have $101_2 \times 111_2$, which can be obtained by employing a shift and add approach, as follows:

$$
\begin{aligned}
N &= 101 \\
2N &= 1010 \quad N \text{ shifted left by one place} \\
4N &= \underline{10100} \quad N \text{ shifted left by two places} \\
7N &= 100011
\end{aligned}
$$

An assembly language program for this operation is

```
XRA A        ;CLEAR A & CARRY
MVI A, 5     ;NUMBER 5 INTO A
MOV B, A     ;A INTO B
RAL          ;ROTATE A LEFT
MOV C, A     ;A INTO C
RAL
ADD C        ;A + C
ADD B        ;A + B
```

The statements following the semicolons are known as *comment statements*; they are used to describe, or document, each instruction. When preceded by a semicolon (*delimiter*), they do not generate any machine code. In this program the first instruction performs $A \oplus A = 0$, which sets A and the CARRY to 0. MVI A, 5 moves 5 into A, and MOV B, A moves 5 to B. The next instruction (RAL) causes $2 \times 5 = 10$ to be generated in A. Thus, MOV C, A causes 10 to be placed in C. Since A still holds 10, the subsequent RAL produces $2 \times 10 = 20$ in A. ADD C causes the contents of C to be added to A, giving $20 + 10 = 30$ in A. Finally, ADD B produces $30 + 5 = 35$ to reside in A.

EXERCISES

12.2.1 Suppose the accumulator contains a BCD digit in the four LSB's and zeros in the four MSB's. Write a program to convert the BCD digit into its ASCII code.

Ans: ORI 30H

12.2.2 Suppose a group of numbers (a *table*) is stored beginning at location 3000H of memory. The D register contains a number less than 256 that represents the number of the element of the table we wish to find. Write a program in hexadecimal machine code to place the requested element of the table into the accumulator.

Ans:				
	LXI B, 3000H	01	00	30H
	MOV A, C	79H		
	ADD D	82H		
	MOV C, A	4FH		
	LDAX B	0AH		

12.2.3 Repeat Ex. 12.2.2 if each element of the table consists of two words. In this case assume that the table has fewer than 128 elements.

Ans:				
	LXI B, 3000H	01	00	30H
	MOV A, C	79H		
	ADD D	82H		
	ADD D	82H		
	MOV C, A	4FH		
	LDAX B	0AH		

12.3 Branch Instructions

The branch instructions allow the programmer to direct the computer to perform the instructions in a nonsequential manner. As we pointed out in Chap. 11, the program counter is normally advanced to the next sequential instruction after each instruction is executed. Using the branch instructions, however, the computer may be directed to perform any instruction in memory by directly affecting the value of the program counter.

The branch instructions contain conditional and unconditional instructions. The conditional branch causes the program counter to be affected only if the condition (indicated by the condition flags) is true. For example, the JC (jump if carry) instruction causes the branch to take place only if the carry flag is set. A great deal of the power of a computer comes from the ability to make branches (i.e., decisions) based on the flag conditions existing at the time the program is executed.

Each of the branch instructions for the 8085 uses direct addressing and therefore requires three words for the instruction. The first word contains the op code, and the second and third words contain the address of the branch. In the 8080 series machines, we recall that the address is always presented in machine code with the least significant byte first, followed by the most significant byte of the address. For example, an unconditional branch instruction to address 4035H takes the form

Mnemonic	Machine Code (in hex)
JMP 4035H	C3 35 40

One prominent use of the conditional branch instruction is in the implementation of loops. We may desire to repeat the execution of a certain set of instructions, or *loop*, until a particular condition is fulfilled (e.g., until the zero flag is set). This technique can be used to repeat a sequence of instructions a fixed number of times.

Suppose we wish to move the data stored in locations 100H through 1FFH to locations 1500H to 15FFH. This means that we desire to move 256 bytes of data. Obviously, we would not want to use 256 pairs of transfer instructions to perform this operation. Instead we will use a loop and repeat the basic move instructions 256 times. The following program uses a loop counter (the L register) to determine when the 256 operations have been completed:

```
                                                      Machine code
          Mnemonics                                   (in binary)
          MVI L, 0                                    00101110
                                                      00000000
          LXI B, 100H                                 00000001
                                                      00000000
                                                      00000001
          LXI D, 1500H                                00010101
                                                      00000000
                                                      00010101
START:    LDAX B      ;    START OF LOOP              00001010
          STAX D                                      00010010
          INX B                                       00000011
          INX D                                       00010011
          DCR L                                       00101101
          JNZ START ;      END OF LOOP                11000010
                                                      ⟨LOWER BYTE OF
                                                      ADDR OF START⟩
                                                      ⟨HIGHER BYTE OF
                                                      ADDR OF START⟩
```

In this program we have introduced a new symbol, START:, which is a *label* used to identify the address of the instruction LDAX B. This label is useful in referring to the instruction, as we have done in the last instruction JNZ START.

The LDAX B and STAX D instructions are the instructions that actually move the data in memory. The remaining instructions are "overhead" required to implement the loop. The loop counter is implemented with the L register. On the first pass through the loop, commencing at START, the DCR L produces FFH in the

L register. On each subsequent pass through the loop, the counter is decremented until L again holds 0. At this time the zero flag is set and the instruction immediately following JNZ START will be executed. This will obviously require 256 passes, which completes the data transfer.

The preceding example illustrates all the parts of a typical assembly language program instruction. Each instruction in assembly language consists of four fields: label, op code, operand, and comment fields. Each of these fields are separated by delimiters, which include the semicolon, the colon, and space. For example, consider the instruction

```
START:    LDAX B    ; START OF LOOP
```

The label field contains a symbol that is used by the assembler to represent the address of the instruction LDAX B. The delimiter colon separates the label field from the op code field. The LDAX symbol represents the op code of the instruction, and B represents the operand for the instruction. The delimiter (space) separates the op code and operand fields. Finally, the delimiter semicolon distinguishes between the operand and the comment fields. Recall that the comment field is ignored in the conversion from assembly language to machine language.

Often in the processing of commands from an external device, such as a keyboard, it is necessary to determine whether the ASCII code for a certain character is contained in the A register. The program will then produce a branch to the appropriate *routine* (sequence of instructions) to execute the command.

Suppose, for example, we have a system that will perform four different functions depending on whether the word contained in the A register is the ASCII code for the letters C, D, E, or F. The following instructions will cause the appropriate routines (ROUC, ROUD, ROUE, ROUF) to be executed:

```
CPI 43H     ; COMPARE ASCII C
JZ  ROUC
CPI 44H     ; COMPARE ASCII D
JZ  ROUD
CPI 45H     ; COMPARE ASCII E          (12.5)
JZ  ROUE
CPI 46H     ; COMPARE ASCII F
JZ  ROUF
    .
    .
    .
```

In this program the ASCII code for C (43H) is first compared to the content of A, and if a match occurs, the zero flag is set. In this case JZ ROUC causes the program to branch to routine ROUC. If no match occurs, the process repeats for ASCII characters D, E, and F. If no match is found, execution will continue at the instruction immediately following JZ ROUF.

In many cases we are interested in branching to perform a routine (subroutine) and then returning to the sequence of instructions. To accomplish this requires that the program counter be stored to enable the computer to return after the subroutine is executed. This is accomplished automatically in the 8085 upon execution of the CALL instruction. The address of the next sequential location of the program is placed on the stack at the time the CALL instruction is executed. Conditional and unconditional CALL instructions are available.

The subroutine structure allows the programmer to invoke a sequence of instructions by the use of one instruction. The basic CALL instruction uses direct addressing and hence is a three-word instruction. The 8085 also has a one-word subroutine call, the RESTART (RST) instruction. This instruction allows one to invoke a subroutine whose *entry point* (first instruction) is located at 0, 8H, 10H, 18H, 20H, 28H, 30H, or 38H.

Using subroutines, the programmer can conveniently divide a program into manageable parts. In fact, in many cases a program may consist of simply a sequence of CALL instructions.

The last instruction in a subroutine is a RETURN (RET) instruction. This causes the return address that is on the stack to be transferred to the program counter. The execution then returns to the calling program.

As another example, let us consider a subroutine to shift the bits in the accumulator four places to the left. The 8085 does not have a shift instruction, but a set of rotate instructions is available. The RAL instruction causes the bits in the A register to be rotated one place to the left, with the carry flag providing the link between the MSB and the LSB. Thus a shift left is accomplished by clearing the carry flag and then rotating left, using the instructions ANA A and RAL. A subroutine for this shift is

```
SHL4:   ANA A   ;CLEAR CARRY
        RAL     ;ROTATE LEFT 1 BIT
        ANA A
        RAL
        ANA A                              (12.6)
        RAL
        ANA A
        RAL
        RET
```

The program in the example would be called from the main program by CALL

SHL4, where SHL4 is a mnemonic symbol that represents the address of the entry point of the subroutine.

Subroutines may be called from a subroutine. This is called *nesting* of the subroutines. For example, we might use a subroutine to perform each of the shift left operations. A subroutine that produces a single shift left of the A register is

```
SHL:  ANA A
      RAL          (12.7)
      RET
```

This subroutine may be used to develop the four-bit left shift by writing

```
SHL4: CALL SHL
      CALL SHL
      CALL SHL
      CALL SHL
      RET
```

Fig. 12.1 illustrates the flow of execution of the subroutines for this example.

Execution of a subroutine generally requires data to be passed from the calling program to the subroutine. This data passage may be accomplished in many ways. In our previous examples the data was passed in and out via the A register. Other registers and memory can also be used to pass data. A complete treatment of this important topic is available in the literature.

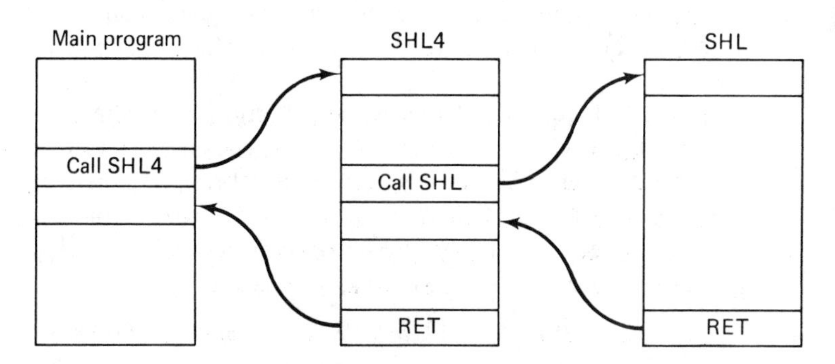

Figure 12.1 *Nesting of subroutines*

EXERCISES

12.3.1 Write a subroutine to provide a 1-millisecond time delay. Assume that the clock cycle of the 8085 is 0.5 microsecond.

			Clock cycles
Ans:	MSEC:	MVI B, 56H	7
	DLY:	NOP	4
		NOP	4
		DCR B	5
		JNZ DLY	10
		RET	10

12.3.2 Using the subroutine of Ex. 12.3.1, write a subroutine to generate a 10 ms delay.

```
Ans:  TMSEC:  MVI C, 0AH
      TSEC:   CALL MSEC
              DCR C
              JNZ TSEC
              RET
```

12.4 Stack and I/O Operations

The stack for the 8085 microprocessor may be used to store data in addition to storing the return addresses for subroutine calls. Sixteen bits of data (a register pair) may be stored using the PUSH instruction and recovered using the POP instruction. In addition to storing register pairs, PUSH PSW allows the programmer to store the accumulator and the condition flags. They may be recovered using the POP PSW instruction.

The execution of subroutines frequently includes operations that affect the contents of the general-purpose registers or the condition flags. It may be desirable to restore these quantities to their original values before returning to the calling program.

In the last example of the previous section we wrote a program to shift the accumulator four bits to the left. In the execution of this subroutine, the value of the CARRY flag and other flags are altered. To restore these flags to their original condition requires that the flags be stored or saved at the beginning of the subroutine and recovered at the end of the subroutine. The stack can be used for this purpose as illustrated by the following program:

```
SHL4:   PUSH PSW    ;SAVE STATUS
        CALL SHL
        CALL SHL
        CALL SHL
        CALL SHL
        MOV B, A
        POP PSW     ;RESTORE STATUS
        MOV A, B
        RET
```

The 8085 instruction set includes one input instruction and one output instruction. Data may be input from an external device by the instruction

IN ⟨INPUT PORT ADDR⟩

where ⟨INPUT PORT ADDR⟩ is the second byte of the instruction and contains the input port address, which can be from 0 to 255. A similar instruction is included for output operations, given by

OUT ⟨OUTPUT PORT ADDR⟩

where ⟨OUTPUT PORT ADDR⟩ is the second byte of the instruction and contains the output port address (also 0 to 255).

There are many techniques for handling input or output operations. We will consider only one, which is perhaps the most frequently used in communicating with terminal devices such as teletypes, CRT terminals, and A/D converters. This approach is known as the *asynchronous I/O*, or the *handshaking* method. In this approach the peripheral device indicates whether it is available to send or receive information by using status bits. Suppose, for instance, that we wish to output a word to a peripheral device connected to OUTPUT PORT 3. The status bit, or flag, that indicates whether the peripheral is BUSY or READY (NOT BUSY) is connected to the LSB of INPUT PORT 4, as shown in Fig. 12.2. A subroutine to output an eight-bit word contained in the C register is given by the following program:

```
OUT8:   PUSH PSW    ;STORE PROCESSOR STATUS
WAIT:   IN 4        ;INPUT PERIPHERAL STATUS
        RAR         ;ROTATE STATUS TO CARRY
        JC WAIT     ;JUMP IF BUSY
        MOV A, C    ;LOAD A
        OUT 3       ;OUTPUT TO PORT 3
        POP PSW     ;RESTORE PROCESSOR STATUS
        RET         ;RETURN TO MAIN PROGRAM
```

The program initially saves the value of the A register and the processor status flags by placing them on the STACK (PUSH PSW). Next a loop, labeled WAIT, is implemented to input the peripheral status (IN 4). The status is then rotated into the CARRY (RAR). If the CARRY is set (peripheral busy), then the loop is repeated (JC WAIT) until the peripheral is ready. Register C is then moved to A (MOV A, C) and A is output to PORT 3 (OUT 3). The value of the A register and the processor status flags are then restored (POP PSW), and program execution returns to the calling program.

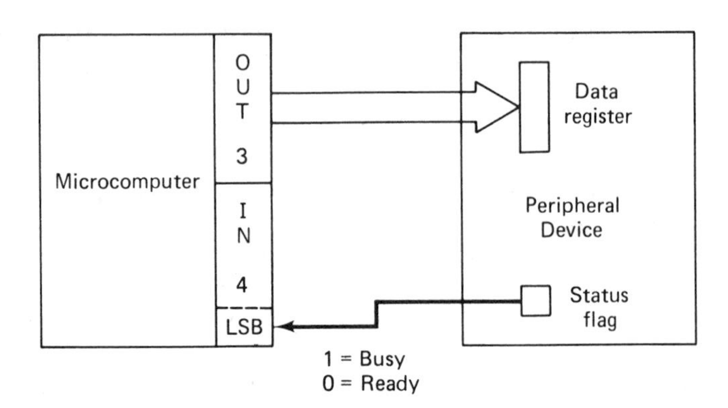

Figure 12.2 *Handshaking I/O*

A similar subroutine can be written to input a character into A. Assume, for example, that the peripheral is to input an eight-bit word into A through INPUT PORT 3 and that the peripheral status flag (DATA READY = 0) is connected to INPUT PORT 4, line 1 (LSB + 1). A program to execute this transfer is the following:

```
IN8:   IN 4       ;INPUT PERIPHERAL STATUS
       RAR        ;ROTATE STATUS TO CARRY
       RAR
       JC IN8     ;JUMP IF BUSY
       IN 3       ;INPUT CHARACTER TO A
       RET        ;RETURN TO MAIN PROGRAM
```

In this example, when execution returns to the calling program, the A register contains the desired character. It should be recognized, however, that the CARRY has not been restored to its value prior to the calling of IN8.

EXERCISES

12.4.1 In the IN8 subroutine above, write an assembly language program sequence to save the A register and processor flags, call IN8, transfer A to C, and restore A and the processor flags.

Ans:

```
PUSH PSW    ;SAVE A AND FLAGS
CALL IN8    ;INPUT CHARACTER TO A
MOV C, A    ;MOVE CHARACTER TO C
POP PSW     ;RESTORE A AND FLAGS
```

12.4.2 It is frequently necessary to save all the registers and the condition flags during execution of a subroutine and to restore these values later. Write two program sequences to accomplish this function using the STACK.

Ans:

```
SAV1:   PUSH PSW    ;SAVE A AND FLAGS
        PUSH B      ;SAVE B AND C
        PUSH D      ;SAVE D AND E
        PUSH H      ;SAVE H AND L
          .
          .
          .
REST1:  POP H       ;RESTORE H AND L
        POP D
        POP B
        POP PSW     ;RESTORE A AND FLAGS
```

12.5 Communication with a Serial Device

In order to illustrate the programming concepts we have described in this chapter, we will develop a system for communicating with a serial device such as a teletypewriter. The teletypewriter (TTY) and CRT were described in Chap. 11. We recall that the TTY operates at 10 characters per second. Each character consists of 11 bits: a start bit, 8 bits for the ASCII code and parity bit, and 2 stop bits (see Fig. 11.25). Let us now develop a program for inputting and outputting serial information to the TTY. The TTY is assumed to be wired in the *full duplex* configuration, shown in Fig. 12.3. "Full duplex" indicates that the input and output sections of the TTY are functionally distinct; that is, an input character can be read from the keyboard without being printed on the TTY. In this configuration it is necessary for the microcomputer to echo the character to the TTY if we desire to print the key that is depressed on the keyboard.

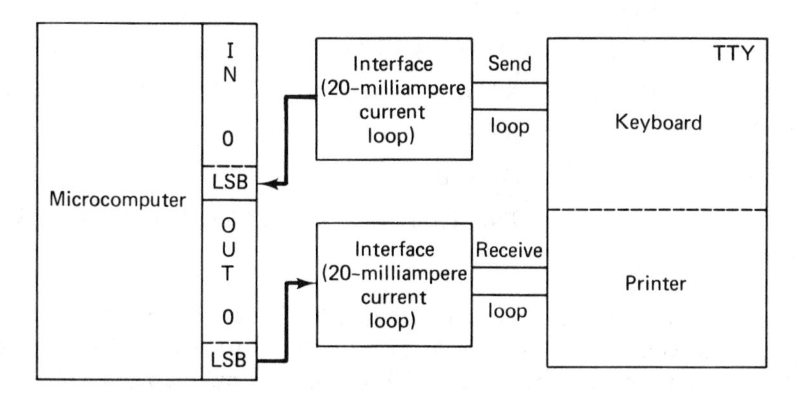

Figure 12.3 *TTY interface to a microcomputer*

The interface shown in Fig. 12.3 consists of the appropriate circuitry necessary to convert the 20-milliampere current loop of the TTY to voltage logic signals that are compatible with the microcomputer input and output ports.

Let us first consider writing a subroutine to read a character from the TTY, store it in C, and echo it to the TTY for printing. The following subroutine will accomplish this task:

```
READ:   IN 0          ;INPUT TTY STATUS
        RAR           ;ROTATE TO CARRY
        JC READ       ;JUMP IF NO START BIT
        CALL HDLY     ;DELAY 4.55 MSEC
        OUT 0         ;SEND START BIT TO TTY
        MVI C, 80H    ;SET MSB OF C
CHAR:   CALL FDLY     ;DELAY 9.09 MSEC
        IN 0          ;INPUT BIT
        OUT 0         ;ECHO BIT
        RAR           ;MOVE BIT TO CARRY
        MOV A, C      ;MOVE C TO A
        RAR           ;ROTATE BIT INTO MSB OF A
        MOV C, A      ;MOVE A TO C
        JNC CHAR      ;JUMP IF CARRY = 0
        CALL FDLY     ;DELAY 9.09 MSEC
        IN 0          ;INPUT STOP BIT
        OUT 0         ;ECHO STOP BIT
        RET
FDLY:   CALL HDLY
HDLY:   MVI B, OBOH
DLY:    XTHL          ;EXCHANGE H, L AND STACK
        XTHL
        DCR B
        JNZ DLY       ;JUMP IF B NOT ZERO
        RET
```

The program begins by reading INPUT PORT 0 into A and loops until a zero is detected in the LSB of A (START BIT = 0). A time delay of 4.55 milliseconds is then generated by CALL HDLY to delay until the middle of the start bit in time. The A register is now echoed to OUTPUT PORT 0 to send a start bit to the TTY.

Next, C is set equal to 80H, which places a 1 in the MSB. In CHAR the first instruction (CALL FDLY) creates a delay of 9.09 milliseconds in order to read in the middle of the first data bit. The bit is echoed to the TTY (OUT 0) and then rotated into the CARRY (RAR). Register C is moved to A, and the first data bit is rotated into the MSB of A (RAR). It should be noted that the 1 originally in the MSB of C (due to the previous MVI C, 80H) is now shifted one position to the right in A and that the remaining bits to the right of this are all zero, including the CARRY. Register A is now moved to C. Since the CARRY = 0, the CHAR loop is again executed to get the second data bit from the TTY. This continues until all eight data bits are read and placed in C (see Ex. 12.5.1). At this time the 1 originally loaded into the C register appears in the CARRY and the CHAR loop is completed. A time delay of 9.09 milliseconds is generated, and a stop bit is read (IN 0) and echoed to the TTY (OUT 0). The program counter is then incremented to the address of FDLY, and another 9.09 millisecond delay is generated. Note that placing FDLY at the end of the READ routine avoided a final call to FDLY, thus saving memory and execution time.

The subroutine FDLY begins by calling HDLY. The B register is loaded with 0BH, and two XTHL are executed to create a small time delay. Clearly, since XTHL exchanges the H and L registers and the stack, the registers and stack are unchanged after the second execution of XTHL. Register B is then decremented and the loop repeated until B is 0. This total operation requires 4.55 milliseconds for a 0.5-microsecond clock cycle. Following the call to HDLY the program counter is incremented to the address of HDLY and HDLY is executed again, giving a total delay of approximately 9.09 milliseconds.

The discussion in this chapter is not intended to be an in-depth treatment of the programming of microcomputers. Nevertheless, it includes many important topics in this area. The interested reader can find these topics more completely developed in the literature.

EXERCISES

12.5.1 Verify that the MVI C, 80H in the READ subroutine causes the program to exit the CHAR loop after eight data bits are placed in register C in the example of this section.

12.5.2 Verify that HDLY generated a time delay of approximately 4.55 milliseconds if the system clock has a period of 0.5 microseconds.

PROBLEMS

12.1 Write three programs of the form of (12.1)–(12.3) to transfer data in memory locations 1000H–100FH to memory locations 1500H–150FH.

12.2 Determine the memory requirement and the execution time for

(a) program (12.4)

(b) program (12.5)

12.3 Write a program to multiply a number contained in the A register by 5.

12.4 Compare the memory and execution time requirements for the program of Prob. 12.3.

12.5 Repeat Prob. 12.3 using loops.

(a) What difficulties do you encounter?

(b) Compare the memory and speed requirements as in Prob. 12.4.

12.6 Write the machine code for program (12.5). Assign addresses to the subroutines ROUC, ROUD, ROUE, and ROUF.

12.7 Compare the memory and time requirements of the programs developed in (12.6) and (12.7).

12.8 In order to multiply two four-bit binary numbers $a_3a_2a_1a_0$ and $b_3b_2b_1b_0$, note that the product is

$$a_0 \times b_3b_2b_1b_0 + 2 \times a_1 \times b_3b_2b_1b_0 + 4 \times a_2 \times b_3b_2b_1b_0 + 8 \times a_3 \times b_3b_2b_1b_0$$

Write a subroutine to multiply a four-bit binary number in the B register by a four-bit binary number in the C register. Place the result in the accumulator.

12.9 List three techniques that may be used to pass data to and from a subroutine.

12.10 Repeat the program illustrated in Sec. 12.5 for a CRT that operates at 9600 baud. Assume that only one stop bit is required.

12.11 Write a subroutine to output a pulse of width 3 milliseconds. Assume the pulse is to be output to the LSB of OUTPUT PORT 1.

12.12 A TTY is connected to a microcomputer with the receive loop connected to the MSB of OUTPUT PORT 1 and the send loop connected to the MSB of INPUT PORT 2 (refer to Fig. 12.3). Write a program to read and echo two sequential characters entered from the TTY keyboard. Store these characters in memory beginning at DATA. End the routine by having the TTY printing carriage positioned for a new line beginning at the left margin. (This is accomplished by a CARRIAGE RETURN followed by a LINE FEED having ASCII codes 0DH and 0AH respectively.)

APPENDIX

THE 8080 INSTRUCTION SET

(Courtesy of Intel Corporation)

A computer, no matter how sophisticated, can only do what it is "told" to do. One "tells" the computer what to do via a series of coded instructions referred to as a **Program**. The realm of the programmer is referred to as **Software**, in contrast to the **Hardware** that comprises the actual computer equipment. A computer's software refers to all of the programs that have been written for that computer.

When a computer is designed, the engineers provide the Central Processing Unit (CPU) with the ability to perform a particular set of operations. The CPU is designed such that a specific operation is performed when the CPU control logic decodes a particular instruction. Consequently, the operations that can be performed by a CPU define the computer's **Instruction Set**.

Each computer instruction allows the programmer to initiate the performance of a specific operation. All computers implement certain arithmetic operations in their instruction set, such as an instruction to add the contents of two registers. Often logical operations (e.g., OR the contents of two registers) and register operate instructions (e.g., increment a register) are included in the instruction set. A computer's instruction set will also have instructions that move data between registers, between a register and memory, and between a register and an I/O device. Most instruction sets also provide **Conditional Instructions.** A conditional instruction specifies an operation to be performed only if certain conditions have been met; for example, jump to a particular instruction if the result of the last operation was zero. Conditional instructions provide a program with a decision-making capability.

By logically organizing a sequence of instructions into a coherent program, the programmer can "tell" the computer to perform a very specific and useful function.

The computer, however, can only execute programs whose instructions are in a binary coded form (i.e., a series of 1's and 0's), that is called **Machine Code**. Because it would be extremely cumbersome to program in machine code, programming languages have been developed. There are programs available which convert the programming language instructions into machine code that can be interpreted by the processor.

One type of programming language is **Assembly Language**. A unique assembly language mnemonic is assigned to each of the computer's instructions. The programmer can write a program (called the **Source Program**) using these mnemonics and certain operands; the source program is then converted into machine instructions (called the **Object Code**). Each assembly language instruction is converted into one machine code instruction (1 or more bytes) by an **Assembler** program. Assembly languages are usually machine dependent (i.e., they are usually able to run on only one type of computer).

THE 8080 INSTRUCTION SET

The 8080 instruction set includes five different types of instructions:

- **Data Transfer Group**—move data between registers or between memory and registers
- **Arithmetic Group** – add, subtract, increment or decrement data in registers or in memory
- **Logical Group** – AND, OR, EXCLUSIVE-OR, compare, rotate or complement data in registers or in memory
- **Branch Group** – conditional and unconditional jump instructions, subroutine call instructions and return instructions
- **Stack, I/O and Machine Control Group** – includes I/O instructions, as well as instructions for maintaining the stack and internal control flags.

Instruction and Data Formats:

Memory for the 8080 is organized into 8-bit quantities, called Bytes. Each byte has a unique 16-bit binary address corresponding to its sequential position in memory.

The 8080 can directly address up to 65,536 bytes of memory, which may consist of both read-only memory (ROM) elements and random-access memory (RAM) elements (read/write memory).

Data in the 8080 is stored in the form of 8-bit binary integers:

DATA WORD

D_7	D_6	D_5	D_4	D_3	D_2	D_1	D_0

MSB LSB

When a register or data word contains a binary number, it is necessary to establish the order in which the bits of the number are written. In the Intel 8080, BIT 0 is referred to as the **Least Significant Bit (LSB)**, and BIT 7 (of an 8 bit number) is referred to as the **Most Significant Bit (MSB)**.

The 8080 program instructions may be one, two or three bytes in length. Multiple byte instructions must be stored in successive memory locations; the address of the first byte is always used as the address of the instructions. The exact instruction format will depend on the particular operation to be executed.

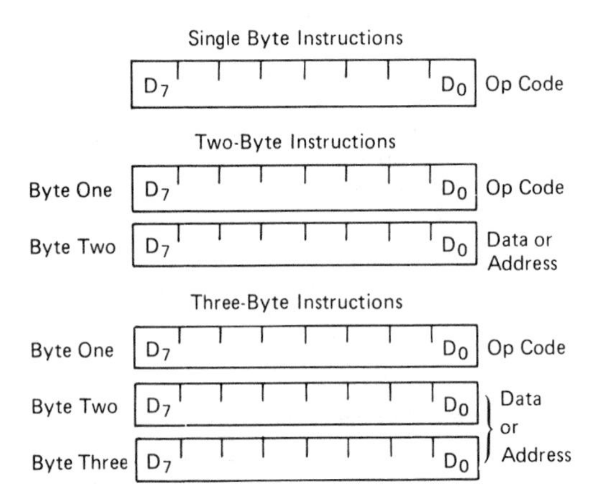

Addressing Modes:

Often the data that is to be operated on is stored in memory. When multi-byte numeric data is used, the data, like instructions, is stored in successive memory locations, with the least significant byte first, followed by increasingly significant bytes. The 8080 has four different modes for addressing data stored in memory or in registers:

- Direct – Bytes 2 and 3 of the instruction contain the exact memory address of the data item (the low-order bits of the address are in byte 2, the high-order bits in byte 3).
- Register – The instruction specifies the register or register-pair in which the data is located.
- Register Indirect – The instruction specifies a register-pair which contains the memory address where the data is located (the high-order bits of the address are in the first register of the pair, the low-order bits in the second).
- Immediate – The instruction contains the data itself. This is either an 8-bit quantity or a 16-bit quantity (least significant byte first, most significant byte second).

Unless directed by an interrupt or branch instruction, the execution of instructions proceeds through consecutively increasing memory locations. A branch instruction can specify the address of the next instruction to be executed in one of two ways:

- Direct – The branch instruction contains the address of the next instruction to be executed. (Except for the 'RST' instruction, byte 2 contains the low-order address and byte 3 the high-order address.)
- Register indirect – The branch instruction indicates a register-pair which contains the address of the next instruction to be executed. (The high-order bits of the address are in the first register of the pair, the low-order bits in the second.)

The RST instruction is a special one-byte call instruction (usually used during interrupt sequences). RST includes a three-bit field; program control is transferred to the instruction whose address is eight times the contents of this three-bit field.

Condition Flags:

There are five condition flags associated with the execution of instructions on the 8080. They are Zero, Sign, Parity, Carry, and Auxiliary Carry, and are each represented by a 1-bit register in the CPU. A flag is "set" by forcing the bit to 1; "reset" by forcing the bit to 0.

Unless indicated otherwise, when an instruction affects a flag, it affects it in the following manner:

Zero: If the result of an instruction has the value 0, this flag is set; otherwise it is reset.

Sign: If the most significant bit of the result of the operation has the value 1, this flag is set; otherwise it is reset.

Parity: If the modulo 2 sum of the bits of the result of the operation is 0, (i.e., if the result has even parity), this flag is set; otherwise it is reset (i.e., if the result has odd parity).

Carry: If the instruction resulted in a carry (from addition), or a borrow (from subtraction or a comparison) out of the high-order bit, this flag is set; otherwise it is reset.

Auxiliary Carry: If the instruction caused a carry out of bit 3 and into bit 4 of the resulting value, the auxiliary carry is set; otherwise it is reset. This flag is affected by single precision additions, subtractions, increments, decrements, comparisons, and logical operations, but is principally used with additions and increments preceding a DAA (Decimal Adjust Accumulator) instruction.

Symbols and Abbreviations:

The following symbols and abbreviations are used in the subsequent description of the 8080 instructions:

SYMBOLS	MEANING
accumulator	Register A
addr	16-bit address quantity
data	8-bit data quantity
data 16	16-bit data quantity
byte 2	The second byte of the instruction
byte 3	The third byte of the instruction
port	8-bit address of an I/O device
r,r1,r2	One of the registers A,B,C,D,E,H,L
DDD,SSS	The bit pattern designating one of the registers A,B,C,D,E,H,L (DDD=destination, SSS=source):

DDD or SSS	REGISTER NAME
111	A
000	B
001	C
010	D
011	E
100	H
101	L

rp — One of the register pairs:

B represents the B,C pair with B as the high-order register and C as the low-order register;

D represents the D,E pair with D as the high-order register and E as the low-order register;

H represents the H,L pair with H as the high-order register and L as the low-order register;

SP represents the 16-bit stack pointer register.

RP — The bit pattern designating one of the register pairs B,D,H,SP:

RP	REGISTER PAIR
00	B-C
01	D-E
10	H-L
11	SP

rh	The first (high-order) register of a designated register pair.
rl	The second (low-order) register of a designated register pair.
PC	16-bit program counter register (PCH and PCL are used to refer to the high-order and low-order 8 bits respectively).
SP	16-bit stack pointer register (SPH and SPL are used to refer to the high-order and low-order 8 bits respectively).
r_m	Bit m of the register r (bits are number 7 through 0 from left to right).
Z,S,P,CY,AC	The condition flags: Zero, Sign, Parity, Carry, and Auxiliary Carry, respectively.
( )	The contents of the memory location or registers enclosed in the parentheses.
$\leftarrow$	"Is transferred to"
$\wedge$	Logical AND
$\forall$	Exclusive OR
$\vee$	Inclusive OR
+	Addition
−	Two's complement subtraction
*	Multiplication
$\leftrightarrow$	"Is exchanged with"
$\overline{}$	The one's complement (e.g., $(\overline{A})$)
n	The restart number 0 through 7
NNN	The binary representation 000 through 111 for restart number 0 through 7 respectively.

Description Format:

The following pages provide a detailed description of the instruction set of the 8080. Each instruction is described in the following manner:

1. The MAC 80 assembler format, consisting of the instruction mnemonic and operand fields, is printed in **BOLDFACE** on the left side of the first line.
2. The name of the instruction is enclosed in parenthesis on the right side of the first line.
3. The next line(s) contain a symbolic description of the operation of the instruction.
4. This is followed by a narative description of the operation of the instruction.
5. The following line(s) contain the binary fields and patterns that comprise the machine instruction.

6. The last four lines contain incidental information about the execution of the instruction. The number of machine cycles and states required to execute the instruction are listed first. If the instruction has two possible execution times, as in a Conditional Jump, both times will be listed, separated by a slash. Next, any significant data addressing modes (see Page 4-2) are listed. The last line lists any of the five Flags that are affected by the execution of the instruction.

Data Transfer Group:

This group of instructions transfers data to and from registers and memory. **Condition flags are not affected** by any instruction in this group.

MOV r1, r2 (Move Register)

(r1) ← (r2)

The content of register r2 is moved to register r1.

Cycles: 1
States: 5
Addressing: register
Flags: none

MOV r, M (Move from memory)

(r) ← ((H) (L))

The content of the memory location, whose address is in registers H and L, is moved to register r.

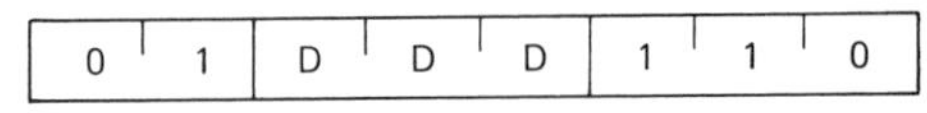

Cycles: 2
States: 7
Addressing: reg. indirect
Flags: none

MOV M, r (Move to memory)

((H) (L)) ← (r)

The content of register r is moved to the memory location whose address is in registers H and L.

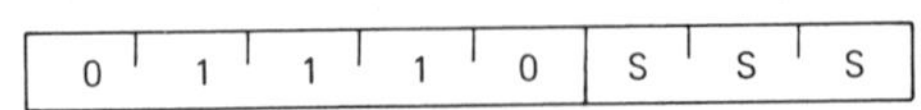

Cycles: 2
States: 7
Addressing: reg. indirect
Flags: none

MVI r, data (Move Immediate)

(r) ← (byte 2)

The content of byte 2 of the instruction is moved to register r.

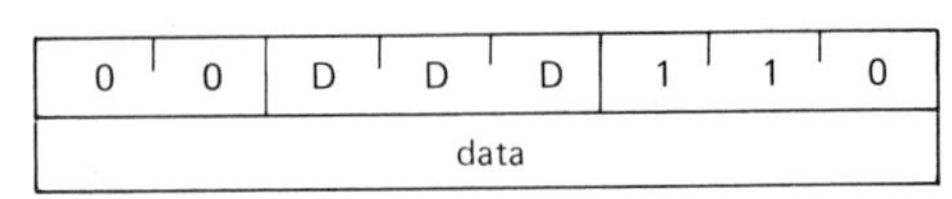

Cycles: 2
States: 7
Addressing: immediate
Flags: none

MVI M, data (Move to memory immediate)

((H) (L)) ← (byte 2)

The content of byte 2 of the instruction is moved to the memory location whose address is in registers H and L.

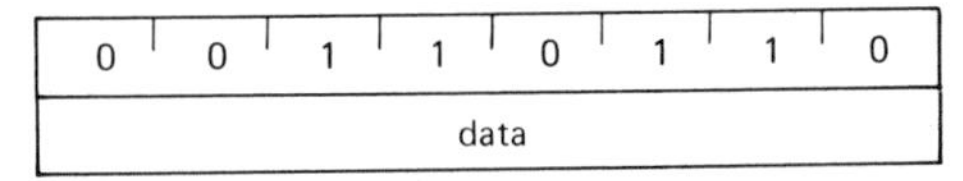

Cycles: 3
States: 10
Addressing: immed./reg. indirect
Flags: none

LXI rp, data 16 (Load register pair immediate)

(rh) ← (byte 3),

(rl) ← (byte 2)

Byte 3 of the instruction is moved into the high-order register (rh) of the register pair rp. Byte 2 of the instruction is moved into the low-order register (rl) of the register pair rp.

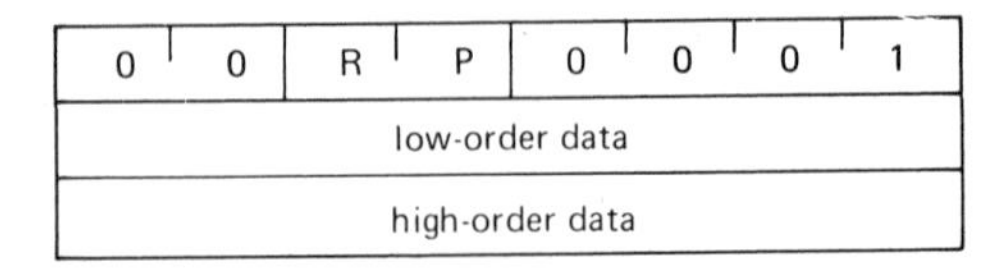

Cycles: 3
States: 10
Addressing: immediate
Flags: none

LDA addr (Load Accumulator direct)

(A) ← ((byte 3)(byte 2))

The content of the memory location, whose address is specified in byte 2 and byte 3 of the instruction, is moved to register A.

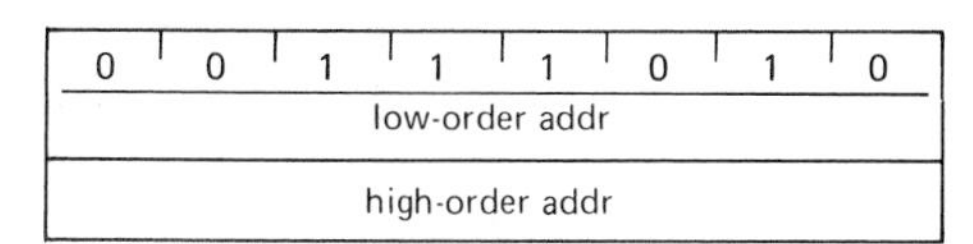

Cycles: 4
States: 13
Addressing: direct
Flags: none

STA addr (Store Accumulator direct)

((byte 3)(byte 2)) ← (A)

The content of the accumulator is moved to the memory location whose address is specified in byte 2 and byte 3 of the instruction.

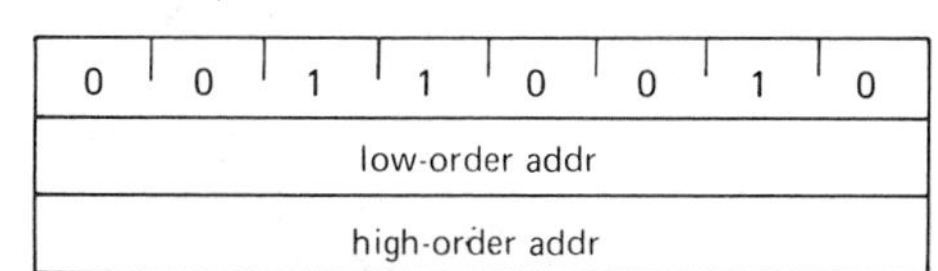

Cycles: 4
States: 13
Addressing: direct
Flags: none

LHLD addr (Load H and L direct)

(L) ← ((byte 3)(byte 2))

(H) ← ((byte 3)(byte 2) + 1)

The content of the memory location, whose address is specified in byte 2 and byte 3 of the instruction, is moved to register L. The content of the memory location at the succeeding address is moved to register H.

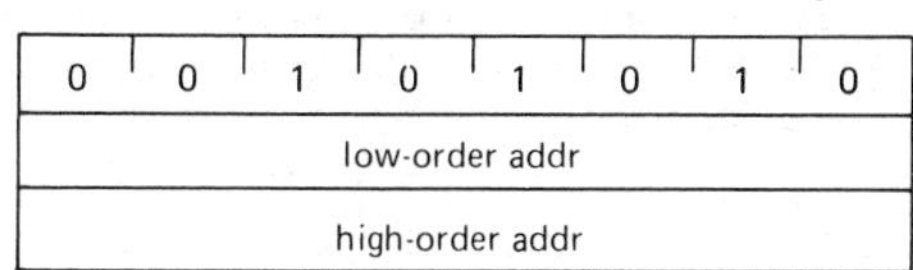

Cycles: 5
States: 16
Addressing: direct
Flags: none

SHLD addr (Store H and L direct)

((byte 3)(byte 2)) ← (L)

((byte 3)(byte 2) + 1) ← (H)

The content of register L is moved to the memory location whose address is specified in byte 2 and byte 3. The content of register H is moved to the succeeding memory location.

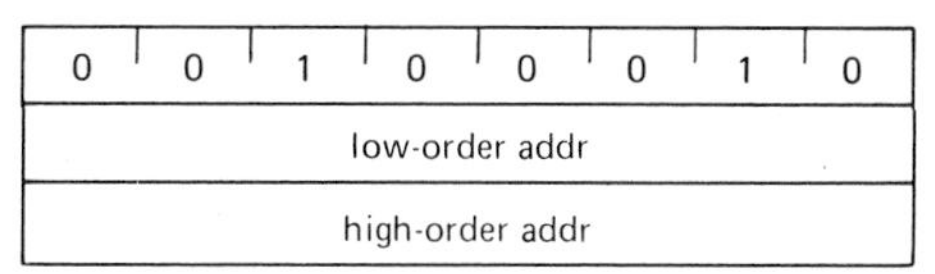

Cycles: 5
States: 16
Addressing: direct
Flags: none

LDAX rp (Load accumulator indirect)

(A) ← ((rp))

The content of the memory location, whose address is in the register pair rp, is moved to register A. Note: only register pairs rp=B (registers B and C) or rp=D (registers D and E) may be specified.

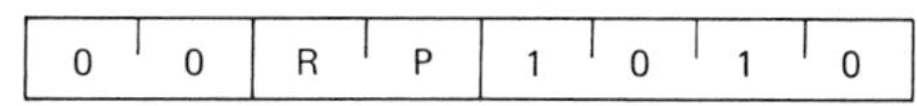

Cycles: 2
States: 7
Addressing: reg. indirect
Flags: none

STAX rp (Store accumulator indirect)

((rp)) ← (A)

The content of register A is moved to the memory location whose address is in the register pair rp. Note: only register pairs rp=B (registers B and C) or rp=D (registers D and E) may be specified.

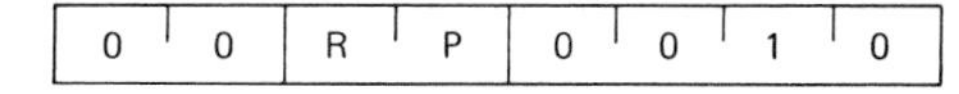

Cycles: 2
States: 7
Addressing: reg. indirect
Flags: none

XCHG (Exchange H and L with D and E)

(H) ↔ (D)

(L) ↔ (E)

The contents of registers H and L are exchanged with the contents of registers D and E.

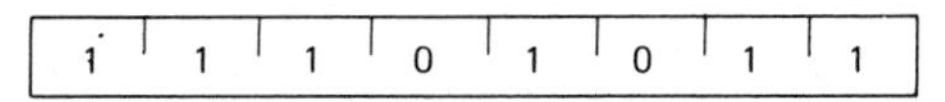

Cycles: 1
States: 4
Addressing: register
Flags: none

Arithmetic Group:

This group of instructions performs arithmetic operations on data in registers and memory.

Unless indicated otherwise, all instructions in this group affect the Zero, Sign, Parity, Carry, and Auxiliary Carry flags according to the standard rules.

All subtraction operations are performed via two's complement arithmetic and set the carry flag to one to indicate a borrow and clear it to indicate no borrow.

ADD r (Add Register)

(A) ← (A) + (r)

The content of register r is added to the content of the accumulator. The result is placed in the accumulator.

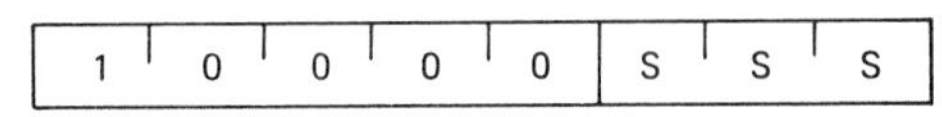

Cycles: 1
States: 4
Addressing: register
Flags: Z,S,P,CY,AC

ADD M (Add memory)

(A) ← (A) + ((H) (L))

The content of the memory location whose address is contained in the H and L registers is added to the content of the accumulator. The result is placed in the accumulator.

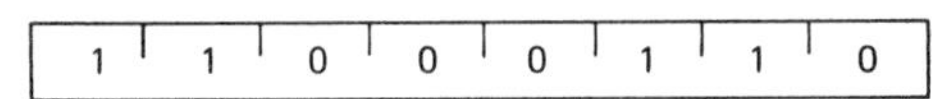

Cycles: 2
States: 7
Addressing: reg. indirect
Flags: Z,S,P,CY,AC

ADI data (Add immediate)

(A) ← (A) + (byte 2)

The content of the second byte of the instruction is added to the content of the accumulator. The result is placed in the accumulator.

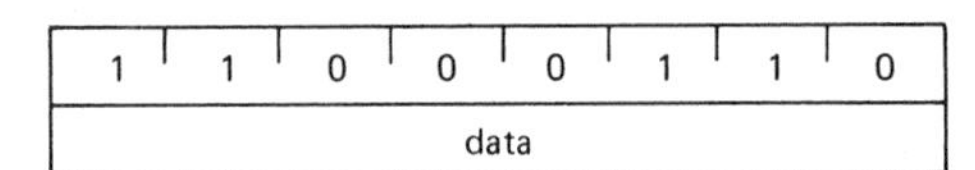

Cycles: 2
States: 7
Addressing: immediate
Flags: Z,S,P,CY,AC

ADC r (Add Register with carry)

(A) ← (A) + (r) + (CY)

The content of register r and the content of the carry bit are added to the content of the accumulator. The result is placed in the accumulator.

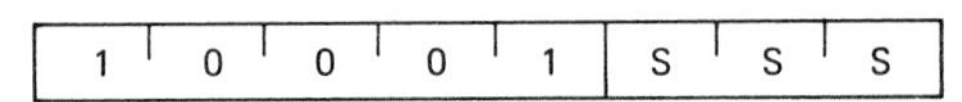

Cycles: 1
States: 4
Addressing: register
Flags: Z,S,P,CY,AC

ADC M (Add memory with carry)

(A) ← (A) + ((H) (L)) + (CY)

The content of the memory location whose address is contained in the H and L registers and the content of the CY flag are added to the accumulator. The result is placed in the accumulator.

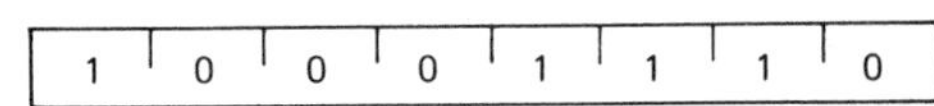

Cycles: 2
States: 7
Addressing: reg. indirect
Flags: Z,S,P,CY,AC

ACI data (Add immediate with carry)

(A) ← (A) + (byte 2) + (CY)

The content of the second byte of the instruction and the content of the CY flag are added to the contents of the accumulator. The result is placed in the accumulator.

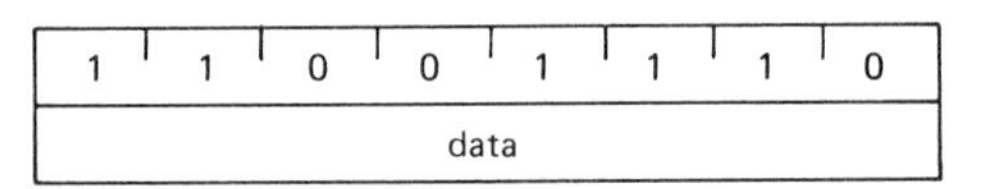

Cycles: 2
States: 7
Addressing: immediate
Flags: Z,S,P,CY,AC

SUB r (Subtract Register)

(A) ← (A) – (r)

The content of register r is subtracted from the content of the accumulator. The result is placed in the accumulator.

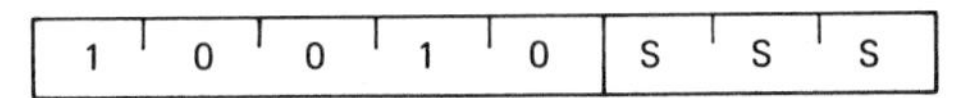

Cycles: 1
States: 4
Addressing: register
Flags: Z,S,P,CY,AC

SUB M (Subtract memory)

(A) ← (A) − ((H) (L))

The content of the memory location whose address is contained in the H and L registers is subtracted from the content of the accumulator. The result is placed in the accumulator.

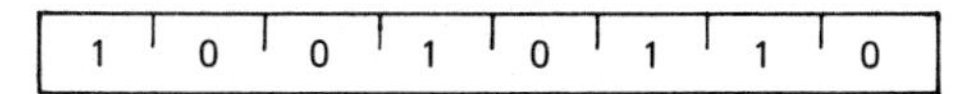

Cycles: 2
States: 7
Addressing: reg. indirect
Flags: Z,S,P,CY,AC

SUI data (Subtract immediate)

(A) ← (A) − (byte 2)

The content of the second byte of the instruction is subtracted from the content of the accumulator. The result is placed in the accumulator.

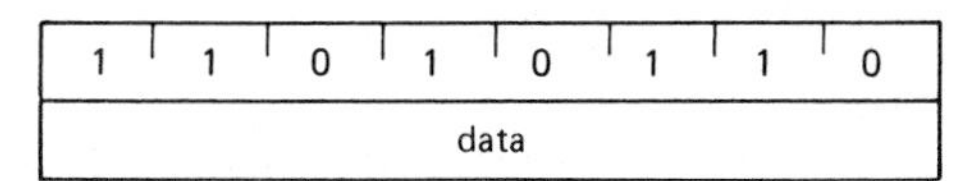

Cycles: 2
States: 7
Addressing: immediate
Flags: Z,S,P,CY,AC

SBB r (Subtract Register with borrow)

(A) ← (A) − (r) − (CY)

The content of register r and the content of the CY flag are both subtracted from the accumulator. The result is placed in the accumulator.

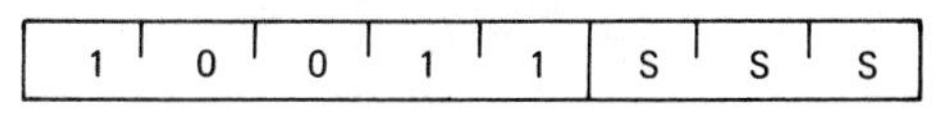

Cycles: 1
States: 4
Addressing: register
Flags: Z,S,P,CY,AC

SBB M (Subtract memory with borrow)

(A) ← (A) − ((H) (L)) − (CY)

The content of the memory location whose address is contained in the H and L registers and the content of the CY flag are both subtracted from the accumulator. The result is placed in the accumulator.

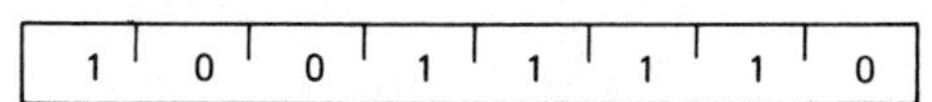

Cycles: 2
States: 7
Addressing: reg. indirect
Flags: Z,S,P,CY,AC

SBI data (Subtract immediate with borrow)

(A) ← (A) − (byte 2) − (CY)

The contents of the second byte of the instruction and the contents of the CY flag are both subtracted from the accumulator. The result is placed in the accumulator.

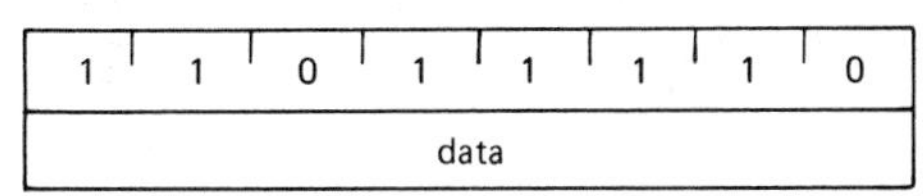

Cycles: 2
States: 7
Addressing: immediate
Flags: Z,S,P,CY,AC

INR r (Increment Register)

(r) ← (r) + 1

The content of register r is incremented by one. Note: All condition flags **except CY** are affected.

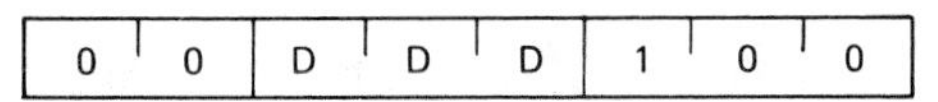

Cycles: 1
States: 5
Addressing: register
Flags: Z,S,P,AC

INR M (Increment memory)

((H) (L)) ← ((H) (L)) + 1

The content of the memory location whose address is contained in the H and L registers is incremented by one. Note: All condition flags **except CY** are affected.

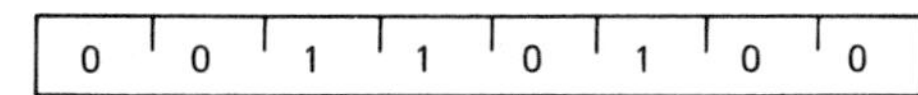

Cycles: 3
States: 10
Addressing: reg. indirect
Flags: Z,S,P,AC

DCR r (Decrement Register)

(r) ← (r) − 1

The content of register r is decremented by one. Note: All condition flags **except CY** are affected.

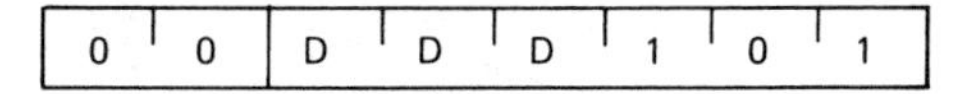

Cycles: 1
States: 5
Addressing: register
Flags: Z,S,P,AC

DCR M (Decrement memory)

((H) (L)) ← ((H) (L)) – 1

The content of the memory location whose address is contained in the H and L registers is decremented by one. Note: All condition flags **except CY** are affected.

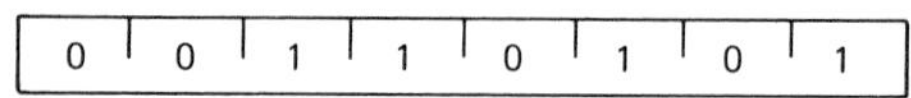

Cycles: 3
States: 10
Addressing: reg. indirect
Flags: Z,S,P,AC

INX rp (Increment register pair)

(rh) (rl) ← (rh) (rl) + 1

The content of the register pair rp is incremented by one. Note: **No condition flags are affected.**

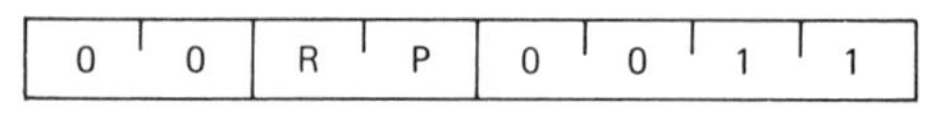

Cycles: 1
States: 5
Addressing: register
Flags: none

DCX rp (Decrement register pair)

(rh) (rl) ← (rh) (rl) – 1

The content of the register pair rp is decremented by one. Note: **No condition flags are affected.**

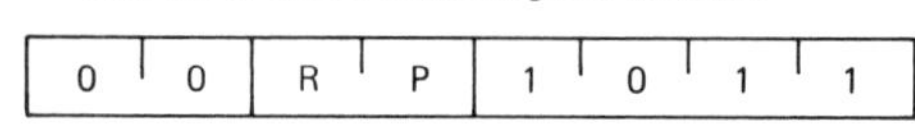

Cycles: 1
States: 5
Addressing: register
Flags: none

DAD rp (Add register pair to H and L)

(H) (L) ← (H) (L) + (rh) (rl)

The content of the register pair rp is added to the content of the register pair H and L. The result is placed in the register pair H and L. Note: **Only the CY flag is affected.** It is set if there is a carry out of the double precision add; otherwise it is reset.

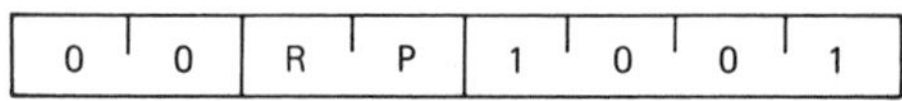

Cycles: 3
States: 10
Addressing: register
Flags: CY

DAA (Decimal Adjust Accumulator)

The eight-bit number in the accumulator is adjusted to form two four-bit Binary-Coded-Decimal digits by the following process:

1. If the value of the least significant 4 bits of the accumulator is greater than 9 **or** if the AC flag is set, 6 is added to the accumulator.
2. If the value of the most significant 4 bits of the accumulator is now greater than 9, **or** if the CY flag is set, 6 is added to the most significant 4 bits of the accumulator.

NOTE: All flags are affected.

0	0	1	0	0	1	1	1

Cycles: 1
States: 4
Flags: Z,S,P,CY,AC

Logical Group:

This group of instructions performs logical (Boolean) operations on data in registers and memory and on condition flags.

Unless indicated otherwise, all instructions in this group affect the Zero, Sign, Parity, Auxiliary Carry, and Carry flags according to the standard rules.

ANA r (AND Register)

(A) ← (A) ∧ (r)

The content of register r is logically anded with the content of the accumulator. The result is placed in the accumulator. **The CY and AC flags are cleared.**

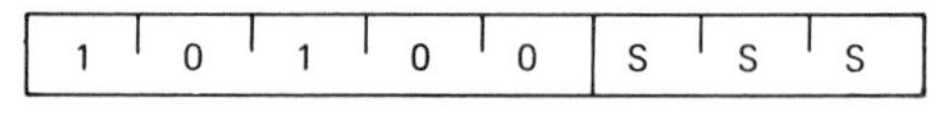

Cycles: 1
States: 4
Addressing: register
Flags: Z,S,P,CY,AC

ANA M (AND memory)

(A) ← (A) ∧ ((H) (L))

The contents of the memory location whose address is contained in the H and L registers is logically anded with the content of the accumulator. The result is placed in the accumulator. **The CY and AC flags are cleared.**

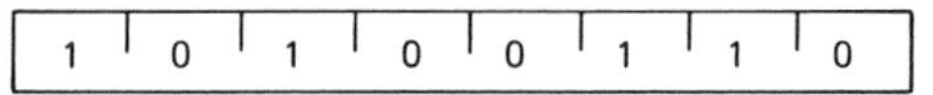

Cycles: 2
States: 7
Addressing: reg. indirect
Flags: Z,S,P,CY,AC

ANI data (AND immediate)

(A) ← (A) ∧ (byte 2)

The content of the second byte of the instruction is logically anded with the contents of the accumulator. The result is placed in the accumulator. **The CY and AC flags are cleared.**

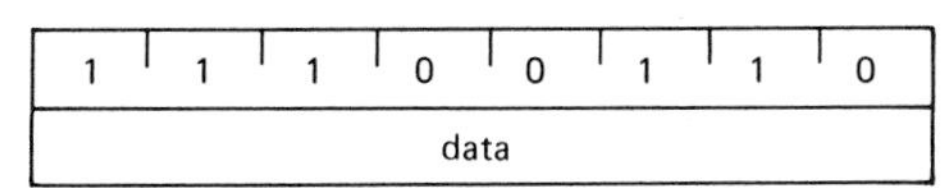

Cycles: 2
States: 7
Addressing: immediate
Flags: Z,S,P,CY,AC

XRA r (Exclusive OR Register)

(A) ← (A) ∀ (r)

The content of register r is exclusive-or'd with the content of the accumulator. The result is placed in the accumulator. **The CY and AC flags are cleared.**

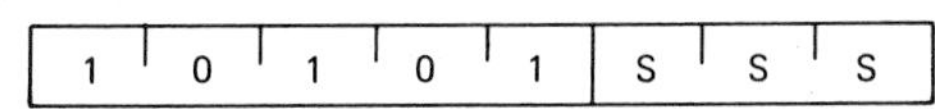

Cycles: 1
States: 4
Addressing: register
Flags: Z,S,P,CY,AC

XRA M (Exclusive OR Memory)

(A) ← (A) ∀ ((H) (L))

The content of the memory location whose address is contained in the H and L registers is exclusive-OR'd with the content of the accumulator. The result is placed in the accumulator. **The CY and AC flags are cleared.**

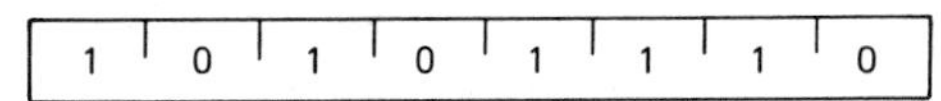

Cycles: 2
States: 7
Addressing: reg. indirect
Flags: Z,S,P,CY,AC

XRI data (Exclusive OR immediate)

(A) ← (A) ∀ (byte 2)

The content of the second byte of the instruction is exclusive-OR'd with the content of the accumulator. The result is placed in the accumulator. **The CY and AC flags are cleared.**

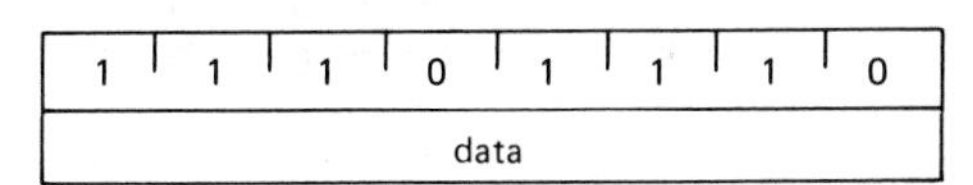

Cycles: 2
States: 7
Addressing: immediate
Flags: Z,S,P,CY,AC

ORA r (OR Register)

(A) ← (A) V (r)

The content of register r is inclusive-OR'd with the content of the accumulator. The result is placed in the accumulator. **The CY and AC flags are cleared.**

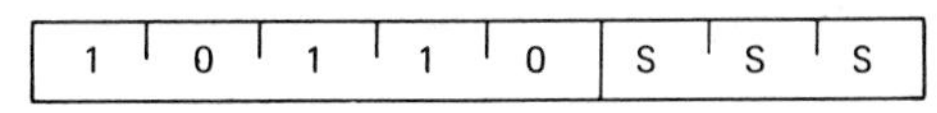

Cycles: 1
States: 4
Addressing: register
Flags: Z,S,P,CY,AC

ORA M (OR memory)

(A) ← (A) V ((H) (L))

The content of the memory location whose address is contained in the H and L registers is inclusive-OR'd with the content of the accumulator. The result is placed in the accumulator. **The CY and AC flags are cleared.**

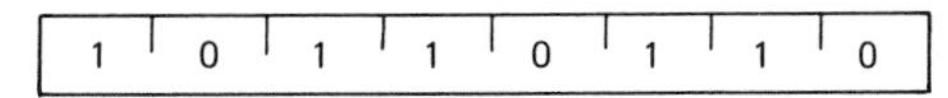

Cycles: 2
States: 7
Addressing: reg. indirect
Flags: Z,S,P,CY,AC

ORI data (OR Immediate)

(A) ← (A) V (byte 2)

The content of the second byte of the instruction is inclusive-OR'd with the content of the accumulator. The result is placed in the accumulator. **The CY and AC flags are cleared.**

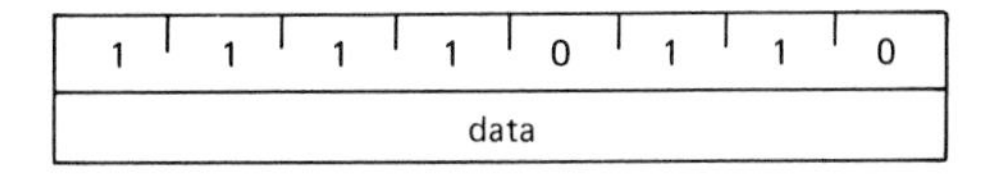

Cycles: 2
States: 7
Addressing: immediate
Flags: Z,S,P,CY,AC

CMP r (Compare Register)

(A) – (r)

The content of register r is subtracted from the accumulator. The accumulator remains unchanged. The condition flags are set as a result of the subtraction. **The Z flag is set to 1 if (A) = (r). The CY flag is set to 1 if (A) < (r).**

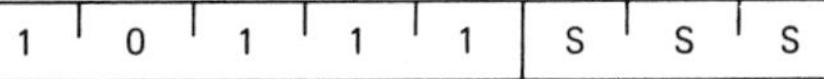

Cycles: 1
States: 4
Addressing: register
Flags: Z,S,P,CY,AC

CMP M (Compare memory)

(A) – ((H) (L))

The content of the memory location whose address is contained in the H and L registers is subtracted from the accumulator. The accumulator remains unchanged. The condition flags are set as a result of the subtraction. The Z flag is set to 1 if (A) = ((H) (L)). The CY flag is set to 1 if (A) < ((H) (L)).

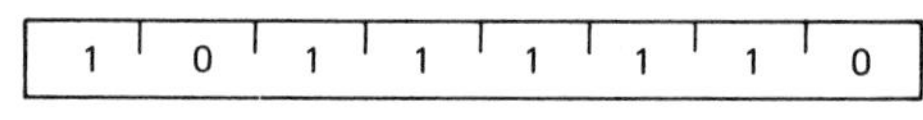

Cycles: 2
States: 7
Addressing: reg. indirect
Flags: Z,S,P,CY,AC

CPI data (Compare immediate)

(A) – (byte 2)

The content of the second byte of the instruction is subtracted from the accumulator. The condition flags are set by the result of the subtraction. The Z flag is set to 1 if (A) = (byte 2). The CY flag is set to 1 if (A) < (byte 2).

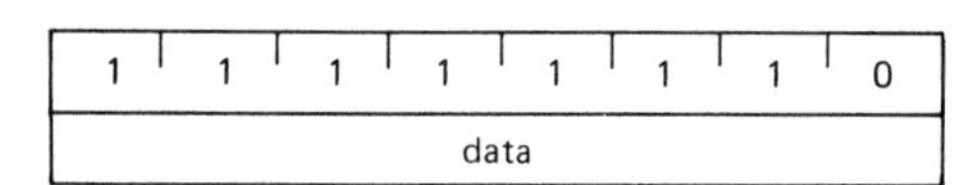

Cycles: 2
States: 7
Addressing: immediate
Flags: Z,S,P,CY,AC

RLC (Rotate left)

$(A_{n+1}) \leftarrow (A_n)$; $(A_0) \leftarrow (A_7)$

$(CY) \leftarrow (A_7)$

The content of the accumulator is rotated left one position. The low order bit and the CY flag are both set to the value shifted out of the high order bit position. **Only the CY flag is affected.**

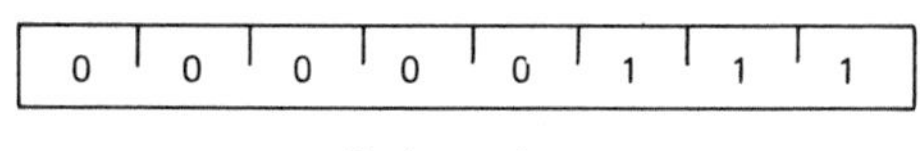

Cycles: 1
States: 1
Flags: CY

RRC (Rotate right)

$(A_n) \leftarrow (A_{n-1})$; $(A_7) \leftarrow (A_0)$

$(CY) \leftarrow (A_0)$

The content of the accumulator is rotated right one position. The high order bit and the CY flag are both set to the value shifted out of the low order bit position. **Only the CY flag is affected.**

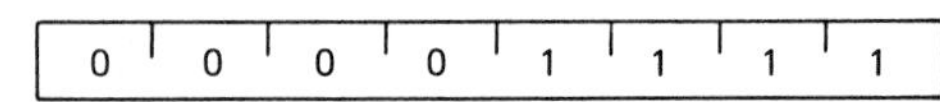

Cycles: 1
States: 4
Flags: CY

RAL (Rotate left through carry)

$(A_{n+1}) \leftarrow (A_n)$; $(CY) \leftarrow (A_7)$

$(A_0) \leftarrow (CY)$

The content of the accumulator is rotated left one position through the CY flag. The low order bit is set equal to the CY flag and the CY flag is set to the value shifted out of the high order bit. **Only the CY flag is affected.**

0	0	0	1	0	1	1	1

Cycles: 1
States: 4
Flags: CY

RAR (Rotate right through carry)

$(A_n) \leftarrow (A_{n+1})$; $(CY) \leftarrow (A_0)$

$(A_7) \leftarrow (CY)$

The content of the accumulator is rotated right one position through the CY flag. The high order bit is set to the CY flag and the CY flag is set to the value shifted out of the low order bit. **Only the CY flag is affected.**

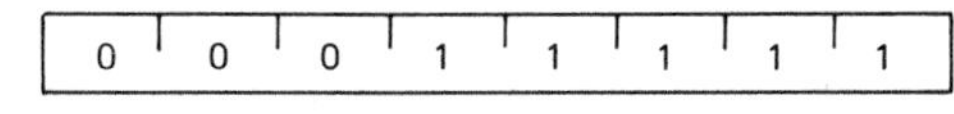

Cycles: 1
States: 4
Flags: CY

CMA (Complement accumulator)

$(A) \leftarrow (\overline{A})$

The contents of the accumulator are complemented (zero bits become 1, one bits become 0). **No flags are affected.**

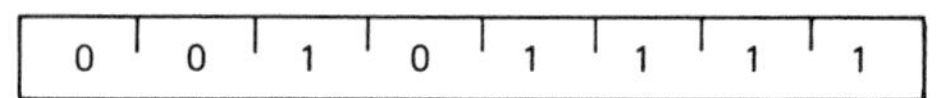

Cycles: 1
States: 4
Flags: none

CMC (Complement carry)

(CY) ← $(\overline{CY})$

The CY flag is complemented. **No other flags are affected.**

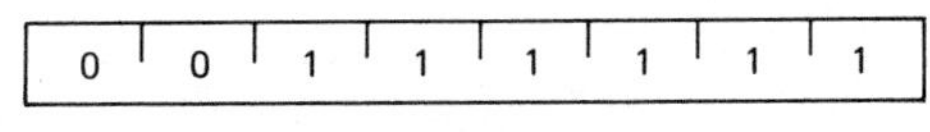

Cycles: 1
States: 4
Flags: CY

STC (Set carry)

(CY) ← 1

The CY flag is set to 1. **No other flags are affected.**

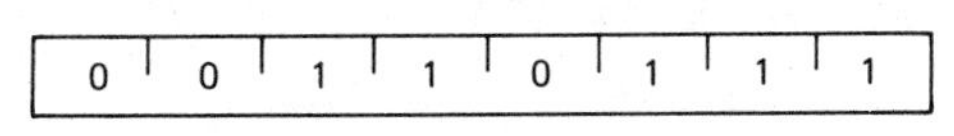

Cycles: 1
States: 4
Flags: CY

Branch Group:

This group of instructions alter normal sequential program flow.

Condition flags are not affected by any instruction in this group.

The two types of branch instructions are unconditional and conditional. Unconditional transfers simply perform the specified operation on register PC (the program counter). Conditional transfers examine the status of one of the four processor flags to determine if the specified branch is to be executed. The conditions that may be specified are as follows:

CONDITION		CCC
NZ	– not zero (Z = 0)	000
Z	– zero (Z = 1)	001
NC	– no carry (CY = 0)	010
C	– carry (CY = 1)	011
PO	– parity odd (P = 0)	100
PE	– parity even (P = 1)	101
P	– plus (S = 0)	110
M	– minus (S = 1)	111

JMP addr (Jump)

(PC) ← (byte 3) (byte 2)

Control is transferred to the instruction whose address is specified in byte 3 and byte 2 of the current instruction.

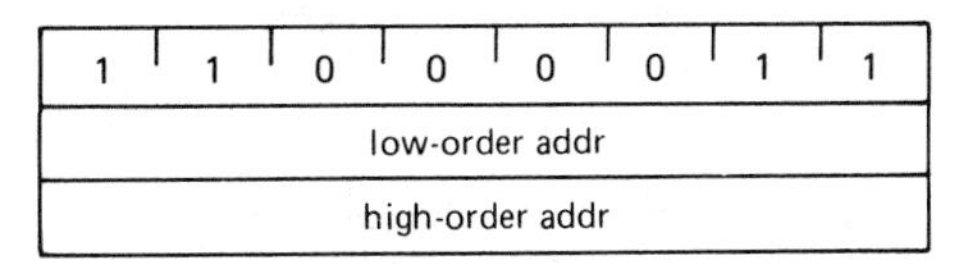

Cycles: 3
States: 10
Addressing: immediate
Flags: none

Jcondition addr (Conditional jump)

If (CCC),

(PC) ← (byte 3) (byte 2)

If the specified condition is true, control is transferred to the instruction whose address is specified in byte 3 and byte 2 of the current instruction; otherwise, control continues sequentially.

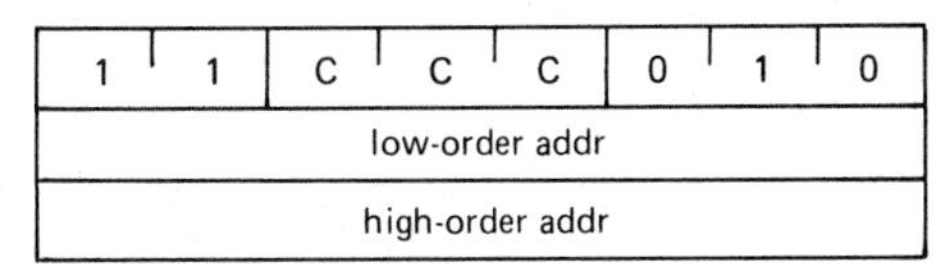

Cycles: 3
States: 10
Addressing: immediate
Flags: none

CALL addr (Call)

((SP) – 1) ← (PCH)
((SP) – 2) ← (PCL)
(SP) ← (SP) – 2
(PC) ← (byte 3) (byte 2)

The high-order eight bits of the next instruction address are moved to the memory location whose address is one less than the content of register SP. The low-order eight bits of the next instruction address are moved to the memory location whose address is two less than the content of register SP. The content of register SP is decremented by 2. Control is transferred to the instruction whose address is specified in byte 3 and byte 2 of the current instruction.

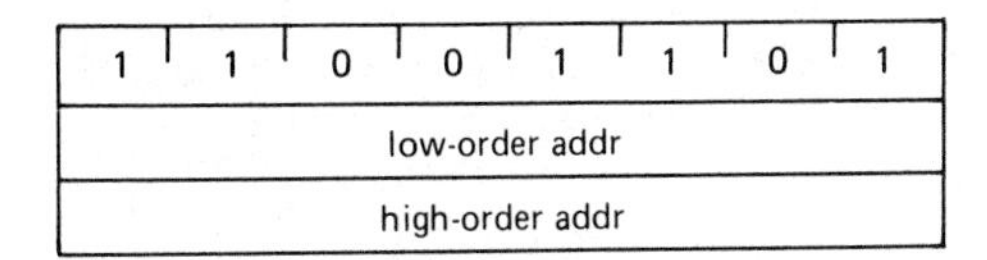

Cycles: 5
States: 17
Addressing: immediate/reg. indirect
Flags: none

Ccondition addr (Condition call)

If (CCC),

((SP) – 1) $\leftarrow$ (PCH)
((SP) – 2) $\leftarrow$ (PCL)
(SP) $\leftarrow$ (SP) – 2
(PC) $\leftarrow$ (byte 3) (byte 2)

If the specified condition is true, the actions specified in the CALL instruction (see above) are performed; otherwise, control continues sequentially.

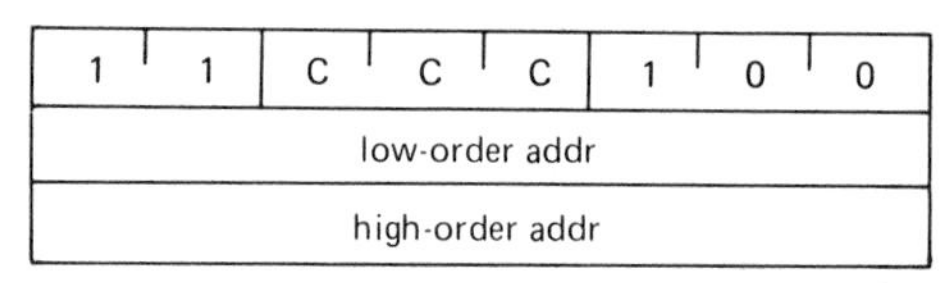

Cycles: 3/5
States: 11/17
Addressing: immediate/reg. indirect
Flags: none

RET (Return)

(PCL) $\leftarrow$ ((SP));
(PCH) $\leftarrow$ ((SP) + 1);
(SP) $\leftarrow$ (SP) + 2;

The content of the memory location whose address is specified in register SP is moved to the low-order eight bits of register PC. The content of the memory location whose address is one more than the content of register SP is moved to the high-order eight bits of register PC. The content of register SP is incremented by 2.

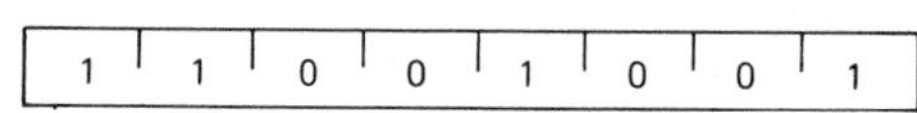

Cycles: 3
States: 10
Addressing: reg. indirect
Flags: none

Rcondition (Conditional return)

If (CCC),

(PCL) $\leftarrow$ ((SP))
(PCH) $\leftarrow$ ((SP) + 1)
(SP) $\leftarrow$ (SP) + 2

If the specified condition is true, the actions specified in the RET instruction (see above) are performed; otherwise, control continues sequentially.

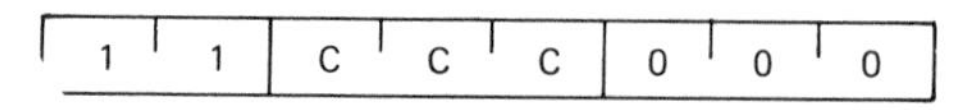

Cycles: 1/3
States: 5/11
Addressing: reg. indirect
Flags: none

RST n (Restart)

((SP) – 1) $\leftarrow$ (PCH)
((SP) – 2) $\leftarrow$ (PCL)
(SP) $\leftarrow$ (SP) – 2
(PC) $\leftarrow$ 8 * (NNN)

The high-order eight bits of the next instruction address are moved to the memory location whose address is one less than the content of register SP. The low-order eight bits of the next instruction address are moved to the memory location whose address is two less than the content of register SP. The content of register SP is decremented by two. Control is transferred to the instruction whose address is eight times the content of NNN.

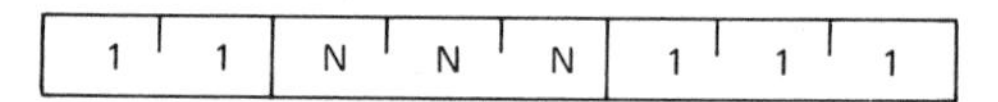

Cycles: 3
States: 11
Addressing: reg. indirect
Flags: none

15	14	13	12	11	10	9	8	7	6	5	4	3	2	1	0
0	0	0	0	0	0	0	0	0	0	N	N	N	0	0	0

Program Counter After Restart

PCHL (Jump H and L indirect – move H and L to PC)

(PCH) $\leftarrow$ (H)
(PCL) $\leftarrow$ (L)

The content of register H is moved to the high-order eight bits of register PC. The content of register L is moved to the low-order eight bits of register PC.

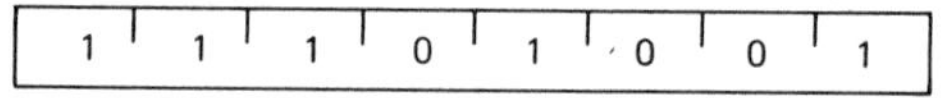

Cycles: 1
States: 5
Addressing: register
Flags: none

Stack, I/O, and Machine Control Group:

This group of instructions performs I/O, manipulates the Stack, and alters internal control flags.

Unless otherwise specified, **condition flags are not affected by any instructions in this group.**

FLAG WORD

D_7	D_6	D_5	D_4	D_3	D_2	D_1	D_0
S	Z	0	AC	0	P	1	CY

PUSH rp (Push)

$((SP) - 1) \leftarrow (rh)$
$((SP) - 2) \leftarrow (rl)$
$(SP) \leftarrow (SP) - 2$

The content of the high-order register of register pair rp is moved to the memory location whose address is one less than the content of register SP. The content of the low-order register of register pair rp is moved to the memory location whose address is two less than the content of register SP. The content of register SP is decremented by 2. **Note: Register pair rp = SP may not be specified.**

1	1	R	P	0	1	0	1

Cycles: 3
States: 11
Addressing: reg. indirect
Flags: none

PUSH PSW (Push processor status word)

$((SP) - 1) \leftarrow (A)$
$((SP) - 2)_0 \leftarrow (CY)$, $((SP) - 2)_1 \leftarrow 1$
$((SP) - 2)_2 \leftarrow (P)$, $((SP) - 2)_3 \leftarrow 0$
$((SP) - 2)_4 \leftarrow (AC)$, $((SP) - 2)_5 \leftarrow 0$
$((SP) - 2)_6 \leftarrow (Z)$, $((SP) - 2)_7 \leftarrow (S)$
$(SP) \leftarrow (SP) - 2$

The content of register A is moved to the memory location whose address is one less than register SP. The contents of the condition flags are assembled into a processor status word and the word is moved to the memory location whose address is two less than the content of register SP. The content of register SP is decremented by two.

1	1	1	1	0	1	0	1

Cycles: 3
States: 11
Addressing: reg. indirect
Flags: none

POP rp (Pop)

$(rl) \leftarrow ((SP))$
$(rh) \leftarrow ((SP) + 1)$
$(SP) \leftarrow (SP) + 2$

The content of the memory location, whose address is specified by the content of register SP, is moved to the low-order register of register pair rp. The content of the memory location, whose address is one more than the content of register SP, is moved to the high-order register of register pair rp. The content of register SP is incremented by 2. **Note: Register pair rp = SP may not be specified.**

1	1	R	P	0	0	0	1

Cycles: 3
States: 10
Addressing: reg. indirect
Flags: none

POP PSW (Pop processor status word)

$(CY) \leftarrow ((SP))_0$
$(P) \leftarrow ((SP))_2$
$(AC) \leftarrow ((SP))_4$
$(Z) \leftarrow ((SP))_6$
$(S) \leftarrow ((SP))_7$
$(A) \leftarrow ((SP) + 1)$
$(SP) \leftarrow (SP) + 2$

The content of the memory location whose address is specified by the content of register SP is used to restore the condition flags. The content of the memory location whose address is one more than the content of register SP is moved to register A. The content of register SP is incremented by 2.

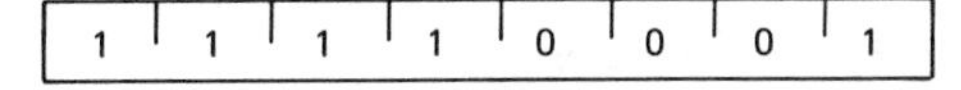

1	1	1	1	0	0	0	1

Cycles: 3
States: 10
Addressing: reg. indirect
Flags: Z,S,P,CY,AC

XTHL (Exchange stack top with H and L)

(L) ↔ ((SP))

(H) ↔ ((SP) + 1)

The content of the L register is exchanged with the content of the memory location whose address is specified by the content of register SP. The content of the H register is exchanged with the content of the memory location whose address is one more than the content of register SP.

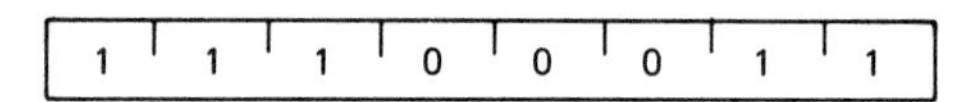

Cycles: 5
States: 18
Addressing: reg. indirect
Flags: none

SPHL (Move HL to SP)

(SP) ← (H) (L)

The contents of registers H and L (16 bits) are moved to register SP.

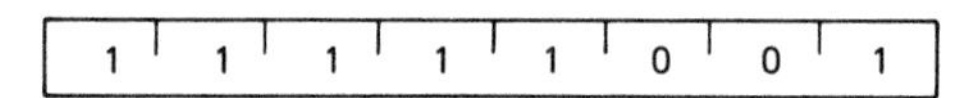

Cycles: 1
States: 5
Addressing: register
Flags: none

IN port (Input)

(A) ← (data)

The data placed on the eight bit bi-directional data bus by the specified port is moved to register A.

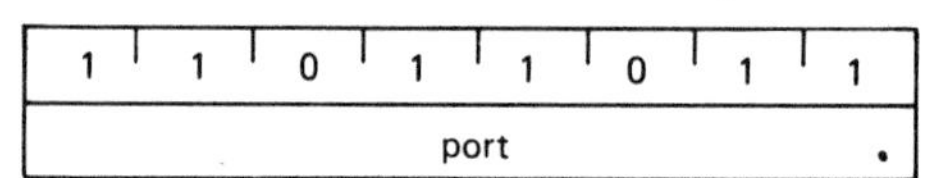

Cycles: 3
States: 10
Addressing: direct
Flags: none

OUT port (Output)

(data) ← (A)

The content of register A is placed on the eight bit bi-directional data bus for transmission to the specified port.

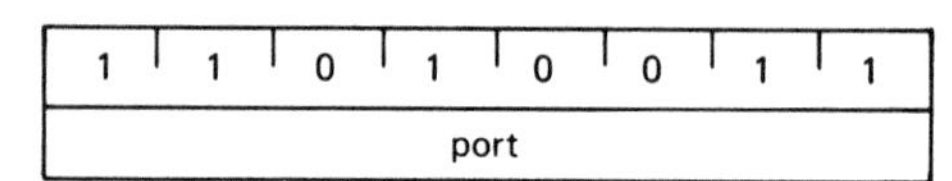

Cycles: 3
States: 10
Addressing: direct
Flags: none

EI (Enable interrupts)

The interrupt system is enabled **following the execution of the next instruction.**

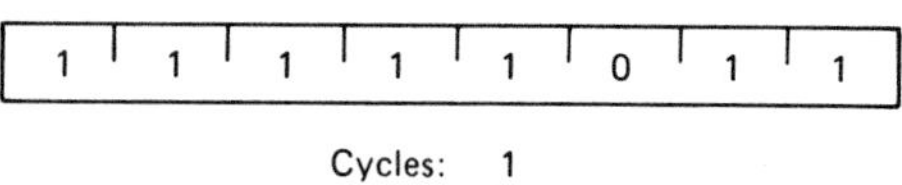

Cycles: 1
States: 4
Flags: none

DI (Disable interrupts)

The interrupt system is disabled **immediately following the execution of the DI instruction.**

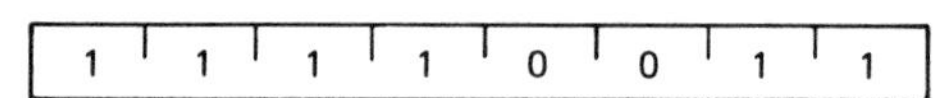

Cycles: 1
States: 4
Flags: none

HLT (Halt)

The processor is stopped. The registers and flags are unaffected.

Cycles: 1
States: 7
Flags: none

NOP (No op)

No operation is performed. The registers and flags are unaffected.

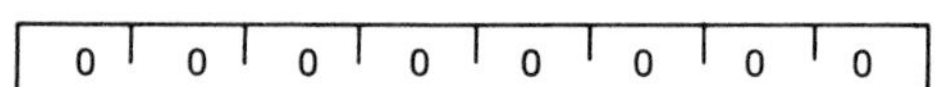

Cycles: 1
States: 4
Flags: none

INSTRUCTION SET

Summary of Processor Instructions

Mnemonic	Description	Instruction Code[1] D_7	D_6	D_5	D_4	D_3	D_2	D_1	D_0	Clock[2] Cycles
$MOV_{r1,r2}$	Move register to register	0	1	D	D	D	S	S	S	5
MOV M, r	Move register to memory	0	1	1	1	0	S	S	S	7
MOV r, M	Move memory to register	0	1	D	D	D	1	1	0	7
HLT	Halt	0	1	1	1	0	1	1	0	7
MVI r	Move immediate register	0	0	D	D	D	1	1	0	7
MVI M	Move immediate memory	0	0	1	1	0	1	1	0	10
INR r	Increment register	0	0	D	D	D	1	0	0	5
DCR r	Decrement register	0	0	D	D	D	1	0	1	5
INR M	Increment memory	0	0	1	1	0	1	0	0	10
DCR M	Decrement memory	0	0	1	1	0	1	0	1	10
ADD r	Add register to A	1	0	0	0	0	S	S	S	4
ADC r	Add register to A with carry	1	0	0	0	1	S	S	S	4
SUB r	Subtract register from A	1	0	0	1	0	S	S	S	4
SBB r	Subtract register from A with borrow	1	0	0	1	1	S	S	S	4
ANA r	And register with A	1	0	1	0	0	S	S	S	4
XRA r	Exclusive Or register with A	1	0	1	0	1	S	S	S	4
ORA r	Or register with A	1	0	1	1	0	S	S	S	4
CMP r	Compare register with A	1	0	1	1	1	S	S	S	4
ADD M	Add memory to A	1	0	0	0	0	1	1	0	7
ADC M	Add memory to A with carry	1	0	0	0	1	1	1	0	7
SUB M	Subtract memory from A	1	0	0	1	0	1	1	0	7
SBB M	Subtract memory from A with borrow	1	0	0	1	1	1	1	0	7
ANA M	And memory with A	1	0	1	0	0	1	1	0	7
XRA M	Exclusive Or memory with A	1	0	1	0	1	1	1	0	7
ORA M	Or memory with A	1	0	1	1	0	1	1	0	7
CMP M	Compare memory with A	1	0	1	1	1	1	1	0	7
ADI	Add immediate to A	1	1	0	0	0	1	1	0	7
ACI	Add immediate to A with carry	1	1	0	0	1	1	1	0	7
SUI	Subtract immediate from A	1	1	0	1	0	1	1	0	7
SBI	Subtract immediate from A with borrow	1	1	0	1	1	1	1	0	7
ANI	And immediate with A	1	1	1	0	0	1	1	0	7
XRI	Exclusive Or immediate with A	1	1	1	0	1	1	1	0	7
ORI	Or immediate with A	1	1	1	1	0	1	1	0	7
CPI	Compare immediate with A	1	1	1	1	1	1	1	0	7
RLC	Rotate A left	0	0	0	0	0	1	1	1	4
RRC	Rotate A right	0	0	0	0	1	1	1	1	4
RAL	Rotate A left through carry	0	0	0	1	0	1	1	1	4
RAR	Rotate A right through carry	0	0	0	1	1	1	1	1	4
JMP	Jump unconditional	1	1	0	0	0	0	1	1	10
JC	Jump on carry	1	1	0	1	1	0	1	0	10
JNC	Jump on no carry	1	1	0	1	0	0	1	0	1'
JZ	Jump on zero	1	1	0	0	1	0	1	0	1ι
JNZ	Jump on no zero	1	1	0	0	0	0	1	0	10
JP	Jump on positive	1	1	1	1	0	0	1	0	10
JM	Jump on minus	1	1	1	1	1	0	1	0	10
JPE	Jump on parity even	1	1	1	0	1	0	1	0	10
JPO	Jump on parity odd	1	1	1	0	0	0	1	0	10
CALL	Call unconditional	1	1	0	0	1	1	0	1	17
CC	Call on carry	1	1	0	1	1	1	0	0	11/17
CNC	Call on no carry	1	1	0	1	0	1	0	0	11/17
CZ	Call on zero	1	1	0	0	1	1	0	0	11/17
CNZ	Call on no zero	1	1	0	0	0	1	0	0	11/17
CP	Call on positive	1	1	1	1	0	1	0	0	11/!7
CM	Call on minus	1	1	1	1	1	1	0	0	11/17
CPE	Call on parity even	1	1	1	0	1	1	0	0	11/17
CPO	Call on parity odd	1	1	1	0	0	1	0	0	11/17
RET	Return	1	1	0	0	1	0	0	1	10
RC	Return on carry	1	1	0	1	1	0	0	(	5/11
RNC	Return on no carry	1	1	0	1	0	0	0	0	5/11
RZ	Return on zero	1	1	0	0	1	0	0	0	5/11
RNZ	Return on no zero	1	1	0	0	0	0	C	0	5/11
RP	Return on positive	1	1	1	1	0	0	0	0	5/11
RM	Return on minus	1	1	1	1	1	0	0	0	5/11
RPE	Return on parity even	1	1	1	0	1	0	0	0	5/11
RPO	Return on parity odd	1	1	1	0	0	0	0	0	5/11
RST	Restart	1	1	A	A	A	1	1	1	11
IN	Input	1	1	0	1	1	0	1	1	10
OUT	Output	1	1	0	1	0	0	1	1	10
LXI B	Load immediate register Pair B & C	0	0	0	0	0	0	0	1	10
LXI D	Load immediate register Pair D & E	0	0	0	1	0	0	0	1	10
LXI H	Load immediate register Pair H & L	0	0	1	0	0	0	0	1	10
LXI SP	Load immediate stack pointer	0	0	1	1	0	0	0	1	10
PUSH B	Push register Pair B & C on stack	1	1	0	0	0	1	0	1	11
PUSH D	Push register Pair D & E on stack	1	1	0	1	0	1	0	1	11
PUSH H	Push register Pair H & L on stack	1	1	1	0	0	1	0	1	11
PUSH PSW	Push A and Flags on stack	1	1	1	1	0	1	0	1	11
POP B	Pop register pair B & C off stack	1	1	0	0	0	0	0	1	10
POP D	Pop register pair D & E off stack	1	1	0	1	0	0	0	1	10
POP H	Pop register pair H & L off stack	1	1	1	0	0	0	0	1	10
POP PSW	Pop A and Flags off stack	1	1	1	1	0	0	0	1	10
STA	Store A direct	0	0	1	1	0	0	1	0	13
LDA	Load A direct	0	0	1	1	1	0	1	0	13
XCHG	Exchange D & E, H & L Registers	1	1	1	0	1	0	1	1	4
XTHL	Exchange top of stack, H & L	1	1	1	0	0	0	1	1	18
SPHL	H & L to stack pointer	1	1	1	1	1	0	0	1	5
PCHL	H & L to program counter	1	1	1	0	1	0	0	1	5
DAD B	Add B & C to H & L	0	0	0	0	1	0	0	1	10
DAD D	Add D & E to H & L	0	0	0	1	1	0	0	1	10
DAD H	Add H & L to H & L	0	0	1	0	1	0	0	1	10
DAD SP	Add stack pointer to H & L	0	0	1	1	1	0	0	1	10
STAX B	Store A indirect	0	0	0	0	0	0	1	0	7
STAX D	Store A indirect	0	0	0	1	0	0	1	0	7
LDAX B	Load A indirect	0	0	0	0	1	0	1	0	7
LDAX D	Load A indirect	0	0	0	1	1	0	1	0	7
INX B	Increment B & C registers	0	0	0	0	0	0	1	1	5
INX D	Increment D & E registers	0	0	0	1	0	0	1	1	5
INX H	Increment H & L registers	0	0	1	0	0	0	1	1	5
INX SP	Increment stack pointer	0	0	1	1	0	0	1	1	5
DCX B	Decrement B & C	0	0	0	0	1	0	1	1	5
DCX D	Decrement D & E	0	0	0	1	1	0	1	1	5
DCX H	Decrement H & L	0	0	1	0	1	0	1	1	5
DCX SP	Decrement stack pointer	0	0	1	1	1	0	1	1	5
CMA	Complement A	0	0	1	0	1	1	1	1	4
STC	Set carry	0	0	1	1	0	1	1	1	4
CMC	Complement carry	0	0	1	1	1	1	1	1	4
DAA	Decimal adjust A	0	0	i	0	0	1	1	1	4
SHLD	Store H & L direct	0	0	1	0	0	0	1	0	16
LHLD	Load H & L direct	0	0	1	0	1	0	1	0	16
EI	Enable Interrupts	1	1	1	1	1	0	1	1	4
DI	Disable interrupt	1	1	1	1	0	0	1	1	4
NOP	No-operation	0	0	0	0	0	0	0	0	4

NOTES: 1. DDD or SSS – 000 B – 001 C – 010 D – 011 E – 100 H – 101 L – 110 Memory – 111 A.
2. Two possible cycle times, (5/11) indicate instruction cycles dependent on condition flags.

INDEX

A

B

M

N

O

P

R

S

T

U

V

W